Human–Bug Encounters in Multispecies Networks

Brill's Series in the History of the Environment

Series Editor

Aleks Pluskowski (*University of Reading*)

VOLUME 8

The titles published in this series are listed at *brill.com/bshe*

Human–Bug Encounters in Multispecies Networks

Edited by

Laura Hollsten
Otto Latva
Sanna Lillbroända-Annala
Suvi Rytty
Tuomas Räsänen

BRILL

LEIDEN | BOSTON

Research Council of Finland

This work was supported by the Research Council of Finland.

Cover illustration: Jan van Kessel the Elder, *Study of Insects* (1660). Musée des Beaux-Arts de Strasbourg, Creative Commons Attribution-Share Alike 2.0 France.

The Library of Congress Cataloging-in-Publication Data is available online at https://catalog.loc.gov

Typeface for the Latin, Greek, and Cyrillic scripts: "Brill". See and download: brill.com/brill-typeface.

ISSN 1876-6595
ISBN 978-90-04-68060-9 (hardback)
ISBN 978-90-04-71544-8 (e-book)
DOI 10.1163/9789004715448

Published by Koninklijke Brill BV, Leiden, The Netherlands.
Koninklijke Brill BV incorporates the imprints Brill, Brill Nijhoff, Brill Schöningh, Brill Fink, Brill mentis, Brill Wageningen Academic, Vandenhoeck & Ruprecht, Böhlau and V&R unipress.

This book is printed on acid-free paper and produced in a sustainable manner.

Contents

Figures

Notes on Contributors

Concepción Cortés Zulueta
is a postdoctoral Fellow. Her interdisciplinary research focuses on the presence and agency of nonhuman animals in contemporary art and visual culture. She has lead a research project on the intertwined perceptions on birds and insects at Universidad de Málaga, Spain, and besides approaching birdsongs as history, she is preparing a book on the conventions and misconceptions on fly eyes and insect vision. She has made temporary research stays at La Sapienza, Università di Roma; the National Art Library (Victoria & Albert Museum); the New Zealand Centre for Human-Animal Studies (University of Canterbury, Christchurch); the Museo Nacional de Ciencias Naturales (MNCN, Madrid), and the Massachusetts Institute of Technology.

Michaela Fenske
is a Professor and Chair of European Ethnology at the University of Würzburg, Germany. Her research interests include historical anthropology, narrative culture, anthropology of the rural, and environmental humanities. Together with her team, she established the field of Multispecies Studies at the University of Würzburg, so far resulting in studies on wolves, ravens, carps, bees, rats, forests, and apple trees. Her current research project "Gardens in transformation" focuses on multispecies communities in gardens. In addition, she engages in exploring the possibilities of socio-ecological sustainability on university campus (REKLINEU TP 10).

Sophie FitzMaurice
is a postdoctoral Research Fellow at the Centre for History and Economics, Magdalene College, University of Cambridge. She received her Ph.D. in History from the University of California, Berkeley, in 2023, where she wrote her dissertation on the material and environmental history of the telegraph system in North America. Her broader research interests lie at the intersection of environmental history and the history of technology and infrastructure.

Laura Hollsten
holds the title of docent in environmental history at the Faculty of Arts, Psychology, and Theology at Åbo Akademi University in Turku, Finland. Her research interests include early modern history, the history knowledge and science, environmental history, and global history. She has been involved in the research project "Humans and Ticks in the Anthropocene" at the University of Eastern Finland, the University of Turku, and Åbo Akademi University.

Karine Aasgaard Jansen
is a senior Researcher in medical anthropology at the Chr. Michelsen Institute (CMI), Norway. She is also the coordinator of the research group "Poverty and Global Health". From 2016–19 she led the research project "Contagion and culture: The 2005–07 chikungunya epidemic in the Western Indian Ocean". The study was funded by the Swedish Research Council and based at Umeå University, Sweden. In addition to her longstanding research interest in arboviral diseases and multispecies ethnography, Karine has worked on pandemic preparedness, vaccine hesitancy and fertility control including abortion.

Otto Latva
is a historian focusing on human–animal and human–plant studies as well as environmental history. He has studied widely the societies and cultures of the early modern period as well as the nineteenth and twentieth centuries. Latva has the title of docent in the field of more-than-human history.

Sanna Lillbroända-Annala
is an ethnologist and senior Lecturer at Åbo Akademi University in Finland. She has conducted research in urban environments from the perspective of gentrification and cultural heritage, as well as in recent years with special interest towards non-humans in urban settings. Lillbroända-Annala holds the title of docent in the field of urban ethnology.

Heta Lähdesmäki
is a historian focusing on environmental history, human–animal relations and wildlife conflicts. She has studied for instance human–wolf relations, bird feeding and rat conflicts in twentieth-century Finland. She works as a postdoctoral researcher at the University of Turku, Finland.

Heidi Mikkola
is a doctoral Researcher in media studies at the University of Turku. Her dissertation centers around the aesthetics and agencies produced by wildlife documentaries, exploring the relationships between humans, nonhuman animals, environments, and technologies.

Marianne Mäkelin
is a doctoral Researcher in science and technology studies and sociology at the University of Helsinki. Her work explores human-insect relations and knowledge practices in biosciences.

Suvi Rytty
is a postdoctoral Researcher in the Department of Finnish History at the University of Turku, Finland. She specializes in the history of medicine and health. In her dissertation, she has studied alternative medicine, vegetarianism, nudism, and anti-vaccinationism, which were practiced in the early 20th century in the context of the Finnish life reform movement. In her current postdoctoral research project, funded by the Kone Foundation, she examines the natural lifestyle in Finland from the late 19th century to the present day.

Tuomas Räsänen
works as a Professor of Environmental History at the University of Eastern Finland. He specializes on the history of human-animal relations, the history of environmentalism, as well as on marine environmental history, particularly in the 20th century. He has been leading a research project, funded by the Research Council of Finland, that explores the relationship between humans and ticks in Finland.

Minna Santaoja
is a multidisciplinary environmental social Researcher with a degree in environmental policy, currently working at the Arctic Centre at the University of Lapland, Finland. Her research focuses on multispecies agency and knowledge practices in different contexts. Human-insect relations have been a continuous thread in Minna's work and she has written previously on e.g. amateur entomology, sustainability of entomophagy, and response-ability in human-wasp cohabitance. She is currently the editor in chief of the Finnish scholarly journal Alue & Ympäristö on regional studies and environmental social research.

Taina Syrjämaa
is Professor of European and World History at the University of Turku. She has led two four-year research projects on animal history, financed by the Academy of Finland. The projects have focused, respectively, on animal agency and on the industrial exploitation of animals. Her current research interests focus on animal mobilities, on the history of cattle and pets, especially cats, and on the visibility vs. invisibility of animals in society.

Emily Webster
is Assistant Professor in the History and Philosophy of Health and Medicine in the Department of Philosophy at Durham University. Her research focuses on the ecology of historical epidemics, drawing on contemporary biology

and ecology alongside traditional historical methods to tell multi-species, multi-scalar histories of infectious disease that ground humans in their physical environments. She is currently working on her book project, "Infectious Ecologies: A Biological History of Epidemics in the British Empire". She received her Ph.D. in History and MSc in Public Health Sciences from the University of Chicago in 2021.

Introduction

Laura Hollsten, Suvi Rytty, Otto Latva, Sanna Lillbroända-Annala and Tuomas Räsänen

"Here, in these small, and [...] so despised creatures, we can find the greatest masterpieces of nature." Thus wrote Carl von Linné in 1739 in a speech where he admired the remarkable talents of insects. Insects have been living on our planet for more than 400 million years, and, according to entomologist Karl Kjer, they did everything first: "They were the first to form social societies, farm, and sing – just about anything you can imagine."[1] Termites and ants, as well as some species of bees and wasps, build societies that exhibit social behavior and a division of labor. Leaf cutter ants carry out agricultural work: they can cut leaves without digesting them but feed the leaves to their fungus garden and eat the fungus. For many insects, such as crickets and cicadas, sound is an important form of communication: some insects squeak, scratch or hiss while others rattle and beep.

Insects are among the most abundant organisms on Earth, with about 1.4 billion for every human. Together with arachnids and other invertebrate animals, they form the biological foundation for all agricultural production, and without them ecosystems would not be able to flourish. Insects are responsible for the pollination of about 80 percent of the trees and bushes on Earth and they play a crucial role in the dispersal of fruit and seeds from plants. Arachnids are equally essential to biodiversity. Spiders are the most important predatory invertebrates living in rural and urban areas, thereby affecting human cultures. The predaceous habits of arachnids are vital to humans, as they can serve as an environmentally-friendly form of pest control by keeping the populations of flies, mosquitoes, and aphids in check. Mites are another group of arachnids, and they live in a variety of ecosystems and play an important role as they decompose animal and plant matter, thereby enabling the release of nutrients that are later utilized for growing plants. Some species of mites also serve as natural pest controllers, in that they manage types of fungus gnats and soil pests.

1 M.D. Trautwein et al., "Advances in Insect Phylogeny at the Dawn of the Postgenomic Era," *Annual Review of Entomology* 57 (2012). See also Brandon Bryn, "Insects Evolved with Earth's First Land Plants," American Association of the Advancement of Science, https://www.aaas.org/news/science-insects-evolved-earths-first-land-plants.

 | DOI:10.1163/9789004715448_002

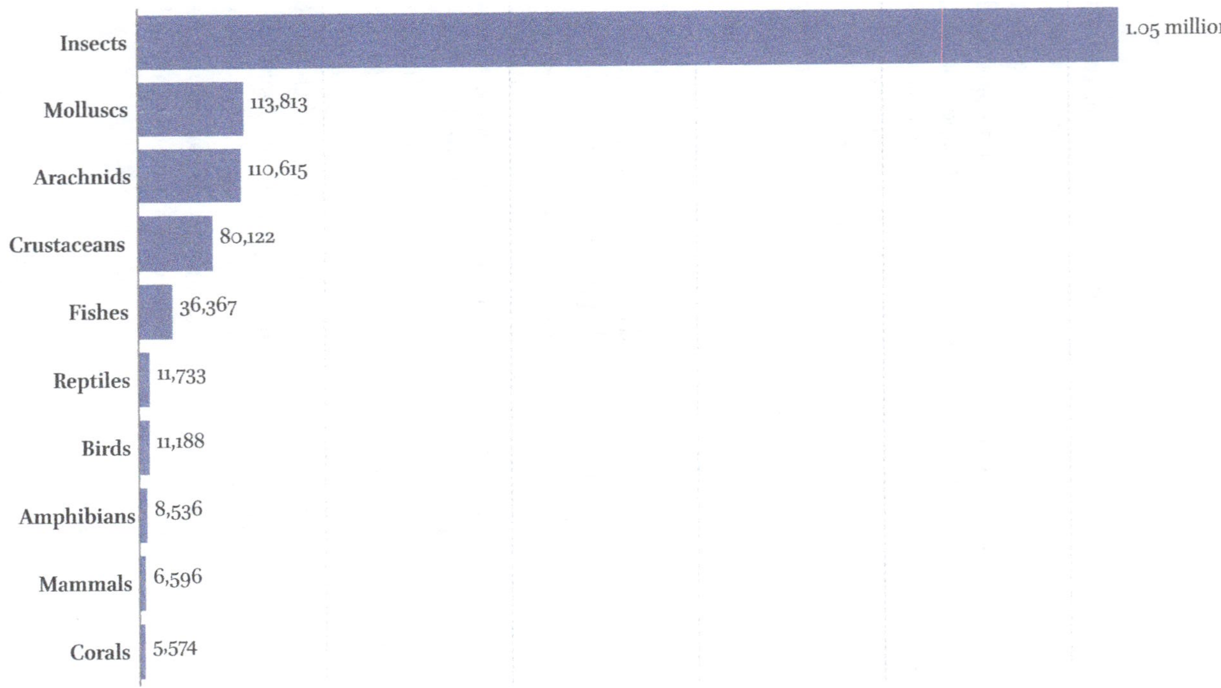

FIGURE 1.1 The vast majority of species on Earth are insects or arachnids. "Data Page: Number of described species", part of the following publication: Hannah Ritchie, Fiona Spooner and Max Roser (2022) – "Biodiversity"
DATA ADAPTED FROM INTERNATIONAL UNION FOR CONSERVATION OF NATURE (IUCN) RED LIST. RETRIEVED FROM HTTPS://OURWORLDINDATA.ORG/GRAPHER/NUMBER-OF-DESCRIBED-SPECIES [ONLINE RESOURCE]

While invertebrates have been sustaining ecosystems for millions of years, humans are latecomers. During our short spell on Earth, we have been dependent on insects and arachnids for food and sustenance. Yet, as our numbers have grown and human ways of life have evolved, we have invaded ecosystems to the point where insect populations are in global decline. The columnist George Monbiot, of *The Guardian*, coined the term "insectageddon" to highlight the implications of the global decline in insect populations.[2]

Until recently, insect biologists and entomologists, who focus on insects and their relationship with humans, have been almost the only group of researchers to be concerned about the ongoing collapse of insect populations. For other scholars, not to mention the general public, the decline in insect populations has been less visible. Often, the absence of bugs that sting, bite and contaminate our bodies, homes, and fields, and terrorize our companion and farm animals, has been a cause of relief rather than alarm. Those of us who are neither biologists nor entomologists do not always appreciate the vast variety of insects on the planet. Our attitude towards them is often characterized by a value judgment based on social, cultural, religious, economic and health reasons, through which we define insects and arachnids as either good or bad, useful or harmful, beautiful or disgusting.

These dichotomies have a long history. Some insects and arachnids have been regarded as sacred beings in certain cultures. The dung beetle was worshiped as the earthly manifestation of the sun, as well as the symbol of rebirth and resurrection in ancient Egypt. The scarab beetle rolls a ball of manure, which created a belief among the Egyptians that a god would reset the sun and roll it out into the sky every morning.[3] Spiders, for their part, have been associated with creation myths as their ability to weave webs can be seen as a metaphor for creating new worlds. Female spider figures feature in many cultures. A spider grandmother, for example, plays an important role in the Navajo creation myth, while the female spider figure of Jorogumo appears in Japanese folklore.[4] Spiders also carry meaning in today's popular culture, both

2 George Monbiot, "Insectageddon: The Scale and Speed of Environmental Collapse is Beyond Imagination," *The Guardian*, October 20, 2017; Jeroen P. van der Sluijs, "Insect Decline, an Emerging Global Environmental Risk," *Current Opinion in Environmental Sustainability* 46 (2020), doi:10.1016/j.cosust.2020.08.012.

3 Mohamed A. Kenawy and Yousrya Abdel-Hamid, "Insects in Ancient (Pharaonic) Egypt: A Review of Fauna, Their Mythological and Religious Significance and Associated Diseases," *Egyptian Academic Journal of Biological Sciences. A, Entomology* 8, no. 1 (2015): 22.

4 Abba A. Abba, "Transcending Cultural Boundary: Renegotiating the Significance of the Spider Folktale," *Journal of Literature, Languages and Linguistics* 27 (2016); Gene Kritsky and Ron H. Cherry, *Insect Mythology* (San Jose, CA: Writers Club Press, 2000); Natalie E. Duffus, Craig R. Christie and Juliano Morimoto, "Insect Cultural Services: How Insects Have

as villains, such as the spider-like monster Shelob in *The Lord of the Rings* and as heroic figures such as Spiderman.[5]

Pollinators, such as bees, have long been perceived as useful. Human interaction with them by robbing wild bees of their honey began in the prehistoric period, possibly even before the appearance of *Homo sapiens*. The oldest evidence of true beekeeping comes from ancient Egypt.[6] Insects and arachnids, as well as products made by and out of them, have played a role in the global economy as they have been traded as useful global commodities, the most famous of which is arguably silk, produced by the larvae of the mulberry silkworm (*Bombyx mori*). Natural historians, merchants, and agriculturalists have been involved in the acquisition and development of other beneficial insects in European countries, engaging in what could be termed as practical entomology.

Medicine was an important area of use in ancient and early modern times. Ants have been used in folk medicine and were recommended in some early modern pharmacopeias.[7] Spanish flies (*Lytta vesicatoria*) are beetles, which the Spanish exported to pharmacies throughout Europe and which thereafter figured in pharmacopeias up until the beginning of the twentieth century. The active substance cantharidin these beetles contain was believed to increase blood circulation and to alleviate toothache, muscle pain, fever, and rheumatism.[8] Crushed spiders and spider oil have been used as amulets in folk medicine. Insects containing carminic acid have been important ingredients in red dyes. Crushed Kermes lice were used until Europeans began importing cochineal from Central and South America, where the lice were farmed on plantations from the sixteenth century.[9] Another red dye imported to Europe from

Changed Our Lives and How Can We Do Better for Them," *Insects* 12, no. 5 (2021), doi:10.3390/insects12050377.

5 Emily A. Geest, Ashley R. Knoch and Andrine A. Shufran, "Villainous Snakes and Heroic Butterflies, the Moral Alignment of Animal-Themed Characters in American Superhero Comic Books," *Journal of Graphic Novels and Comics* 13, no. 5 (2022).

6 Gene Kritsky, "Beekeeping from Antiquity Through the Middle Ages," *Annual Review of Entomology* 62, no. 1 (2017), doi:10.1146/annurev-ento-031616-035115.

7 I. Svanberg and Å. Berggren, "Ant Schnapps for Health and Pleasure: The Use of Formica rufa L. (Hymenoptera: Formicidae) to Flavour Aquavit," *Journal of Ethnobiology and Ethnomedicine* 15, no. 1 (2019), doi:10.1186/s13002-019-0347-7.

8 Aleksander Karol Smakosz, "Bug as a Drug. Lytta vesicatoria L. Applications in Nineteenth Century Official Medicine," *Pharmacognosy Reviews* 16, no. 31 (2022).

9 Abraham Rowe, "Global Cochineal Production: Scale, Welfare Concerns, and Potential Interventions," (2020), doi:10.31219/osf.io/t57w2; Carlos Marichal, "Mexican Cochineal, Local Technologies and the Rise of Global Trade from the Sixteenth to the Nineteenth Centuries," in *Global History and New Polycentric Approaches Europe, Asia and the Americas in a World*

Southeast Asia was shellac, a resin that is formed in the bark of trees where the female lac bug (*Kerria lacca* or *Lacifer lacca*) secretes it. Shellack was used as a dye, a binding agent in pigments and for various types of polish.[10]

Humans and social insects have both developed divisions of labor, but the human farming of silkworms, bees and edible insects create a relation of insect subordination, where insects can be viewed as slaves, laborers, or domestic animals, depending on our perspective.[11] Farming insects for dye, fabric, medicine, and food has in recent years raised questions of insect welfare and ethical concern. Recent research shows that adult dipterans, cockroaches, bees, grasshoppers, butterflies, and some beetles feel pain. As for arachnids, the research on the subject is so far inconclusive.[12]

The raising and killing of insects in food production is less regulated than that of conventional livestock. Insect farmers and consumers of edible insects now face questions about what insect welfare means, and how it might differ from vertebrate welfare. A further question concerns the treatment of insects in entomology and using insects in experiments, which often means causing them pain or killing them. Generally speaking, the discussion about human ethical obligations towards insects is nothing new, for some religions, such as Jainism or Buddhism, have for centuries forbidden harming or killing insects.

Ethical thinking extended to insects is complicated by the fact that many of them are harmful from a human perspective. Various pests, lice, and bedbugs have annoyed and troubled people since antiquity and biblical times. Locusts, grasshoppers, and crickets, for example, have been perceived as threats to agricultural productivity. Some insects and arachnids are harmful as they pose a threat to human and animal health, although the awareness that insects can spread diseases only grew with the development of bacteriology. For example, it was not until the end of the nineteenth century that the transmission of the causative bacterium of plague from rodents by flea bites was shown.[13]

Network System, ed. Manuel Perez Garcia and Lucio De Sousa (Singapore: Palgrave Macmillan, 2018).

10 Edward D. Melillo, "Global Entomologies: Insects, Empires, and the 'Synthetic Age' in World History," *Past & Present* 223, no. 1 (2014).

11 Jennifer Bonnell, "Occupational Hazards: Honeybee Labour as an Interpretive Device in Animal History" in *Traces of the Animal Past: Methodological Challenges in Animal History*, ed. Jennifer Bonnell and Sean Kheraj (Calgary: University of Calgary Press, 2022).

12 S. Kralj-Fišer and M. Gregorič, "Spider Welfare," in *The Welfare of Invertebrate Animals*, ed. Claudio Carere and Jennifer Mather (Cham: Springer, 2019).

13 T. Butler, "Plague History: Yersin's Discovery of the Causative Bacterium in 1894 Enabled, in the Subsequent Century, Scientific Progress in Understanding the Disease and the Development of Treatments and Vaccines," *Clinical Microbiology and Infection* 20, no. 3 (2014): 202.

Today, it is known that bloodsucking insects and arachnids, such as mosquitoes, ticks, and fleas, act as vectors transmitting infectious pathogens, such as parasites, bacteria, or viruses, between humans, or from animals to humans. Vector-borne diseases, the most well-known of which include malaria, dengue, chikungunya fever, yellow fever, the Zika virus, West Nile fever, Japanese encephalitis, tick-borne encephalitis and Lyme borreliosis, are a real threat for millions of people throughout the world. Vector-borne diseases account for more than 17% of all infectious diseases and cause more than 700,000 deaths annually. The burden of them is highest in tropical and subtropical areas, and they tend to affect the poorest populations.[14]

Climate change has increased the geographic spread of vector-borne diseases since vectors, such as mosquitoes and ticks, are climate-sensitive and benefit from a warmer climate and changing rainfall patterns. Ticks (*Ixodes*), which transmit a wide variety of pathogens not only in the tropics but also in temperate North America, Europe, and Asia keep proliferating and moving into northern areas that were once too cold for them. As a result, the problem of tick-borne diseases has extended to new areas such as Canada and northern Europe, but ticks have also spread to higher altitudes in Central Europe and in the Greater Alpine Region.[15]

Another example of a species that has spread as a result of anthropogenic environmental change is the spruce bark beetle (*Ips typographus*), which is causing considerable damage to European and North American forests because it benefits from higher temperatures and storm damage. It also thrives in monocultural spruce forests favored by the wood processing industry. Hitherto, human attitudes toward insects and arachnids that are considered to be harmful to humans and their economical endeavors have been marked by hostility. The perceived necessity to annihilate insects has even been expressed in terms of warfare.

Pesticides have been used for centuries to fight off harmful insects. The increase in commerce and transport, together with the expansion of monoculture, contributed to the global spread of insect pests. In turn, this led to the use of pesticides, such as arsenic compounds and nicotine, and, after the Second World War, synthetic pesticides. Their use intensified after the second half of

14 World Health Organization, "Vector-borne diseases," https://www.who.int/news-room/fact-sheets/detail/vector-borne-diseases.

15 Shlomit Paz, "Climate Change Impacts on Vector-Borne Diseases in Europe: Risks, Predictions and Actions," *The Lancet Regional Health – Europe* 1 (2021): 1, doi:10.1016/j.lanepe.2020.100017; Patricia A. Nuttall, "Climate Change Impacts on Ticks and Tick-Borne Infections," *Biologia* 77 (2022): 1503, doi:10.1007/s11756-021-00927-2.

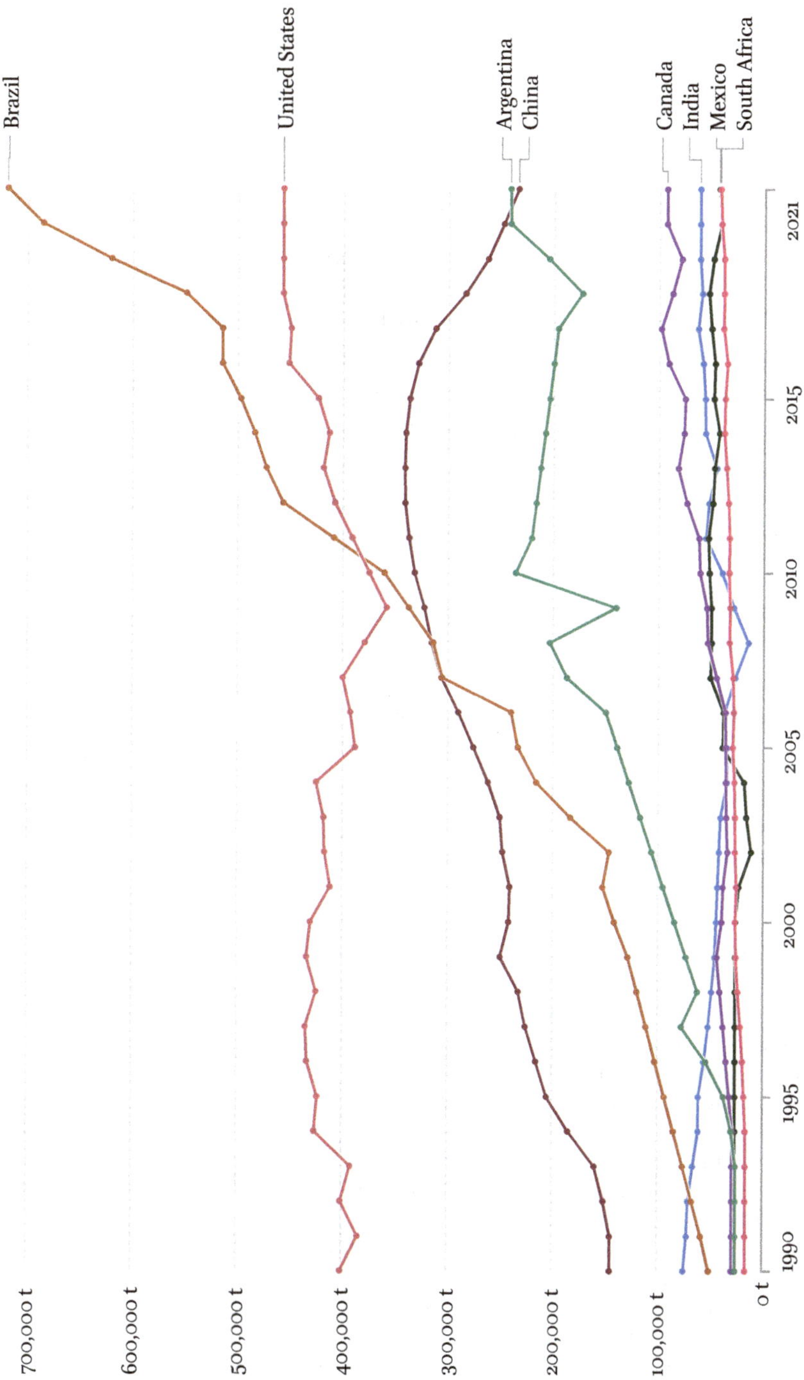

FIGURE 1.2 Graph showing pesticide use in Europe, Asia, Africa, Australia and North and South America from 1990 to 2020. "Data Page: Total pesticide use", part of the following publication: Hannah Ritchie, Pablo Rosado and Max Roser (2023) – "Agricultural Production" DATA ADAPTED FROM FOOD AND AGRICULTURE ORGANIZATION OF THE UNITED NATIONS. RETRIEVED FROM HTTPS://OURWORLDINDATA.ORG/GRAPHER/PESTICIDE-USE-TONNES [ONLINE RESOURCE]

the nineteenth century, reaching new heights in the 1990s (see Figure 1.2).[16] According to José Ramón Bertomeu-Sánchez, factors contributing to the global circulation of DDT in the 1950s and 1960s included intensive monoculture farming, the involvement of chemical industry, and experts who advocated its use in public health and pest control, but also the geopolitical strategies of the Cold War and transnational institutions.[17] The consequences of DDT are now well-known: in 1948 D.E. Howell discovered DDT in human fat, and soon biologists found that fish feeding on insects had also been killed.[18] In 1962, Rachel Carson brought attention to the lethal impact of DDT on birds – and possibly on humans. These interrelated developments began to demonstrate the workings of multispecies ecological networks and their fragility.

Insects and arachnids often represent an aspect of the natural world we wish to keep at a safe distance. Nevertheless, they invade our homes, both as unwanted vermin and as harmful guests buzzing around our windows. The idea that home is a human place excluding nature is fundamental to Western conceptualizations of domestic space. The Australian scholar Emma R. Power writes that pests appear as nonhuman species that disrupt people's experiences of home and their sense of security and purity.[19] Human border-making and narratives that depict human habitat as an inappropriate abode for pests are significant in these discussions.

Bugs not only occupy our homes, but they also invade our bodies. Besides various bacteria, worms and lice, intrusive creatures, such as Deer fly and ticks attach themselves to human bodies. Consequently, they cross, intersect and break the boundaries between humans and animals and humans and nature. An alternate view of pests emerges through a stance whereby pests become

16 José Ramón Bertomeu-Sánchez, "Introduction. Pesticides: Past and Present," *HoST-Journal of History of Science and Technology* 13, no. 1 (2019). For general overviews on the history of pesticides, see Frederick Rowe Davis, *Banned: A History of Pesticides and the Science of Toxicology* (New Haven: Yale University Press, 2014); Graham A. Matthews, *A History of Pesticides* (Boston, MA: Cabi, 2018).

17 Bertomeu-Sánchez, "Pesticides: Past and Present". See also Edmund Russell, *War and Nature: Fighting Humans and Insects with Chemicals from World War I to Silent Spring* (Cambridge: Cambridge University Press, 2001).

18 Frank N. Egerton, "Understanding Food Chains and Food Webs, 1700–1970," *Bulletin of the Ecological Society of America* 88, no. 1 (2007).

19 Emma R. Power, "Nature in the Home," in *International Encyclopedia of Housing and Home*, ed. Susan J. Smith (San Diego: Elsevier Science & Technology, 2012); Emma R. Power, "Pests and Home-Making: Depictions of Pests in Homemaker Magazines," *Home Cultures* 4, no. 3 (2007); Emma R. Power, "Kill'em dead!: The Ordinary Practices of Pest Control in the Home," *Antennae: The Journal of Nature in Visual Culture* 23 (2012).

part of human space and human experiences of space. As researchers Riikka Hohti and Maggie MacLure have observed, insects, like viruses, "oblige us to rethink the boundaries between inside and outside, proximity and distance, the one and the many, life and death." They argue that thinking *with* insects can reorient thought and action, open up theoretical and practical directions, and inform the development of methodologies.[20]

Eventually, how people interpret the insect worlds is influenced by the articulation of human difference due to categories such as gender, social and economic status, ethnic identity or culture. In Western culture, for example, an interest in bugs has traditionally been seen as a masculine trait, while femininity has been emphasized by a dislike of bugs.[21] The history of colonialism, meanwhile, provides an example of how political-economic interests influenced the attitudes and actions of colonial officials towards insects and how it highlighted racial inequalities. Especially in Sub-Saharan Africa, the tremendous malaria mortality figures of white settlers were the largest obstacle to colonization. Hence, the study of malaria and the development of tropical medicine in the late nineteenth century were intended to protect colonial interests, rather than to help indigenous people or to defeat malaria more broadly.[22]

Sometimes images of bugs as disgusting, pestilent or invasive intruders have been projected onto particular groups of people, such as migrants and minorities, to enhance differences and to justify inequality, discrimination or even persecution. As Miriam Ticktin has noted, the imagery about invasive others takes many forms.[23] Already in 1964, Alex Bein studied the history of the denunciation of Jews as parasites, increasingly understood not merely as metaphor but as a biological term.[24] Similarly, historian Clapperton Mavhunga has shown how freedom fighters were likened to subhuman "vermin beings" to be exterminated during the Zimbabwean War of Liberation.[25] Samuel Dolbee

20 Riikka Hohti and Maggie MacLure, "Insect-Thinking as Resistance to Education's Human Exceptionalism: Relationality and Cuts in More-Than-Human Childhoods," *Qualitative Inquiry* 28, no. 3–4 (2022).

21 See for example Michaela Fenske's chapter in this volume.

22 Jesse B. Bump and Ifeyinwa Aniebo, "Colonialism, Malaria, and the Decolonization of Global Health," *PLOS Global Public Health* 2, no. 9 (2022), doi:10.1371/journal.pgph.0000936.

23 Miriam Ticktin, "Invasive Others: Toward a Contaminated World," *Social Research: An International Quarterly, Special Issue on the Invasive Other* 84, no. 1 (2017).

24 Alex Bein, "The Jewish Parasite: Notes on the Semantics of the Jewish Problem, with Special Reference to Germany," *The Leo Baeck Institute Year Book* 9, no. 1 (1964), doi:10.1093/leobaeck/9.1.3.

25 Clapperton Chakanetsa Mavhunga, "Vermin Beings: On Pestiferous Animals and Human Game," *Social Text* 29, no. 1 (2011). See also Russell, *War and Nature*.

presents a striking example of how locusts shaped people's imagination and impacted the lives of deportees in the Ottoman region during the Armenian genocide. Along the road of deportation Armenians were not only symbolically linked with swarms of locusts, but the deportations also made the locust invasions worse, as fallow and unplowed lands formed excellent breeding grounds for these bugs. Because the locusts had consumed everything on their way, the deportees had to eat locusts in turn, which also served as a means of survival.[26]

1 Thinking about Insects and Arachnids in the Humanities and Social Sciences

Traditionally, insects and arachnids, as well as the diseases they spread have been studied in the context of natural sciences. Although bugs of all sorts have preceded, co-evolved and lived with humans in a variety of ways, and affected societal and historical developments as co-agents, human relations to them have remained rather scantily investigated from the perspective of the humanities and social sciences – except perhaps when insects have been examined as the vectors of devastating diseases. As the burden of arthropod-borne diseases has traditionally been the heaviest in the tropics, historians of Africa have since the 1970s examined human responses to disease-carrying insects and arachnids, such as tsetse flies and ticks.[27] Human attitudes to and actions against mosquitoes as the deadliest human killers in history have also been the subject of much research, particularly in the context of colonial history.[28]

26 Samuel Dolbee, *Locusts of Power: Borders, Empire, and Environment in the Modern Middle East* (Cambridge: Cambridge University Press, 2023). See also Jeannie N. Shinozuka, *Biotic Borders: Transpacific Plant and Insect Migration and the Rise of Anti-Asian Racism in America, 1890–1950* (Chicago: University of Chicago Press, 2022), which builds connections between transpacific plant and insect migration and the rise of anti-Asian racism in America during the early decades of the 20th century.

27 See for example Helge Kjekshus, *Ecology Control and Economic Development in East African History: The Case of Tanganyika 1850–1950* (Berkeley: University of California Press, 2022); James L. Giblin, "East Coast Fever in Socio-Historical Context: A Case Study from Tanzania," *The International Journal of African Historical Studies* 23, no. 3 (1990), doi:10.2307/219597; Paul F. Cranefield, *Science and Empire: East Coast Fever in Rhodesia and the Transvaal* (Cambridge: Cambridge University Press, 2002).

28 John R. McNeill, *Mosquito Empires: Ecology and War in the Greater Caribbean, 1620–1914* (Cambridge: Cambridge University Press, 2010); Rohan Deb Roy, *Malarial Subjects: Empire, Medicine and Nonhumans in British India, 1820–1909* (Cambridge: Cambridge University Press, 2017); Timothy C. Winegard, *The Mosquitoes: A Human History of Our Deadliest*

In the global North, where the role of disease vectors and pests has been less pronounced than in tropical regions, the interest in studying insects and other bugs from the perspective of humanities seems to be more recent. In addition to the history of medicine,[29] bugs have attracted attention in the history of science and ideas when studying for instance the history of entomology and knowledge production in natural history.[30] Climate change and the related proliferation of ticks in the Western countries have generated social debate on tick-borne diseases, which has lately drawn also the attention of sociologists and historians in Western Europe and North America.[31] As for northern Europe, by contrast, the ways in which the proliferation of ticks carrying diseases has affected human behavior and societies is still a rather unexplored subject.

Due to scientific research and ecological thinking, the value of insects and arachnids as useful parts of the ecosystem has been increasingly recognized in recent decades. There is a growing awareness of their role in the survival of the living world and human culture.[32] The global decline of insect populations and the spreading of certain disease vectors to new areas reflect how bugs are

---

Predator (New York, NY: Dutton, 2019). On empires and insects, see, Pierre-Etienne Stockland, "Statecraft and Insect Oeconomies in the Global French Enlightenment (1670–1815)" (Ph.D. diss., Columbia University, 2018).

29 Plague is perhaps one of the most studied zoonotic disease in the Western history of medicine, see for example Ole Jørgen Benedictow, *The Complete History of the Black Death* (Woodbridge, Suffolk: The Boydell Press, 2021).

30 One of the earliest historical researches on human reactions to ectoparasites was James Ronald Busvine's *Insects, Hygiene and History* (London: Athlone Press, 1976). For more recent research on "vermin" as creatures and category in the early modern literary, scientific, and political imagination, see Lucinda Cole, *Imperfect Creatures: Vermin, Literature, and the Sciences of Life, 1600–1740* (Ann Arbor: University of Michigan Press, 2016). For research on the scientific production of knowledge about insects, see for example Brian W. Ogilvie, "Order of Insects: Insect Species and Metamorphosis between Renaissance and Enlightenment," in *The Life Sciences in Early Modern Philosophy*, ed. Ohad Nachtomy and Justin E.H. Smith (Oxford: Oxford University Press, 2014); Mary Terrall, "Following Insects Around: Tools and Techniques of Eighteenth-Century Natural History," *The British Journal for the History of Science* 43, no. 4 (2010); Mary Terrall, *Catching Nature in the Act: Réaumur and the Practice of Natural History in the Eighteenth Century* (Chicago: University of Chicago Press, 2014).

31 See for example Robert A. Aronowitz, "Lyme Disease: The Social Construction of a New Disease and Its Social Consequences," *The Milbank Quarterly* 69, no. 1 (1991); P. Peretti-Watel, J. Ward, R. Lutaud and V. Seror, "Lyme Disease: Insight from Social Sciences," *Médecine et Maladies Infectieuses* 49 (2019).

32 Timothy D. Schowalter, *Insects and Society* (Boca Raton, FL: CRC Press, 2019); Uli Beisel, Ann H. Kelly and Noémi Tousignant, "Knowing Insects: Hosts, Vectors and Companions of Science," *Science as Culture* 22, no. 1 (2013), doi:10.1080/09505431.2013.776367.

affected by and entangled with the most threatening and debated global environmental changes, often referred to with the catchphrase the Anthropocene. These include climate change as well as the complex issues of biodiversity crisis and human-animal relations. These are problems that science alone cannot solve, because they are dependent on human behavior directed by social and cultural values and conventions, political and religious ideas as well as how people see themselves in relationship to the world around them.[33]

Consequently, investigating interactions between humans and insects/arachnids is essential within the humanities and social sciences. A fitting perspective is provided by the environmental humanities, which approaches ecological crises by taking seriously cultural differences and divergent histories, values and ethical frameworks, as well as questions of socioeconomic inequality,[34] but also by the human-animal studies investigating the relations and interactions between humans, nonhuman animals, society and culture. In the West, there is a burgeoning interest in studying insects from the perspectives offered by these research trends,[35] but as researchers within the area of human-animal studies have noted, the field has been characterized by a certain bias toward vertebrates, in particular mammals.[36]

Humans tend to be more likely to feel an affinity or kinship with other animals that are perceived as similar, whereas insects are perceived as different and difficult to understand or sympathize with. Nevertheless, we do share several characteristics with insects. For example, 60 percent of the DNA code of fruit flies and humans is identical,[37] most of the enzymes made by insects

33 Sverker Sörlin, "Environmental Humanities: Why Should Biologists Interested in the Environment Take the Humanities Seriously?," *BioScience* 62, no. 9 (2012), doi:10.1525/bio.2012.62.9.2.

34 Ursula K. Heise, "Introduction: Planet, Species, Justice – and the Stories We Tell About Them," in *The Routledge Companion to the Environmental Humanities,* 1st ed., ed. Ursula Heise, Jon Christensen and Michelle Niemann (London: Routledge, 2017), 2.

35 For example how poison and honey draw bee and beekeeper together in uneven gift relations, and how beekeepers make their bodies and their selves vulnerable to bees have been examined by Kelsey Green and Franklin Ginn, "The Smell of Selfless Love: Sharing Vulnerability with Bees in Alternative Apiculture," *Environmental Humanities* 4 (2014). Brian Morris's book *Insects and Human Life* (London: Routledge, 2006) studies how differently people respond to, make use of, and relate to insects in non-Western cultures. In *A Philosophy of the Insect* (New York: Columbia University Press, 2019) Jean-Marc Drouin has explored the question of what insects are as well as their scientific, aesthetic, ethical, and historical relationship with humanity.

36 Lisa Jean Moore and Rhoda M. Wilkie, "Introduction to *The Silent Majority: Invertebrates in Human-Animal Studies*," *Society and Animals* 27, no. 7 (2019), doi:10.1163/15685306-00001903.

37 University of Cambridge, "How Close Are You to a Fruit Fly."

and humans are very similar, both humans and insects possess brains, hearts, digestive tracts and reproductive organs, and human muscles and nerve cells work in a similar way to those of many insects. There are also similarities in behavior: many insects give gifts or sing to potential mates.[38]

Due to their unquestionable significance for humans and the global ecosystem, it is time to counteract the bias towards mammals and to bring insects into the humanities and social sciences. We urgently need deeper and more nuanced insights into insects and arachnids as parts of natureculture,[39] a concept illuminating the inseparability between nature and culture in ecological relationships, which are both biophysically and socially formed. This anthology aspires to provide this by examining the numerous ways in which human lives intersect with, influence, and are influenced by insects and arachnids as part of a wider network of species, but also of social and cultural interpretations and perceptions.

The overarching theme of the volume is the various ways insects and arachnids have acted as agents of change in diverse eco-social networks, consisting of humans, other classes of non-humans and inanimate nature. In this vein, the authors in this book investigate how the agency and actions of insects and arachnids have been perceived and interpreted, seen or remained unseen, from the early modern period to the present day. This includes how these perceptions and interpretations have been intertwined or collided with, as well as influenced by, ecological and social factors, such as climate change and the population dynamics of wildlife, financial and political interests, and different cultural conceptions.

The authors in this volume also examine various forms of knowledge, ranging from scientific studies to everyday awareness and experiences. Another point of focus is how these conceptions have been narrated and negotiated, and how they have affected our everyday practices and interactions with insects, arachnids, and their host animals, as well as the wider natural environment. Finally, the volume sheds light on questions concerning how to live together with insects and arachnids and how to nurture a new ethics that incorporates insects into the sphere of moral evaluation.

•••

38 Arizona State University, "Ask a Biologist."

39 Donna J. Haraway, *When Species Meet* (Minneapolis, MN: University of Minnesota Press, 2008).

The starting point for this publication has been a research project entitled *Humans and Ticks in the Anthropocene* (330762),[40] funded by the Research Council of Finland between 2020–2024. The research team of this project consists of historians, ethnologists and social scientists from the University of Eastern Finland, Åbo Akademi University and the University of Turku. The project aims to examine the multi-faceted relationship between humans and ticks, and seeks to provide a comprehensive analysis of the discussions about ticks in the Finnish society from the late nineteenth century to the present. Therefore, several chapters in this volume examine human relations to ticks in the Finnish context, starting from the observation that human attitudes towards ticks have been largely determined by the awareness that ticks can spread many harmful pathogens.

In Finland, it was realized already in the latter part of the 19th century that these previously harmless pests could actually be spreading diseases to domestic animals. However, it was the knowledge that ticks could also spread serious diseases to humans that changed people's attitudes towards them. Tick-borne diseases of public health importance throughout Europe including Finland, are borreliosis (Lyme disease)[41] and tick-borne encephalitis (TBE).[42]

40 "Humans and Ticks in the Anthropocene."

41 Borreliosis or Lyme disease, caused by the bacterium *Borrelia burgdorferi*, is the most common vector-borne disease in Europe and in North America. The infection can affect the skin, nervous system, joints, and heart. No vaccine against borreliosis is currently available, but most cases can be treated successfully with a few weeks of antibiotics. An estimated 65,000–85,000 Lyme borreliosis cases per year were reported in Europe in 2006, although the disease is suspected to be significantly underreported. Based on clinical symptoms, more than 5000 Lyme disease cases were reported in Finland in 2022. See Leah Burn et al., "Incidence of Lyme Borreliosis in Europe from National Surveillance Systems (2005–2020)," *Vector Borne Zoonotic Dis.* 23, no. 4 (2023), doi:10.1089/vbz.2022.0071; The Finnish Institute for Health and Welfare, "Borrelioosin seuranta ja esiintyvyys Suomessa", last modified December 15, 2023, https://thl.fi/fi/web/infektiotaudit-ja-rokotukset/taudit-ja-torjunta/taudit-ja-taudinaiheuttajat-a-o/borrelia/borrelioosin-seuranta-ja-esiintyvyys-suomessa.

42 TBE is a viral infectious disease that attacks the central nervous system and can even lead to death. It can, however, be prevented with a vaccine. The disease is widespread from central Europe through Siberia to Japan. It is caused by a flavivirus which includes three subtypes: European, Far-eastern, and Siberian. In 2020, there were over 3700 confirmed TBE cases reported in Europe. In 2022 the number of confirmed cases in Finland was 123. European Centre for Disease Prevention and Control, "Tick-Borne Encephalitis," in *ECDC, Annual epidemiological report for 2020* (Stockholm: ECDC, 2022); The Finnish Institute for Health and Welfare, "Puutiaisaivotulehduksen esiintyvyys Suomessa," last modified January15,2024,https://thl.fi/fi/web/infektiotaudit-ja-rokotukset/taudit-ja-torjunta/taudit-ja-taudinaiheuttajat-a-o/puutiaisaivotulehdus/puutiaisaivotulehduksen-esiintyvyys-suomessa.

TBE – initially known by the Finns as Kumlinge disease – was discovered already in the 1950s,[43] but it was the prevalence of Lyme disease which became apparent in the 1990s that marked a watershed period after which people's attitudes towards ticks began to be determined by fear and loathing.[44]

In order to broaden the picture of multispecies networks affecting human relations to bugs, the Finnish research team organized in March 2022 an international seminar "Humans, Ticks and Insects in Multispecies Networks" that resulted in this anthology. The chapters in this volume widen the perspective from ticks to other arachnids and insects such as spiders, violet carpenter bees, wasps, flies, ants, cockroaches, beetles, grasshoppers, mosquitoes, and rat-fleas, while the geographical scope expands from Finland to Western Europe, and from North America to India, as well as to Réunion and Mauritius.

In the chapters the humans interacting with insects and arachnids include natural historians and scientists, journalists, wildlife documentarists, public health officials, colonial administrators, physicians, and hunters, but also informants both in Réunion and Mauritius discussing their encounters with mosquitoes, and in Finland describing their experiences and feelings regarding ticks. They all represent categories of human difference, which shapes how humans interpret bug worlds.

Although the authors in this volume represent different scholarly traditions, such as history, ethnology, anthropology, media studies, sociology and environmental policy, the chapters are linked to each other through themes that intertwine and intersect throughout the volume. These are the ethics of human-arthropod relations; the formation, transmission, and circulation of knowledge about insects and arachnids; different ways of seeing, perceiving and representing insects and arachnids; ways of understanding the diseases caused by insects and arachnids and efforts to manage the health risks they pose; and eco-social networks in which insects and arachnids form a part.

In the opening chapter, Michaela Fenske observes that a different, more mindful attitude toward insect rights supports the development of new human practices and new aesthetics as well as new ethics. She proposes an "ethics of situatedness," which takes into account specific circumstances, and questions binary thinking in which the arguments are either in favor of or against all insects. Minna Santaoja notes in her chapter that insects tend to be placed differently on human moral scales: the dichotomous view of the benign bee, for example, and the aggressive wasp has a long history, based on anthropomorphic constructions. By deconstructing the story of wasps as aggressive and

43 See for example Suvi Rytty's chapter in this volume.
44 See for example Otto Latva's chapter in this volume.

exploring narratives for wasp-human conviviality she proposes rehearsing multispecies narratives of animacy in order to allow us to be better prepared in human-wasp encounters.

Laura Hollsten examines scholarly knowledge, traditional knowledge and practical knowledge about spiders in connection with the use of spiders in medicine, natural history, and silk production in eighteenth-century Europe. One of the objectives of her chapter is to bring spiders into the study, not simply as objects of knowledge, but as crucial parts of knowledge networks. In a chapter about the development and decline of economic entomology, Sophie FitzMaurice discusses the intricate relationship between scientific practice, agriculture, religion, and ethics in nineteenth- and early twentieth-century America, noting that the knowledge propagated by American entomologists drew from farmers' practical experience. In a comparison between economic entomology and economic ornithology, FitzMaurice shows that representatives of both disciplines sorted the natural world into the dichotomous categories of beneficial and injurious species.

The debates about gene drive mosquitoes and the ambition to eradicate mosquitoes from the point of view of more-than-human biopolitics are discussed by Marianne Mäkelin. The development of genetically modified, so-called gene drive mosquitoes marks a new chapter in human-mosquito relations, evoking much positive interest. Mäkelin explores these ethical and political debates about the development and conditions of the possible use of gene drive mosquitoes, thereby participating in discussions on how insects shape networks of global biosciences and environmental interventions.

Knowledge about insects and arachnids has been circulated not only in the context of natural history and science, but also in various media. Otto Latva examines the relationship between ticks and people in Finland between 1889 and 1990 by studying Finnish public discussion in newspapers and magazines. His findings show that the habit of perceiving ticks as a harmless nuisance persisted in Finnish culture for a surprisingly long time. Although red water fever, which kills cattle, and tick-borne encephalitis, which causes severe symptoms in humans, were known to be tick-borne diseases, they did not cause a significant change in the way ticks were discussed in the press. It was only with the discovery of the link between Lyme disease and ticks that these arachnids began to be portrayed almost exclusively as "frightening enemies" in the public debate.

If animals are perceived as "others," invertebrates have been considered as the ultimate others, despite their many similarities to humans. The otherness of insects in cinematic representations is discussed by Concepción Cortés Zulueta and Heidi Mikkola respectively. Cortés Zulueta shows how the cliché about insects as monstrous others was spread through representations in magic lantern slides, scientific microphotographs, and films, mixing various kinds of misconceptions about common houseflies. These appear intimately entangled with negative emotions and are often related to insect vision and to the alleged dangers they pose for humans.

The long tradition of portraying insects as monstrous others continued in horror films featuring giant or invasive insects. In her chapter, Mikkola notes how insects are associated with attributes like "monstrous" and "creepy," and how wildlife documentarists have managed to make insects appear either frightening or more relatable. One of her examples is a montage of a cockroach scurrying through pipes under a hotel building; a scene referencing film genres, such as horror thrillers of the 1990s and film noir. Both chapters exemplify the ways in which art and fiction can be used as powerful tools and mediators to imagine and relate to insects.

Although human attitudes towards insects and arachnids are greatly influenced by our understanding of the diseases they cause, these understandings are relatively young. Many of these diseases are old, even ancient, but the role of insects and arachnids as vectors has only been discovered much more recently. In her chapter, Taina Syrjämaa analyzes the slow process of uncovering the agency of ticks in spreading Babesia protozoans that caused Redwater fever in cattle from the 1860s to the 1930s. During this process, human perceptions of ticks changed in close connection with practices and daily actions in a shared and lived multispecies environment.

When it comes to tick-borne diseases, experts' risk assessments and the public's fears do not always coincide. For although diseases are essentially biological conditions, the public understanding about them is also influenced by social and cultural meanings attached to them. In her chapter, Suvi Rytty analyzes how public perceptions of TBE and Lyme borreliosis as scary diseases have been socially constructed through the image conveyed by newspapers since the 1950s. She shows how these perceptions have not only been influenced by scientific evidence, but also by changes in the disease spectrum, the health policy priorities of different eras, individual experiences and the way these diseases have been featured in the news at different times.

With the growing awareness of a health risk, tick encounters and imagined encounters with ticks have brought about protective and preventive measures. In her chapter, Sanna Lillbroända-Annala discusses tick-related practices and materiality, evoked by strong emotions, that cause people to act defensively so as to avoid tick bites and tick-borne diseases. These practices and materializations become control mechanisms for avoiding diseases and managing health risks.

Multispecifity is in some ways ingrained in all the chapters of this book, as the authors analyze how insects and arachnids have influenced the human world. Sometimes their effect takes a direct form, but often, however, insects and arachnids have an impact on the larger web of life and their actions determine how humans encounter and relate to other species. In her chapter, Emily Webster concentrates on animal experiments in the early twentieth-century Bombay (Mumbai) colonized by the British. These experiments were designed to examine the network of humans, rats, fleas and bacteria in terms of the transmission of the plague. She discusses how the experiments relied on multispecies relationships and ecologies within the city to develop and test the mechanisms of the rat-flea theory noting also how colonial sites have served as "living laboratories" for the western knowledge production.

In these constellations of the transmission of severe diseases, insects and arachnids are typically portrayed as unidirectional vectors. Karine Aasgard Jansen argues, however, that we need to understand disease-carrying insects, in her case mosquitoes, as co-participants. Thus, a multispecies approach that not only takes into account different species, but also objects and spaces, would show more respect to the research subjects. She also stresses that standardized global public health measures in the fight against arboviral diseases often fail if they do not take into consideration, how local cultural customs and practices impact on uses and management of mosquito contact zones.

Multispecies approaches are able not only to demonstrate that human attitudes toward insects and arachnids are prone to change across time and space, but also that changes in one part of the network lead to changes in the whole. In their chapter, Heta Lähdesmäki and Tuomas Räsänen demonstrate how the proliferation of ticks and tick-borne diseases in Finland has conflicted the relationship between humans and White-tailed deer, which is an important host species for ticks, and have also engendered tension between people with different views on the uses of nature.

Common to all the contributions in this anthology is the ambition to overcome dichotomies: between humans and animals, between arthropods and mammals and between those insects and arachnids that are considered benign and those that are perceived as harmful. These divisions derive from an anthropocentric perspective that has characterized – and continues to do so – relations between humans and arthropods. Still, violet carpenter bees, wasps, mosquitoes, flies, ticks, and spiders are living their lives in gardens, forests, swamps, and human homes, searching for food, making sounds, pollinating and decomposing plants, killing their prey and being killed by other species. These activities may have negative consequences for us: insects and arachnids may bite or sting humans, which may lead to an allergic reaction or a serious illness for the human, and, in some cases, to death for both the human and the insect.

However, today we have more biological knowledge about these animals than ever before. Therefore, we know that what attracts insects and arachnids to humans in the first place is the way we smell. An entomologist cited by Minna Santaoja describes wasps as a misunderstood species. We interpret wasps as aggressive, but in fact they approach humans not in order to attack, but because they are attracted by and interested in the human smell that they sense through their sensitive olfactory system. Therefore, by simply stepping back and holding our breath we may well cause the wasp to lose interest in us. Not so in the case of the tick, which is drawn to the smell of human skin. Once it burrows into human skin, it will not let go, so we simply have to learn to avoid human – tick encounters altogether.

In order to learn to live with insects and arachnids, we all have to become entomologists and arachnologists, at least to the extent that we learn what motivates the species we loathe, fear, and wish to avoid. Whether we like it or not, our lives are entangled with, and dependent on insects and arachnids, which all play their role in ecosystems. In accordance with Donna Haraway's idea of staying with the trouble, we need to learn to co-live with insects and arachnids with as much intelligence and compassion as we can while navigating our way among our fellow species.[45] In the thick of it, with ticks, mosquitoes and the rest of our multispecies world.

Bibliography

Abba, Abba A. "Transcending Cultural Boundary: Renegotiating the Significance of the Spider Folktale." *Journal of Literature, Languages and Linguistics* 27 (2016): 22–26.

Arizona State University, "Ask a Biologist." Accessed November 20, 2023. https://askabiologist.asu.edu/insect-and-human-biology.

Aronowitz, Robert A. "Lyme Disease: The Social Construction of a New Disease and Its Social Consequences." *The Milbank Quarterly* 69, no. 1 (1991): 79–112.

Bein, Alex. "The Jewish Parasite: Notes on the Semantics of the Jewish Problem, with special Reference to Germany." *The Leo Baeck Institute Year Book* 9, no. 1 (1964): 3–40. doi:10.1093/leobaeck/9.1.3.

45 Donna J. Haraway, *Staying with the Trouble: Making Kin in the Chthulucene* (Durham, NC: Duke University Press, 2016).

Beisel, Uli, Ann H. Kelly and Noémi Tousignant. "Knowing Insects: Hosts, Vectors and Companions of Science." *Science as Culture* 22, no. 1 (2013): 1–15. doi:10.1080/09505431.2013.776367.

Benedictow, Ole Jørgen. *The Complete History of the Black Death.* Woodbridge, Suffolk: The Boydell Press, 2021.

Bertomeu-Sánchez, José Ramón. "Introduction. Pesticides: Past and Present." *HoST-Journal of History of Science and Technology* 13, no. 1 (2019): 1–27.

Bonnell, Jennifer. "Occupational Hazards: Honeybee Labour as an Interpretive Device in Animal History." In *Traces of the Animal Past: Methodological Challenges in Animal History*, edited by Jennifer Bonnell and Sean Kheraj, 49–72. Calgary: University of Calgary Press, 2022.

Bryn, Brandon. "Insects Evolved with Earth's First Land Plants." American Association of the Advancement of Science. https://www.aaas.org/news/science-insects-evolved-earths-first-land-plants.

Bump, Jesse B., and Ifeyinwa Aniebo. "Colonialism, Malaria, and the Decolonization of Global Health." *PLOS Global Public Health* 2, no. 9 (2022): 1–12. doi:10.1371/journal.pgph.0000936.

Burn, Leah, Thao Mai Phuong Tran, Andreas Pilz, Andrew Vyse, Mark A. Fletcher, Frederick J. Angulo, Bradford D. Gessner, Jennifer C. Moïsi, Luis Jodar and James H. Stark. "Incidence of Lyme Borreliosis in Europe from National Surveillance Systems (2005–2020)." *Vector Borne Zoonotic Dis.* 23, no. 4 (2023): 156–171. doi:10.1089/vbz.2022.0071.

Busvine, James Ronald. *Insects, Hygiene and History*. London: Athlone Press, 1976.

Butler, T. "Plague History: Yersin's Discovery of the Causative Bacterium in 1894 Enabled, in the Subsequent Century, Scientific Progress in Understanding the Disease and the Development of Treatments and Vaccines." *Clinical Microbiology and Infection* 20, no. 3 (2014): 202–209. doi:10.1111/1469-0691.12540.

Cole, Lucinda. *Imperfect Creatures: Vermin, Literature, and the Sciences of Life, 1600–1740*. Ann Arbor: University of Michigan Press, 2016.

Cranefield, Paul F. *Science and Empire: East Coast Fever in Rhodesia and the Transvaal.* Cambridge: Cambridge University Press, 2002.

Davis, Frederick Rowe. *Banned: A History of Pesticides and the Science of Toxicology.* New Haven: Yale University Press, 2014.

Deb Roy, Rohan. *Malarial Subjects: Empire, Medicine and Nonhumans in British India, 1820–1909*. Cambridge: Cambridge University Press, 2017.

Dolbee, Samuel. *Locusts of Power: Borders, Empire, and Environment in the Modern Middle East.* Cambridge: Cambridge University Press, 2023.

Drouin, Jean-Marc. *A Philosophy of the Insect.* New York: Columbia University Press, 2019.

Duffus, Natalie E., Craig R. Christie and Juliano Morimoto. "Insect Cultural Services: How Insects Have Changed Our Lives and how Can We Do Better for Them." *Insects* 12, no. 5 (2021). doi:10.3390/insects12050377.

Egerton, Frank N. "Understanding Food Chains and Food Webs, 1700–1970." *Bulletin of the Ecological Society of America* 88, no. 1 (2007): 50–69. doi:10.1890/0012-9623(2007)88[50:UFCAFW]2.0.CO;2.

European Centre for Disease Prevention and Control. "Tick-Borne Encephalitis." In *ECDC. Annual epidemiological report for 2020*. Stockholm: ECDC, 2022.

The Finnish Institute for Health and Welfare. "Borrelioosin seuranta ja esiintyvyys Suomessa." Last modified December 15, 2023. https://thl.fi/fi/web/infektiotaudit-ja-rokotukset/taudit-ja-torjunta/taudit-ja-taudinaiheuttajat-a-o/borrelia/borrelioosin-seuranta-ja-esiintyvyys-suomessa.

The Finnish Institute for Health and Welfare. "Puutiaisaivotulehduksen esiintyvyys Suomessa." Last modified January 15, 2024. https://thl.fi/fi/web/infektiotaudit-ja-rokotukset/taudit-ja-torjunta/taudit-ja-taudinaiheuttajat-a-o/puutiaisaivotulehdus/puutiaisaivotulehduksen-esiintyvyys-suomessa.

Geest, Emily A., Ashley R. Knoch and Andrine A. Shufran. "Villainous Snakes and Heroic Butterflies, the Moral Alignment of Animal-Themed Characters in American Superhero Comic Books." *Journal of Graphic Novels and Comics* 13, no. 5 (2022): 735–750.

Giblin, James L. "East Coast Fever in Socio-Historical Context: A Case Study from Tanzania." *The International Journal of African Historical Studies* 23, no. 3 (1990): 401–21. doi:10.2307/219597.

Green, Kelsey and Franklin Ginn. "The Smell of Selfless Love: Sharing Vulnerability with Bees in Alternative Apiculture." *Environmental Humanities* 4 (2014): 149–170.

Haraway, Donna J. *Staying with the Trouble: Making Kin in the Chthulucene*. Durham, NC: Duke University Press, 2016.

Haraway, Donna J. *When Species Meet*. Minneapolis, MN: University of Minnesota Press, 2008.

Heise, Ursula K. "Introduction: Planet, Species, Justice – and the Stories We Tell About Them." In *The Routledge Companion to the Environmental Humanities*, 1st ed., edited by Ursula Heise, Jon Christensen and Michelle Niemann, 1–10. London: Routledge, 2017.

Hohti, Riikka, and Maggie MacLure. "Insect-Thinking as Resistance to Education's Human Exceptionalism: Relationality and Cuts in More-Than-Human Childhoods." *Qualitative Inquiry* 28, no. 3–4 (2022): 322–332.

Kenawy, Mohamed A. and Yousrya Abdel-Hamid. "Insects in Ancient (Pharaonic) Egypt: A Review of Fauna, Their Mythological and Religious Significance and Associated Diseases." *Egyptian Academic Journal of Biological Sciences. A, Entomology* 8, no. 1 (2015): 15–32.

Kjekshus, Helge. *Ecology Control and Economic Development in East African History: The Case of Tanganyika 1850–1950*. Berkeley: University of California Press, 2022.

Kralj-Fišer, S. and M. Gregorič, "Spider Welfare." In *The Welfare of Invertebrate Animals*, edited by Claudio Carere and Jennifer Mather, 105–122. Cham: Springer, 2019.

Kritsky, Gene. "Beekeeping from Antiquity Through the Middle Ages." *Annual Review of Entomology* 62, no. 1 (2017): 249–264. doi:10.1146/annurev-ento-031616-035115.

Kritsky, Gene and Ron H. Cherry. *Insect Mythology*. San Jose, CA: Writers Club Press, 2000.

Marichal, Carlos. "Mexican Cochineal, Local Technologies and the Rise of Global Trade from the Sixteenth to the Nineteenth Centuries." In *Global History and New Polycentric Approaches: Europe, Asia and the Americas in a World Network System*, edited by Manuel Perez Garcia and Lucio De Sousa, 255–273. Singapore: Palgrave Macmillan, 2018. doi:10.1007/978-981-10-4053-5_12.

Matthews, Graham A. *A History of Pesticides*. Boston, MA: Cabi, 2018.

Mavhunga, Clapperton Chakanetsa. "Vermin beings: On Pestiferous Animals and Human Game." *Social Text* 29, no. 1 (2011): 151–176.

McNeill, John R. *Mosquito Empires: Ecology and War in the Greater Caribbean, 1620–1914*. Cambridge: Cambridge University Press, 2010.

Melillo, Edward D. "Global Entomologies: Insects, Empires, and the 'Synthetic Age' in World History." *Past & Present* 223, no. 1 (2014): 233–270.

Monbiot, George. "Insectageddon: The Scale and Speed of Environmental Collapse is Beyond Imagination." *The Guardian*, October 20, 2017.

Moore, Lisa Jean and Rhoda M. Wilkie. "Introduction to *The Silent Majority: Invertebrates in Human-Animal Studies*." *Society & Animals* 27, no. 7 (2019): 653–655. doi:10.1163/15685306- 00001903.

Morris, Brian. *Insects and Human Life*. London: Routledge, 2006.

Nuttall, Patricia A. "Climate Change Impacts on Ticks and Tick-Borne Infections." *Biologia* 77 (2022): 1503–1512. doi:10.1007/s11756-021-00927-2.

Ogilvie, Brian W. "Order of Insects: Insect Species and Metamorphosis Between Renaissance and Enlightenment." In *The Life Sciences in Early Modern Philosophy*, edited by Ohad Nachtomy and Justin E.H. Smith, 222–245. Oxford: Oxford University Press, 2013. doi:10.1093/acprof:oso9780199987313.003.0012.

Paz, Shlomit. "Climate Change Impacts on Vector-Borne Diseases in Europe: Risks, Predictions and Actions." *The Lancet Regional Health – Europe* 1 (2021): 1. doi:10.1016/j.lanepe.2020.100017.

Peretti-Watel, P., J. Ward, R. Lutaud and V. Seror. "Lyme Disease: Insight from Social Sciences." *Médecine et Maladies Infectieuses* 49 (2019): 133–139. https://doi.org/10.1016/j.medmal.2018.12.005.

Power, Emma R. "Nature in the Home." In *International Encyclopedia of Housing and Home*, edited by Susan J. Smith, 6–11. San Diego, CA: Elsevier Science & Technology, 2012.

Power, Emma R. "Kill'em Dead!: The Ordinary Practices of Pest Control in the Home." *Antennae: The Journal of Nature in Visual Culture* 23 (2012): 136–148.

Power, Emma R. "Pests and Home-Making: Depictions of Pests in Homemaker Magazines." *Home Cultures* 4, no. 3 (2007): 213–236.

Rowe, Abraham. "Global Cochineal Production: Scale, Welfare Concerns, and Potential Interventions." (2020). doi:10.31219/osf.io/t57w2.

Russell, Edmund. *War and Nature: Fighting Humans and Insects with Chemicals from World War I to Silent Spring*. Cambridge: Cambridge University Press, 2001.

Schowalter, Timothy D. *Insects and Society*. Boca Raton, FL: CRC Press, 2019.

Shinozuka, Jeannie N. *Biotic Borders: Transpacific Plant and Insect Migration and the Rise of Anti-Asian Racism in America, 1890–1950*. Chicago: University of Chicago Press, 2022.

Sluijs, Jeroen P. van der. "Insect Decline, an Emerging Global Environmental Risk." *Current Opinion in Environmental Sustainability* 46 (2020): 39–42. doi:10.1016/j.cosust.2020.08.012.

Smakosz, Aleksander Karol. "Bug as a Drug. Lytta vesicatoria L. Applications in Nineteenth Century Official Medicine." *Pharmacognosy Reviews* 16, no. 31 (2022): 27–33.

Stockland, Pierre-Etienne. "Statecraft and Insect Oeconomies in the Global French Enlightenment (1670–1815)." Ph.D. diss., Columbia University, 2018.

Svanberg, I. and Å. Berggren. "Ant Schnapps for Health and Pleasure: The Use of Formica rufa L. (Hymenoptera: Formicidae) to Flavour Aquavit." *Journal of Ethnobiology and Ethnomedicine* 15, no. 1 (2019): 68. doi:10.1186/s13002-019-0347-7.

Sörlin, Sverker. "Environmental Humanities: Why Should Biologists Interested in the Environment Take the Humanities Seriously?" *BioScience* 62, no. 9 (2012): 788–789. doi:10.1525/bio.2012.62.9.2.

Terrall, Mary. "Following Insects Around: Tools and Techniques of Eighteenth-Century Natural History." *The British Journal for the History of Science* 43, no. 4 (2010): 573–588.

Terrall, Mary. *Catching Nature in the Act: Réaumur and the Practice of Natural History in the Eighteenth Century*. Chicago: University of Chicago Press, 2014.

Ticktin, Miriam. "Invasive Others: Toward a Contaminated World." *Social Research: An International Quarterly, Special Issue on the Invasive Other* 84, no.1 (2017): XXI–XXXIV.

Trautwein M.D., B.M. Wiegmann, R. Beutel, K.M. Kjer and D.K. Yeates. "Advances in Insect Phylogeny at the Dawn of the Postgenomic Era." *Annual Review of Entomology* 57 (2012): 449–468.

University of Cambridge. "How Close Are You to a Fruit Fly." Accessed November 20, 2023. https://www.cam.ac.uk/research/features/how-close-are-you-to-a-fruit-fly.

University of Turku. "Humans and Ticks in the Anthropocene." Accessed November 20, 2023. https://sites.utu.fi/huti/.

Winegard, Timothy C. *The Mosquitoes: A Human History of Our Deadliest Predator.* New York, NY: Dutton, 2019.

World Health Organization. "Vector-borne diseases." Accessed on 17.11.2023. https://www.who.int/news-room/fact-sheets/detail/vector-borne-diseases.

PART 1

Ethics

CHAPTER 1

Becoming Aware of Insects: Dangers and Endangerments in the Anthropocene

Michaela Fenske

1 Unidentified Flying Objects in the Anthropocene

Strange unidentified flying objects appeared in an orchard in Franconia, a region in the north of the federal state of Bavaria, Germany, in the spring of 2006. The gardeners in the idyllic countryside location were still panicking even after having successfully struggled against the UFOs. The scene, as described by entomologist Heinz Bußler, is horrible: twelve "aliens" were killed, but two "Ufos" were still alive.[1] The report of the event demonstrates that the first representatives of the violet carpenter bee (Xylocopa violacea) that arrived in Germany after migrating from the warm river valleys into other regions were not welcome. Gardeners who were unfamiliar with this predominantly Mediterranean wild bee felt threatened. The unidentified flying objects were conceived as potentially dangerous without any chance of integrating into the regional biospheres. Therefore, entomologist Bußler worried about the survival chances this insect might have in Germany.

The story of the violet carpenter is representative of the experience of many other insects in our time.[2] Due to climate change, violet carpenter bees, similar to other primarily Mediterranean insects, migrate.[3] Their migration might be

1 Heinz Bußler, "Mediterrane Holzbienen entdecken Bayern. Die hummelgroße Violette Holzbiene erschreckt so manchen Bürger," *LWF aktuell* 58 (2007).

2 The present contribution is based on explorative research in the form of informal talks with experts from different fields such as the practice of nature conservation, the biosciences, and horticultural advisory work, as well as the evaluation of media representations and participant observation as a passionate gardener; see Michaela Fenske, Fieldnotes in Franconia, 2020–2022. I particularly thank Prof. Dr. Andrea Holzschuh, Isolde Keil-Vierheilig and Dr. Klaus Mandery for their cooperation. Preliminary results of my research are published in Michaela Fenske, "Mit der Blauen Holzbiene denken: Neue Perspektiven auf die Verflechtungen von Menschen und Insekten," *Bayerisches Jahrbuch für Volkskunde* (2023).

3 For information about the violet carpenter bee, see Klaus Mandery, *Die Bienen und Wespen Frankens: Ein historischer Vergleich über neue Erhebungen und alte Sammlungen* (Nürnberg: Bund Naturschutz in Bayern, 2001); Paul Westrich, *Die Wildbienen Deutschlands*, 2nd edition (Stuttgart: Ulmer, 2nd ed., 2019), 207–211, 667–668.

because of unfavorable conditions in their original biospheres or the development of favorable conditions in other regions. At any rate, area shifts and desynchronized developments in space and time, like that of the flowering of plants, the maturing of dusting insects and other effects of today's polycrisis, effect insects worldwide.[4] Consequently, a lot of insects have enforced their mobility, and migrant insects arrive in cities, villages and landscapes inhabited by humans (and other living beings). An extreme mode of mobility (or rather immobility) is the insects' disappearance by dying: we are facing a worldwide mass extinction of insects,[5] an "insectageddon."[6] Wild or solitary bees, such as the violet carpenter bee, are particularly affected by this disappearance.[7]

In this respect, the Anthropocene,[8] the age when humans are drastically influencing the living conditions on planet earth,[9] is also an age of insects. This is probably one of the reasons why the societies of the Global North are

---

4 For a general survey, see Frauke Fischer and Hilke Oberhansberg, *Wal macht Wetter: Warum biologische Vielfalt unser Klima rettet* (München: oekom, 2023), 15–55; for insects, see Werner Gnatzy and Jürgen Tautz, *Insekten–Erfolgsmodelle der Evolution: Faszinierend und bedroht* (Berlin: Springer, 2023), 225–235.

5 For Germany, see Caspar A. Hallmann et al., "More than 75 Percent Decline over 27 Years in Total Flying Insect Biomass in Protected Areas," *PLoS One* 12, no. 10 (2017): e0185809, doi:10.1371/journal.pone.0185809; for agrarian landscapes, see Deutsche Akademie der Naturforscher Leopoldina et al., *Biodiversität und Management von Agrarlandschaften* (Halle: Köthen, 2020); Marcel Robischon, *Planet der Insekten: Von duftenden Ameisen, betrügerischen Leuchtkäfern und gespenstischen Faltern* (Bern, Stuttgart, Wien: Haupt, 2011), 173. From a global perspective e.g. Eric Himmel, *Extinct and Endangered: Insects in Peril* (New York: Abrams 2023).

6 George Monbiot, "Insectageddon: The Scale and Speed of Environmental Collapse is beyond Imagination," *The Guardian*, October 23, 2017.

7 For Germany, see Westrich, *Die Wildbienen Deutschlands*, 7–10.

8 Although the concept of the Anthropocene is adopted in the present chapter, the author is aware of both the critique directed at this concept (such as its blindness for place, time and responsibilities) and the suggestions for alternative concepts. The latter include Capitalocene, Plantationocene and Anglocene to mark a special set of ethics and practices of exploitation, colonialism, sexism and racism; see Jane Davis, Alex A. Moulton, Levi van Sant and Brian Williams, "Anthropocene, Capitalocene, … Plantationocene?: A Manifesto for Ecological Justice in an Age of Global Crises," *Geography Compass* 13, no. 5 (2019): e12438, doi:10.1111/gec3.12438; Novocene to emphasize the world forming power of hyperintelligence; see James Lovelock, *Novocene: The Coming Age of Hyperintelligence* (London: Penguin Books, 2019); Chthulucene, to underline both the becoming of humans with other beings and to emphasize the differences of this becoming in space and time; see Donna J. Haraway, *Staying with the Trouble: Making Kin in the Chthulucene* (Durham and London: Duke University Press, 2016).

9 For the concept of Anthropocene within the humanities, see Eva Horn and Hannes Bergthaller, *The Anthropocene: Key Issues for the Humanities* (London and New York: Routledge, 2020).

currently "rediscovering" insects (and, together with them, the arachnids, such as spiders and ticks, that are often erroneously regarded as insects in popular perception).[10] For a long time overseen and neglected, insects and other invertebrates are increasingly gaining attention in societies and sciences.[11] Today's societal debates in the Global North focus primarily on growing potential dangers posed by insects to humans and on the increasing endangerment of insects through mass extinction. But both, dangers from and endangerment of insects, are inseparably connected insofar as the endangerment of insects leads to dangers for humans, since the existence of insects is of fundamental relevance for human survival.[12]

Insects have long played a rather subordinate role in the humanities and social sciences as in societies at large. A prominent exception concerns the well-established research on honeybees.[13] The latter are a sort of "domestic pig among insects."[14] Humans and honeybees share a long period of becoming together, mainly for the benefit of humans. Today, numerous studies focus

10 In addition to the increased discussion of insects in the context of species and nature conservation, insects are also a prominent subject in art, design and everyday culture; see Thomas Ruhland, "Insekten und die Regeneration des Lebendigen in der Naturgeschichte des 18. Jahrhunderts," in *Vielfältig verflochten. Interdisziplinäre Beiträge zur Mensch-Tier-Relationalität*, ed. Forschungsschwerpunkt "Tier – Mensch – Gesellschaft" (Bielefeld: transcript, 2017), 103–104. This new love of insects sometimes takes on strange forms: one wonders, for example, why dogs should be excited about toys shaped as insects, such as those which have been offered by the Hunter company in branches of the Freßnapf retail chain in Germany since summer 2022. The range of toys of the brand "Florenz" includes ladybirds, honeybees, wild bees, a caterpillar and politically correct bags of flower seeds for the gardening dog; see Hunter, "Hundespielzeug Florenz", accessed August 19, 2023, https://www.wirliebenhunter.de/hunde/hundespielzeug/3234/hundespielzeug-florenz.

11 This attention often focuses on the fact that invertebrates are particularly threatened by extinction. Therefore, they also play a role in the context of extinction studies; see Thom van Dooren, *A World in a Shell: Snail Stories for a Time of Extinction* (Cambridge, MA, and London: MIT Press, 2022).

12 See Petro Cardoso, "Insects, Spiders, Snails and Empathy: Representing Invertebrate Extinctions in Natural History Museums," in *Animals, Plants and Afterimages: The Art and Science of Representing Extinction*, ed. Valérie Bienvenue and Nicholas Chare (New York and Oxford: Berghahn, 2022); Robischon, *Planet der Insekten*.

13 Summarized by Michaela Fenske, "Reduktion als Herausforderung: Kulturwissenschaftliche Annäherungen an Tiere in ländlichen Ökonomien," *Jahrbuch für die Geschichte des ländlichen Raumes* 13 (2016). Under the pressure of the ecological crises, even the swarming of honeybees, for a long time not appreciated by humans, is now seen as positive; see Michaela Fenske, "Narrating the Swarm: Changing Metanarratives in Times of Crisis," *Narrative Culture* 4, no. 2 (2017).

14 For this wording, I am indebted to my life science proofreader and colleague Prof. Jörg Müller; see Fenske, "Mit der Blauen Holzbiene denken" (2023).

primarily on the potential dangers insects (and arachnids such as ticks) constitute for humans. A prominent example of these dangers are diseases transmitted by the bite of the female mosquito, particularly malaria.[15]

In the present chapter, the violet carpenter bee serves as a guide and partner for my reflections.[16] Thinking with the wild bee opens up new perspectives: it allows one to break away from the traditional heritage of European modernity, from adopted attitudes, values, patterns of behavior and practices in order to develop more empathy towards insects and other invertebrates. Becoming aware of insects is the first step towards an "art of inclusion,"[17] or learning to live together, as an indispensable practice to stop mass extinction. We will follow this wild bee on its way through different regions of Europe, into our gardens, and we will search for it in our encyclopedias and children's books to understand the role of insects within Western societies. These observations and reflections are made within the theoretical frame of multispecies studies.[18] Based on the concept of "local distinctiveness,"[19] my arguments derive from the experiences of special regional multispecies assemblages or communities.[20]

---

15 See Uli Beisel and Carsten Wergin, "Understanding Multispecies Mobilities: From Mosquito Eradication to Coexistence," in *Mosquitopia: The Place of Pests in a Healthy World*, ed. Marcus Hall and Dan Tamïr (New York: Routledge, 2022); Clare I.R. Chandler and Uli Beisel, "The Anthropology of Malaria. Locating the Social," *Medical Anthropology* 36, no. 5 (2017); Mike Wolf and Kevin Hall, "Asian Tiger Mosquitos as Undesirable Cross-Border Commuters: Invasive Species and the Regulation of (Bio-)insecurities in Europe," *Journal for European Ethnology and Cultural Analysis* 5, no. 1 (2020).

16 In this manner, I follow the methodological consequence of multispecies studies that considers other living beings not as research objects but as research partners. The topic of "thinking with animals" has been repeatedly addressed in social anthropology in the wake of Claude Lévi-Strauss' *The End of Totemism*; see the German translation: Claude Lévi-Strauss, *Das Ende des Totemismus* (Frankfurt a.M: Suhrkamp, 1965), 116.

17 Anna Tsing, "Arts of Inclusion, or How to Love a Mushroom," *Mānoa* 22, no. 2 (2010).

18 In this perspective, insects are seen as co-creators of natureculture. This perspective relies mainly on Eben Kirksey and Stefan Helmreich, "The Emergence of Multispecies Ethnography," *Cultural Anthropology* 25, no. 4 (2010). It has become part of the Environmental Humanities, see *The Routledge Companion to the Environmental Humanities*, ed. Ursula K. Heise, Jon Christensen and Michelle Niemann (London and New York: Routledge, 2017).

19 See Owain Jones and Paul Cloke, *Tree Cultures: The Place of Trees and Trees in Their Place* (Oxford and New York: Berg, 2002), 9.

20 Heiko Paeth, climate expert of the University of Würzburg in Franconia, and his colleagues argue that the temperature in Lower Franconia is already higher than in other regions of Germany; see Daniel Schönbein, Luzia Keupp, Felix Pollinger and Heiko Paeth, *Klimabericht für Unterfranken: Die Region unter dem Einfluss der globalen Erwärmung* (Würzburg: Universität Würzburg, 2020); see also Ralf Wege, "Klimawandel mit dramatischen Folgen für Unterfranken," *BR24*, October 20, 2021, https://www.br.de/nachrichten/bayern/klimawandel-mit-dramatischen-folgen-fuer-unterfranken,SmMYrrN. The living assemblages in this region are already facing the challenge to manage hitherto unknown effects, such as increased aridity, disruptures of the vegetation cycle or species mobility.

FIGURE 2.1 "Affective Ecology": Violet Carpenter Bee and Clary Sage
© FENSKE 2021

The region chosen as an example is that of Franconia, in the north of the federal state of Bavaria, Germany. Concerning climate change, Lower Franconia is a model region for what we are likely to witness in the rest of Germany in ten or twenty years. The violet carpenter bee has become a relevant actor here.

As an initial step of learning to live together, we need to face the cultural heritage concerning insects in Europe where insects have often been conceived by humans as 'suspicious strangers', somehow expressing a special kind of xenophobia.

Although the region shares similar challenges with other regions in Germany, Lower Franconia has to deal with various effects of the socio-ecological crises sooner than other regions

2 Insects as 'Suspicious Strangers'

From the perspective of our cultural heritage, the situation in the Franconian orchard in the Summer of 2006 described initially is understandable. The gardeners were faced by previously unknown large, dark flying insects. These strangers were so strong that they were able to drill deep corridors into the wood of dead apple trees. The violet carpenter bee, with its length of 23 to 28 millimeters, is the largest solitary bee living in Europe. Its color, a mixture of blue and violet, has a metallic reflection in the sunlight, and its buzzing sounds loud and deep like an engine. The bee takes its English name from the fact that, similar to a carpenter, the female bee bores oblong holes into rotting wood. The bee lays eggs in separate chambers in these holes, supplying the future larvae with a store of pollen. In the perspective of the Franconian gardeners, these beings could not possibly be terrestrial, they seemed to originate directly from human nightmares.

Narrating is a pivotal practice not only of understanding, but also of shaping the world. Therefore, folklore and narrative culture[21] are essential constituents of cultural heritage. Narratives concerned with insects are fairly unilateral in Western cultures. Insects primarily incorporate and illustrate human nightmares and horror visions in stories, horror movies, comics, games and other media.[22] According to recent research in the field of social anthropology, insects are also seen as particularly "plastic" or "viscous" and thus as shape-shifting, since they are ultimately not completely controllable.[23] This feature adds to making them suitable agents in human nightmares.

Scholars of insects and other invertebrates lament the fact that these beings are faced with a strong human bias: either they are simply overseen and neglected or, with very few exceptions, they have an extraordinarily negative image.[24] As a window into our cultural heritage, one may check the entries of

---

21 For the relevant concept of narrative culture, see Ulrich Marzolph and Regina Bendix, "Introduction: Narrative Culture: A Concept and Its Scope," *Narrative Culture* 1, no. 1 (2014).

22 See Hugh Raffles, *Insectopedia* (New York: Vintage, 2011).

23 See Renisa Mawani, "Insect Wars. Bees, Bedbugs and Biopolitics," in *Routledge Handbook of Law and Theory*, ed. Andreas Philippopoulos-Mihalopoulos (London and New York: Routledge, 2019); Uli Beisel, "Unbequeme Mobilitäten. Stechmücken, Menschen und sich verschiebende Multispecies Encounters," in *Herausgefordertes Leben: Seuchen bei Menschen, Tieren, Pflanzen. Forschungskolloquium im Sommersemester 2022 am Lehrstuhl für Europäische Ethnologie der Universität Würzburg*, organized by Michaela Fenske and Pearl-Sue Carper, University of Würzburg, June 23, 2022.

24 See Cardoso, "Insects, Spiders, Snails," 234–235.

the German *Enzyklopädie des Märchens* (Encyclopedia of Folk and Fairy Tales), a concise dictionary of folk narrative research covering the past two centuries.[25] Only 11 of the more than 3,000 entries deal with insects. The insects discussed are ants, crickets, dung beetles, honeybees, flies, fleas, lice, locusts and mosquitos. Only ants and honeybees enjoy a positive image, as their extreme diligence, hard work and cleverness are praised. This positive image has a long tradition: honeybees and, to a lesser extent, ants have been regarded as model animals since antiquity. They are often employed to narrate virtues, such as the organization of good citizenship and prosperous economics.[26] Contrasting with this positive imagery, the other insects treated in the Encyclopedia are mainly categorized as vermin, representing the world's annoying, dangerous and menacing aspects. Only a few other than the listed insects have been popular in Western narrative culture since premodern times, such as bugs, butterflies, May beetles and wasps.[27] Except for butterflies (and here only the adult insects, not their intermediary stage of caterpillars), all of them are more or less narrated as unpleasant pests.

Both the knowledge about and the perception of insects in the societies of the Global North are extremely reduced. The researcher of narrative culture Rudolf Schenda was the first to address one of the essential problems of this reduced world view regarding animals in general and insects in particular: such a restricted world view limits the perception of the animals in question.[28] The persistent transmission of traditional stereotypes and prejudices up to the very present, thus, prevents an unbiased encounter between humans and animals, and especially between man and insects. In recent studies, cultural scientists point out that these "dominant discourses"[29] and the ideas manifest in them have disastrous consequences in times of mass species extinction. The worldwide focus on the honeybee makes other species, many of which are endangered, "invisible and mute."[30] The legacy of modernity, thus, incorporates a

25 *Enzyklopädie des Märchens: Handwörterbuch zur historischen und vergleichenden Erzählforschung*, vols. 1–15, ed. Kurt Ranke et.al. (Berlin and New York: De Gruyter, 1977–2015).

26 See Ulrike Kruse, "Von Bienen und Menschen," in *Tiere, Bilder, Ökonomien: Aktuelle Forschungsfragen der Human-Animal-Studies*, ed. Chimaira (Bielefeld: transcript, 2013); Niels Werber, *Ameisengesellschaft: Eine Faszinationsgeschichte* (Frankfurt a.M.: S. Fischer, 2013).

27 Rudolf Schenda, *Das ABC der Tiere: Märchen, Mythen und Geschichten* (München: Beck, 1995).

28 Schenda, *Das ABC der Tiere*.

29 Katharina Alsen and Eline D. Tabak, "Charismatische Insekten? Maja Lundes Bienes historie (2015) und die Narrative der Honigbiene auf der soziozoologischen Skala," *Tierstudien* 21 (2022).

30 Alsen and Tabak, "Charismatische Insekten?".

reduction in perspectives and possibilities of perception. The consequence is a reduction in thinking, as the depletion of species goes together with a curtailment in words.[31]

In the present situation that is ruled by the effects of limited awareness (and, as a consequence, also of a lack of knowledge) concerning insects, we are confronted with increasing numbers of unfamiliar insects, newcomers from all over the world. Other insect species, for example, that were previously not "native" have recently been sighted in central Europe, in addition to the violet carpenter bee. The amber wood cockroach, the European praying mantis, the Asian tiger mosquito or the box tree borer are among them. Not all these insects are as friendly and cooperative within their new communities as the violet carpenter bee. The Asian tiger mosquito, for example, from the perspective of health care is a potential vector of dangerous diseases. At the same time, insects perceived as "native," for example, ticks or bark beetles, profit from the effects of climate change. Traditional practices of diminishing or exterminating these (and other) insects, such as spraying insecticides, are becoming increasingly ineffective. They are particularly unsuccessful if executed only within national borders.[32] Although they are small and, at first sight, seemingly harmless, the agency of insects is powerful, and the effects of their practices include the killing of trees as well as humans.

In this particular moment of history, when the effects of insect mobility have become a challenge for humans, it is imperative to change our perspectives. Overcoming the binary orders of the modern world view in Europe is an essential step in this context. But before thinking about alternative perspectives and practices, let us ask ourselves: How did we actually arrive at this historic moment of mass extinction? One episode to understand the disappearance of awareness, species and words can be located early in the nineteenth century when children were introduced to the modern world view.

3 Enculturation into the Modern World View

One early mention of the violet carpenter bee in Franconia is contained in a German book of natural history for children published in 1830. The book bears the title "Entertaining Fables and Narratives of the Natural History of Fishes, Bugs, Insects, Plants and Minerals for Smaller Children Who Want to

31 Fenske, "Reduktion als Herausforderung".

32 Wolf and Hall, "Asian Tiger Mosquitos".

Prepare Themselves for This Science."[33] It mentions the violet carpenter bee together with other solitary bees. From a present perspective both features are remarkable: the treatment of solitary bees in an early nineteenth-century book for children as well as the explicit mention of the violet carpenter bee. At the beginning of the nineteenth century, the violet carpenter bee was still comparatively rare in Germany. The fact that the book's female author, Luise Hölder, then in her sixties, had lived in Franconia for some time might explain her awareness of the violet carpenter bee, because the bee had already been encountered, although rarely, in the warm river valleys in that region.[34] Scrolling through the book, one is fascinated to see how many insects the author discusses, not only widespread bugs, such as the rose chafer or the May beetle, but also nowadays comparatively rare species, such as the stag beetle.

Luise Hölder wrote her book at a time which current biologists locate as the peak of biodiversity in Europe, including that of insects.[35] Scholars from literary studies have documented a fair variety of formerly existing words, names as well as terms, for contemporary other-than-humans since the beginning of the 19th century in an interdisciplinary research project; the loss of species went together with a disappearance of these words.[36] As nature and culture are deeply interwoven, their research illustrates a dramatic loss of natureculture.

Insects exerted a considerable fascination in European societies at the peak of biodiversity in the early nineteenth century. The discipline of entomology was institutionalized, and hundreds of thousands of insects ended their lives as objects in showcases of museums and private collections. Luise Hölder's children's book was part of this widespread interest in insects. The book illustrates the state of popular knowledge about different insect species at the time and how children were introduced to this sphere of natureculture s.[37]

33 Luise Hölder, *Unterhaltende Fabeln und Erzählungen aus der Naturgeschichte der Fische, Käfer, Insekten, des Gewächs- und Mineralreichs für kleinere Kinder die sich zu dieser Wissenschaft vorbereiten wollen* (Leipzig: Carl Cnobloch, 1830), digitized collection of the Staatsbibliothek Preußischer Kulturbesitz zu Berlin.

34 See e.g. Mandery, *Die Bienen und Wespen*; Westrich, *Die Wildbienen Deutschlands*, 667–668.

35 See Vera Müller, "Direkte und indirekte Treiber. Landwirtschaft und Biodiversitätsverlust: [Im Gespräch mit Tobias Plieninger]," *Forschung & Lehre* 5 (2021).

36 Lars Langer et al., "The Rise and Fall of Biodiversity in Literature: A Comprehensive Quantification of Historical Changes in the Use of Vernacular Labels for Biological Taxa in Western Creative Literature," *People and Nature* 3, no. 5 (2021).

37 See the exhibition Internationale Jugendbibliothek, "Ich weiß etwas, was du nicht weißt: Weltwissen in Kinder- und Jugendsachbüchern aus fünf Jahrhunderten," May 25, 2022,

Children were also encultured into how to deal with insects, including disrespect towards and the license to kill some of them. The author introduces two boys visiting the countryside in the chapter "Honeyfeast." The boys meet a beekeeper and his honeybees, and also observe various solitary bees. One bee was, as the author says, "luckily killed" by the boys when the insect was "audaciously" tasting their sweets on the table.[38] The species killed was named exactly: it was a northern white-tailed bumblebee – a solitary bee that is currently highly protected in Germany.

As a matter of fact, the practice of killing insects often remains unquestioned to this day in the everyday lives of many people as well as in the sciences, where it is considered a prerequisite for the insects' determination and classification.[39] However, the killing of wild bees is now prohibited by German law and requires special permission even in the context of scientific research. Hölder's book, with its matter-of-fact narratives, supplies meaningful insights into the establishment of modern world views, orders and practices. Children were encultured into the world's matters of course. This feature makes books for children such an important source for cultural analysis. The unquestionable right to kill insects is not the only introduction into the binary world of the adults, as children are also introduced to a binary opposition of gender and other forms of separation. In this context, insects were generally declared "a boy's thing," a formula that adds weight to Hölder's female authorship. Girls and women were not supposed to be introduced to the insect world. Similar to the human world, the world of insects was categorized in a binary order: that of harmful and that of useful insects. This simplifying order of insects continues until today to be recommended for use in our everyday spaces, such as gardens.[40]

The modern world order's numerous binary oppositions, for example, nature/culture, female/male, harmful/useful, as well as native/invasive and common/rare, simplify complex realities. According to French philosopher Bruno Latour, such a purified and unambiguous world never corresponded to the people's historical everyday experiences.[41] However, the ideal of a binary world view proved to be extremely effective for a long time, and was, therefore, also particularly long-lasting. Today, in the Anthropocene, it becomes obvious

---

to April 25, 2023. Accessed August 20, 2023, https://www.ijb.de/ausstellungen/single/ich-weiss-etwas-was-du-nicht-weisst.

38 Hölder, *Unterhaltende Fabeln und Erzählungen*, 226.

39 Hans Malicky, *Vom Handwerk der Entomologie* (Berlin: Springer, 2019).

40 *Gartennützlinge, Gartenschädlinge*, 9th ed. (München: Obst- und Gartenbauverlag des Bayerischen Landesverbandes für Gartenbau und Landespflege, 2015).

41 Bruno Latour, *We Have Never Been Modern* (Cambridge, MA: Harvard University Press, 1993).

that any strictly binary perspective is disastrously narrow. As for insects, this perspective is largely responsible for both mass extinction and mass migration. The good news in the present context is that many migrant insects are not only harmless for most of the other living beings but also enrich their new multispecies assemblages with their special competences. Thus, there are chances that the ever-changing local multispecies communities may learn to live together.

4 Living within Emergent Ecologies

Eben Kirksey, Australian anthropologist and well-known scholar of multispecies studies, proposed the concept of "emergent ecologies" to understand and deal with developments currently experienced all over the contemporary world, including such phenomena as mass extinction and climate change.[42] Kirksey calls for a radically new perspective towards the effects of today's crises: instead of apocalyptic projections of the future or nostalgic restorations of the past, he argues for an appreciation of new opportunities. This also implies the effort to understand the new multispecies communities "that have been formed and transformed by chance."[43]

The multispecies assemblages with violet carpenter bees give an insight into the chances this perspective offers. The violet carpenter bee was endemic mainly in the Mediterranean region until the twenty-first century and was rarely encountered in the region of Franconia, where gardeners were shocked at their first encounter with the bee. Due to changing climatic conditions, the violet carpenter bee began to migrate and encountered adequate living conditions in various regions of Franconia. The bee has now become common in Franconia and further migrated to other regions in Germany and northern Europe, particularly since the turn of the millennium.[44] Biologists describe the violet carpenter bee as polylectic, implying that it is able to collect its harvest from a large variety of different flowers. From the perspective of humans, violet carpenter bees are hard and effective workers in gardens.[45] They are highly

42 Eben Kirksey, *Emergent Ecologies* (Durham and London: Duke University Press, 2015).

43 Kirksey, *Emergent Ecologies*, 1.

44 There are also reports from Poland; see Józef Banaszak, Weronika Banaszak-Cibicka and Lucyna Twerd, "Possible Expansion of the Range of Xylocopa violacea L. (Hymenoptera, Apiformes, Apidae) in Europe," *Turkish Journal of Zoology* 43 (2019).

45 Based on the concept of animal labor by Jocelyne Porcher and Jean Estebanez, "Animal Labor: At the Forefront of Innovative Research," in *Animal Labor: A New Perspective on Human-Animal Relations*, ed. Jocelyne Porcher and Jean Estebanez (Bielefeld: transcript, 2019).

efficient for the pollination of fruit and vegetables; and they not only pollinate a broad range of plants, but as wild bees, also perform this task more effectively than honeybees.[46]

Even so, the violet carpenter bee is particularly attached to certain plants, for example, the papilionacea, i.e. the plants of the pea or bean families. Gardeners know that the outstanding passion of the violet carpenter bee for clary sage, a thermophilic Mediterranean plant, is more than a side note.[47] This intimate plant-bee symbiosis demonstrates the garden-shaping power of the carpenter bee. The interpretation of this close relationship as an "affective ecology"[48] allows us to better understand the related interdependences as well as creativity as an essential constituent of the agency of species: due to the agency of clary sage, the plant will be found and pollinated effectively especially by violet carpenter bees. Together with the insects, new plants enrich or disturb local multispecies assemblages and vice versa. Not every newcomer is bound to enrich local multispecies communities. Particularly from a human perspective, the effects of ticks, tiger mosquitos, apple loopers, box tree borers or amber wood cockroaches are hard to integrate. But since they are not going to disappear, the multispecies communities and especially humans will have to learn to coexist with them. Therefore, it is useful to include this broad perspective in our research.

So, the question remains, how we are to differentiate between the various species, and how should we deal with those species with which we feel uncomfortable? How to live within emergent ecologies that are also inhabited by tiger mosquitos? The obvious answer to these questions is that we need a thorough environmental education for children, young adults and adults of any age in the gardens, forests, fields, moors and so on. Societies need a more detailed knowledge to cope with their everyday encounters with insects, particularly in

46 For this reason, in the context of food production, the services of specific varieties of solitary bees are at times preferred. In Israel, the carpenter bee is of interest; see Adi Sadeh, Avi Shmida and Tamar Kaesar, "The Carpenter Bee Xylocopa pubescens as an Agricultural Pollinator in Greenhouses," *Apidologie* 38 (2007); see also Andrea Holzschuh, Jan-Hendrik Dudenhöffer and Teja Tscharntke, "Landscapes with Wild Bee Habitats Enhance Pollination: Fruit Set and Yield of Sweet Cherry," *Biological Conservation* 153 (2012).

47 Natural scientist Wilhelm Schuster, for example, describes that the insect can smell the powerful scent of the clary sage over long distances, cf. Wilhelm Schuster, "Warum, wie und wann ist die stahlblauflügelige grosse Holzbiene (Xylocopa violacea) bei uns im Untermaintal eingewandert?," *Societas entomologica* 23, no. 12 (1908).

48 Carla Hustak, and Natasha Meyers, "Involutionary Momentum: Affective Ecologies and the Sciences of Plant/Insect Encounters," *differences: A Journal of Feminist Cultural Studies* 23, no. 2 (2012), doi:10.1215/10407391-1892907.

the Global North. And altogether, the solid preparation for these encounters involves new ethics, new practices and new aesthetics.

5 Looking for New Ethics, New Practices and New Aesthetics

On the way to the growing uncertainties, the twenty-first century witnesses intense discussions about social ethics. In addition, the traditional ethical attitude towards insects seems increasingly questionable: in the Anthropocene, killing those insects which we do not know or that only scare us may be just as delicate as welcoming all insects with open arms. Therefore, the debate in multispecies studies concerning ethical attitudes towards insects is controversial. Honeybee researchers plead for an "intra-species mindfulness" – a mindfulness that aims to facilitate the coexistence of species.[49] Reflecting on interspecies justice, philosophers generally integrate insects as deserving protection, with the exception of those that constitute an imminent danger to humans.[50] An approach from the field of geography is particularly well-suited to the complex situation in the current multispecies assemblages, as it suggests ethics to be differentiated according to relevant spaces and depending on diverging conditions. An "ethics of situatedness" takes into account specific circumstances, as it breaks through the binary thinking and action of being either in favor of or against all insects.[51]

A different, more mindful attitude towards insect rights supports the development of new human practices that, in their turn, involve new challenges. A departure from customary practices is needed precisely because insects in rural economies, as in forest and garden management, can, at times, cause existentially harmful damage for the farmers. Regional studies on forests under climate stress give insights into these new attitudes and practices.[52] Confronted with the explosive increase of bark beetles during the progressively frequent drought years, foresters in the south of Germany are starting new forest management practices: they are planting other species of trees and looking for cooperating partners in fauna and flora that will help to make the forest community more resilient. Incidentally, the new practices also include a new

49 Lisa Jean Moore and Mary Kosut, "Among the Colony: Ethnographic Fieldwork, Urban Bees and Intra-species Mindfulness," *Ethnography* 15 (2014).

50 See Anna Wienhues, "Zwischenartliche Gerechtigkeit und das Insektensterben: Wann handelt es sich beim Artentod um eine Ungerechtigkeit?," *Tierstudien* 20 (2021).

51 Jones and Cloke, *Tree Cultures*, 10–11.

52 Laura Hoss, *Fränkische Waldgemeinschaften im Wandel: Eine kulturanthropologische Annäherung* (Würzburg: Universität Würzburg, 2022).

economic modesty, such as refraining from overexploiting the land through annual harvests.

Other management practices concern gardening. This involves, for example, refraining from using pesticides or chemical fertilizers, practicing new mowing regimes, and leaving foliage in the garden as a natural ground cover and fertilizer. Those who garden with the violet carpenter bee recognize the need for sunlit areas and deadwood in the garden. Tidy gardens (as well as forests and other habitats) from which all deadwood and foliage has been removed are also counterproductive for other creatures. Once again, conventional practices are challenged by the violet carpenter bee, now also incorporating aesthetics, i.e. visual habits and ideals of beauty. Insect protection, similar to the protection of many other species, requires people to accept and implement new aesthetics. In this manner, the violet carpenter bee reminds us that the new

FIGURE 2.2 The aesthetics of the other, gardening with Violet Carpenter Bees
© FENSKE 2021

age needs new aesthetics, such as those communicated and practiced by the contemporary natural garden initiatives and the urban gardening movement.

What the violet carpenter bee shows, if we think with it, is the art of managing to live together with insects, whether they are newly arriving or familiar to us.

6 The Art of Living Together

The violet carpenter bee illustrates the diversity of the challenges that humans face in the Anthropocene, which may also be seen as the age of insects. The effects of humans as new geological forces on earth is putting huge pressure on insects and other invertebrates, which are, simultaneously, both forced to look for new habitats and invited to spread because of earth warming. In the face of both the danger posed by and the endangerment of insects, it seems particularly helpful to be open-minded and flexible in our attitudes, ethics and practices towards insects. In a dramatic reduction of natureculture, insects have been reduced in terms of species, numbers and words. The currently growing awareness of the close connections between humans and other living beings also enables a more unprejudiced encounter with insects and other living beings in our everyday spaces, such as gardens, parks or playgrounds, where insects and humans are bound to meet. Both the danger from and endangerment of insects require a flexible attitude and practice of coexistence on the part of their human fellow creatures that is appropriate to the situation and the place equally, specifically in local multispecies assemblages. New concepts of managing the world, such as "emergent ecologies," rely on the integrative ability of local multispecies assemblages as well as their responsiveness and flexibility.

The story of the violet carpenter bee in Germany can also be told as a story of positive moments concerning the living together of human and insect gardeners. Thanks to the effort of entomologist Heinz Bußler and the endeavors of many of his colleagues, conservationists, open-minded gardeners and others, the violet carpenter bee is increasingly well-known and integrated in Germany: the violet carpenter bee was awarded the status of "Garden Animal of the Year" in the summer of 2022 by a German institution engaging in nature conservation.[53] In some regions, the new member of the local multispecies

53 Heinz Sielmann Stiftung, *Gartentierwahl 2022*, accessed August 20, 2023, https://www.sielmann-stiftung.de/news/detail/blauschwarze-holzbiene-ist-gartentier-des-jahres-2022.

assemblages functions as a general symbol for insect protection.[54] And in German everyday culture, the violet carpenter bee is now encountered as a motif on greeting cards or an animal heroine in children's books.[55] It is to be hoped that gardeners in Franconia and elsewhere will learn that sustainable gardening does not mean subduing their gardens to human expectations, but rather that gardening always implies working with other living species, such as insects. This may or may not always be pleasant, but understanding and respecting the complexity of living with insects and their aesthetics is a promising and highly rewarding goal.

Bibliography

Alsen, Katharina, and Eline D. Tabak. "Charismatische Insekten? Maja Lundes Bienes historie (2015) und die Narrative der Honigbiene auf der soziozoologischen Skala." *Tierstudien* 21 (2022): 110–121.

Banaszak, Józef, Weronika Banaszak-Cibicka and Lucyna Twerd. "Possible Expansion of the Range of Xylocopa violacea L. (Hymenoptera, Apiformes, Apidae) in Europe." *Turkish Journal of Zoology* 43 (2019): 650–656.

Beisel, Uli. "Unbequeme Mobilitäten. Stechmücken, Menschen und sich verschiebende Multispecies Encounters." In *Herausgefordertes Leben: Seuchen bei Menschen, Tieren, Pflanzen. Forschungskolloquium im Sommersemester 2022 am Lehrstuhl für Europäische Ethnologie der Universität Würzburg*, organized by Michaela Fenske and Pearl-Sue Carper. University of Würzburg, June 23, 2022.

Beisel, Uli, and Carsten Wergin. "Understanding Multispecies Mobilities: From Mosquito Eradication to Coexistence." In *Mosquitopia: The Place of Pests in a Healthy World*, edited by Marcus Hall and Dan Tamïr, Ch. 3. New York: Routledge, 2022. doi:10.4324/9781003056034-3.

Bußler, Heinz. "Mediterrane Holzbienen entdecken Bayern. Die hummelgroße Violette Holzbiene erschreckt so manchen Bürger." *LWF aktuell* 58 (2007): 50–51.

Cardoso, Petro. "Insects, Spiders, Snails and Empathy: Representing Invertebrate Extinctions in Natural History Museums." In *Animals, Plants and Afterimages: The Art and Science of Representing Extinction*, edited by Valérie Bienvenue and Nicholas Chare, 233–236. New York and Oxford: Berghahn, 2022.

54 See the map with the number KLK 13397 from Grätz Verlag (illustrated by Daniela Drescher), accessed August 22, 2023, https://www.graetz-verlag.de/doppelkarte-bienen_1.

55 Anke Simon and Claudia Botz, *Mein Wildbienen-Buch: Die abenteuerliche Reise der kleinen Wildbiene Mia* (Augsburg: Wißner, 2020).

Chandler, Clare I.R., and Uli Beisel. "The Anthropology of Malaria. Locating the Social." *Medical Anthropology* 36, no. 5 (2017): 411–421.

Davis, Jane, Alex A. Moulton, Levi van Sant and Brian Williams. "Anthropocene, Capitalocene, ... Plantationocene?: A Manifesto for Ecological Justice in an Age of Global Crises." *Geography Compass* 13, no. 5 (2019): e12438. doi:10.1111/gec3.12438.

Deutsche Akademie der Naturforscher Leopoldina et al. *Biodiversität und Management von Agrarlandschaften*. Halle: Köthen, 2020.

van Dooren, Thom. *A World in a Shell: Snail Stories for a Time of Extinction*. Cambridge, MA, and London: MIT Press, 2022.

Enzyklopädie des Märchens: Handwörterbuch zur historischen und vergleichenden Erzählforschung vols. 1–15, edited by Kurt Ranke et al. Berlin and New York: De Gruyter, 1977–2015.

Fenske, Michaela. "Reduktion als Herausforderung: Kulturwissenschaftliche Annäherungen an Tiere in ländlichen Ökonomien." *Jahrbuch für die Geschichte des ländlichen Raumes* 13 (2016): 15–32.

Fenske, Michaela. "Narrating the Swarm: Changing Metanarratives in Times of Crisis." *Narrative Culture* 4, no. 2 (2017): 130–152.

Fenske, Michaela. *Fieldnotes in Franconia*. 2020–2022.

Fenske, Michaela. "Mit der Blauen Holzbiene denken. Neue Perspektiven auf die Verflechtungen von Menschen und Insekten." Bayerisches Jahrbuch für Volkskunde (2023): 25–42.

Fischer, Frauke, and Hilke Oberhansberg. *Wal macht Wetter: Warum biologische Vielfalt unser Klima rettet*. München: oekom, 2023.

Gartennützlinge, Gartenschädlinge, 9th edition. München: Obst- und Gartenbauverlag des Bayerischen Landesverbandes für Gartenbau und Landespflege, 2015.

Gnatzy, Werner, and Jürgen Tautz. *Insekten – Erfolgsmodelle der Evolution: Faszinierend und bedroht*. Berlin: Springer, 2023.

Grätz Verlag, illustr. Daniela Drescher. Accessed August 22, 2023. https://www.graetz-verlag.de/doppelkarte-bienen_1.

Hallmann, Caspar A., Martin Sorg, Eelke Jongejans, Henk Siepel, Nick Hofland, Heinz Schwan, Werner Stenmans et al. "More than 75 Percent Decline over 27 Years in Total Flying Insect Biomass in Protected Areas." *PLoS ONE* 12, no. 10 (2017), e0185809. doi:10.1371/journal.pone.0185809.

Haraway, Donna J. *Staying with the Trouble: Making Kin in the Chthulucene*. Durham and London: Duke University Press, 2016.

Heinz Sielmann Stiftung, *Gartentierwahl* 2022. Accessed August 20, 2023. https://www.sielmann-stiftung.de/news/detail/blauschwarze-holzbiene-ist-gartentier-des-jahres-2022.

Himmel, Eric. *Extinct and Endangered: Insects in Peril*. New York: Abrams, 2023.

Hölder, Luise. *Unterhaltende Fabeln und Erzählungen aus der Naturgeschichte der Fische, Käfer, Insekten, des Gewächs- und Mineralreichs für kleinere Kinder die sich zu dieser Wissenschaft vorbereiten wollen.* Leipzig: Carl Cnobloch, 1830. Digitized collection of the Staatsbibliothek Preußischer Kulturbesitz zu Berlin.

Holzschuh, Andrea, Jan-Hendrik Dudenhöffer and Teja Tscharntke. "Landscapes with Wild Bee Habitats Enhance Pollination: Fruit Set and Yield of Sweet Cherry." *Biological Conservation* 153 (2012): 101–106.

Horn, Eva, and Hannes Bergthaller. *The Anthropocene: Key Issues for the Humanities.* London and New York: Routledge, 2020.

Hoss, Laura. *Fränkische Waldgemeinschaften im Wandel: Eine kulturanthropologische Annäherung.* Würzburg: Universität Würzburg, 2022.

Hunter. "Hundespielzeug Florenz." Accessed August 19, 2023. https://www.wirlieben hunter.de/hunde/hundespielzeug/3234/hundespielzeug-florenz.

Hustak, Carla, and Natasha Meyers. "Involutionary Momentum: Affective Ecologies and the Sciences of Plant/Insect Encounters." *differences: A Journal of Feminist Cultural Studies* 23, no. 2 (2012): 74–118. doi:10.1215/10407391-1892907.

Internationale Jugendbibliothek. "Ich weiß etwas, was du nicht weißt: Weltwissen in Kinder- und Jugendsachbüchern aus fünf Jahrhunderten." May 25, 2022, to April 25, 2023. Accessed August 20, 2023. https://www.ijb.de/ausstellungen/single/ich-weiss -etwas-was-du-nicht-weisst.

Jones, Owain, and Paul Cloke. *Tree Cultures: The Place of Trees and Trees in Their Place.* Oxford and New York: Berg, 2002.

Kirksey, Eben. *Emergent Ecologies.* Durham and London: Duke University Press, 2015.

Kirksey, Eben, and Stefan Helmreich. "The Emergence of Multispecies Ethnography." *Cultural Anthropology* 25, no. 4 (2010): 545–576.

Kruse, Ulrike. "Von Bienen und Menschen." In *Tiere, Bilder, Ökonomien: Aktuelle Forschungsfragen der Human-Animal-Studies,* edited by Chimaira, 63–85. Bielefeld: transcript, 2013.

Langer, Lars, Manuel Burghardt, Roland Borgards, Katrin Böhning-Gaese, Ralf Seppelt and Christian Wirth. "The Rise and Fall of Biodiversity in Literature: A Comprehensive Quantification of Historical Changes in the Use of Vernacular Labels for Biological Taxa in Western Creative Literature." *People and Nature* 3, no. 5 (2021): 1093–1109.

Latour, Bruno. *We Have Never Been Modern.* Cambridge, MA: Harvard University Press, 1993.

Lévi-Strauss, Claude. *Das Ende des Totemismus.* Frankfurt a.M: Suhrkamp, 1965.

Lovelock, James. *Novocene: The Coming Age of Hyperintelligence.* London: Penguin Books, 2019.

Malicky, Hans. *Vom Handwerk der Entomologie.* Berlin: Springer, 2019.

Mandery, Klaus. *Die Bienen und Wespen Frankens: Ein historischer Vergleich über neue Erhebungen und alte Sammlungen*. Nürnberg: Bund Naturschutz in Bayern, 2001.

Marzolph, Ulrich, and Regina Bendix. "Introduction: Narrative Culture: A Concept and Its Scope." *Narrative Culture* 1, no. 1 (2014): 1–7.

Mawani, Renisa. "Insect Wars. Bees, Bedbugs and Biopolitics." In *Routledge Handbook of Law and Theory*, edited by Andreas Philippopoulos-Mihalopoulos, 279–295. London and New York: Routledge, 2019.

Monbiot, George. "Insectageddon: The Scale and Speed of Environmental Collapse Is Beyond Imagination." *The Guardian*. October 23, 2017. https://www.monbiot.com/2017/10/23/insectageddon/.

Moore, Lisa Jean, and Mary Kosut. "Among the Colony. Ethnographic Fieldwork, Urban Bees and Intra-species Mindfulness." *Ethnography* 15 (2014): 516–539.

Müller, Vera. "Direkte und indirekte Treiber. Landwirtschaft und Biodiversitätsverlust: [Im Gespräch mit Tobias Plieninger]." *Forschung & Lehre* 5 (2021): 364–366.

Porcher, Jocelyne, and Jean Estebanez. "Animal Labour: At the Forefront of Innovative Research." In *Animal Labor: A New Perspective on Human-Animal Relations*, edited by Jocelyne Porcher and Jean Estebanez, 11–34. Bielefeld: transcript, 2019.

Raffles, Hugh. *Insectopedia*. New York: Vintage, 2011.

Robischon, Marcel. *Planet der Insekten: Von duftenden Ameisen, betrügerischen Leuchtkäfern und gespenstischen Faltern*. Bern, Stuttgart, Wien: Haupt, 2011.

The Routledge Companion to the Environmental Humanities, edited by Ursula K. Heise, Jon Christensen and Michelle Niemann. London and New York: Routledge, 2017.

Ruhland, Thomas. "Insekten und die Regeneration des Lebendigen in der Naturgeschichte des 18. Jahrhunderts." In *Vielfältig verflochten: Interdisziplinäre Beiträge zur Tier-Mensch-Relationalität*, edited by Forschungsschwerpunkt "Tier – Mensch – Gesellschaft", 103–122. Bielefeld: transcript, 2017.

Sadeh, Adi, Avi Shmida, and Tamar Kaesar. "The Carpenter Bee Xylocopa pubescens as an Agricultural Pollinator in Greenhouses." *Apidologie* 38 (2007): 508–517.

Schenda, Rudolf. *Das ABC der Tiere: Märchen, Mythen und Geschichten*. München: Beck, 1995.

Schönbein, Daniel, Luzia Keupp, Felix Pollinger and Heiko Paeth. *Klimabericht für Unterfranken: Die Region unter dem Einfluss der globalen Erwärmung*. Würzburg: Universität Würzburg, 2020.

Schuster, Wilhelm. "Warum, wie und wann ist die stahlblauflügelige grosse Holzbiene (Xylocopa violacea) bei uns im Untermaintal eingewandert?" *Societas entomologica* 23, no. 12 (1908): 89–90.

Simon, Anke, and Claudia Botz. *Mein Wildbienen-Buch: Die abenteuerliche Reise der kleinen Wildbiene Mia*. Augsburg: Wißner, 2020.

Tsing, Anna. "Arts of Inclusion, or How to Love a Mushroom." *Mānoa* 22, no. 2 (2010): 191–203.

Wege, Ralf. "Klimawandel mit dramatischen Folgen für Unterfranken." *BR24*. October 20, 2021. Accessed August 19, 2023. https://www.br.de/nachrichten/bayern/klimawandel-mit-dramatischen-folgen-fuer-unterfranken,SmMYrrN.

Werber, Niels. *Ameisengesellschaft: Eine Faszinationsgeschichte*. Frankfurt a.M.: S. Fischer, 2013.

Westrich, Paul. *Die Wildbienen Deutschlands*, 2nd edition. Stuttgart: Ulmer, 2019.

Wienhues, Anna. "Zwischenartliche Gerechtigkeit und das Insektensterben: Wann handelt es sich beim Artentod um eine Ungerechtigkeit?" *Tierstudien* 20 (2021): 89–99.

Wolf, Mike, and Kevin Hall. "Asian Tiger Mosquitos as Undesirable Cross-Border Commuters: Invasive Species and the Regulation of (Bio-)insecurities in Europe." *Journal for European Ethnology and Cultural Analysis* 5, no. 1 (2020): 64–76.

CHAPTER 2

Deconstructing Wasp Aggression: Proposing a Critically Anthropomorphic Narrative of Shared Vulnerability

Minna Santaoja

1 Introduction

Wasps[1] are generally perceived as aggressive, annoying, or even evil insects. Media is in a key position in creating multispecies narratives, representing various nonhuman species, and constructing human-wasp relations. Wasps seem to receive only bad press[2] – they are commonly unloved insects. Due to the negative perception, wasps are considered in multispecies ethics even less than other insects. Wasps are often juxtaposed with their black-and-yellow -striped "cousins," the bees. The negative perception of wasps and comparison with hard-working bees seem to have long roots. Going back over two millennia, Aristotle is often quoted to highlight the bias. In the third book of his *On the Generation of Animals,* from the 4th century BC, Aristotle wrote:

> Concerning the generation of animals akin to [bees], as hornets and wasps, the facts in all cases are similar to a certain extent, but are devoid of the extraordinary features which characterize bees; this we should expect, for they have nothing divine about them as the bees have.[3]

The quote is often presented in a shortened form, further emphasizing the difference in attitudes towards wasps and bees. The quote seems to be taken as evidence of culturally long history of wasp dislike but that is a misrepresentation of Aristotle's work. Aristotle was in fact a wasp *aficionado*, testified by

1 With "wasps" I refer here to yellowjackets and hornets, i.e. stinging social wasps. Globally, there are over 100,000 species of wasps. Most are solitary and do not come into conflict with humans. From the 33,000 stinging wasp species, approximately one thousand are social, forming colonies. In Finland, there are twelve species of social vespine wasps.

2 Seirian Sumner, Georgia Law, and Alessandro Cini, "Why We Love Bees and Hate Wasps," *Ecological Entomology* 43 (2018): 836, doi:10.1111/een.12676.

3 Aristotle, *On the Generation of Animals,* trans. Arthur Platt, Book III, Chapter 10 (Infomotions Inc., 2001; original 305 BC), 66.

his extensive writings on wasp natural history and behavior in several books of *Historia Animalium*.[4] The quote speaks more of contemporary bias towards wasps rather than its history.

Another historical example reflecting wasp dislike is a fable *The bee and the wasp* from 1859.[5] The story features a humble bee, working hard to provide for his family. One day, he comes across Mr. Wasp, a slick fellow, who manages to persuade him to go see the sights of the famed London city. Eventually, the vile wasp leaves the bee to drown in a pool of nectar,[6] intrudes the bee's home, and has the bee's children for supper. The insect behaviors portrayed in the story are to an extent based on ecological facts – some wasp species are known to invade beehives and use bee larvae and pupae as food.[7] But the moral of the story is of humans, not of insects: "While pleasure's path pursuing, [...] has sought his undoing." The story is anthropomorphic – assigning human qualities to the insects. Still, that it is precisely wasp that was chosen to portray the deceitful, lazy, pleasure-seeking character tells something about the wasp perceptions of the time.

While the historical examples portray wasps as "having nothing divine" or as creatures of questionable morals and lifestyles, the contemporary narratives emphasize wasp aggression.[8] Human-wasp relations and human fear towards wasps are centered around the possibility of a sting.[9] Wasps are often perceived to sting without any apparent reason, out of sheer evilness, and they are seen to threat humans' bodily integrity. A wasp sting is always unexpected, painful and for people allergic to wasp venom even dangerous – there is a legitimate concern embedded in human-wasp encounters. Still, it is questionable whether wasp behavior can be labeled aggressive – or what is meant by "aggressive wasps." Wasp numbers vary yearly depending on environmental conditions.

4 Seirian Sumner, *Endless Forms. The Secret World of Wasps* (London: William Collins, 2022), 197.

5 Richard Frankum and George Cruikshank, *The Bee and the Wasp. A Fable in Verse* (London: Basil Montagu Pickering, 1859).

6 Curiously, the fable shows that the idea of an "insect bar," now popular for feeding pollinators and relocating uninvited guests such as wasps, existed already in 1889. In the story it was placed "to keep wasps, flies, and ear-wigs, – a mischievous crew, – From devouring some fruit that was ripe, and in view." In the wasp's view, the "bar" was placed there by "a noted Wasp-lover."

7 Makoto Matsuura and Seiki Yamane, *Biology of the Vespine Wasps* (Berlin: Springer Verlag, 1990).

8 In the 102 articles discussing wasps that I collected from Finnish news media for the period 2015–2019, there were 33 references to wasp aggression.

9 Minna Santaoja, Jyrki Torniainen, and Atte Komonen, "Developing Response-ability in Human-Wasp Encounters," *Trace: Journal for Human-Animal Studies* 9 (2023): 120.

The abundance of wasps is not cheered by humans, which is reflected in media headlines such as "Feisty wasps are now tormenting people on beaches and terraces."[10] The "wasp season" overlaps with the time of the year when human domesticity extends outdoors – to gardens, terraces, parks, and beaches – and the boundaries of domesticity become more permeable through open doors and windows. Wasps' intrusion to these spaces perceived as human territory may be experienced particularly annoying as wasps – unlike most wildlife – do not shy away from humans.

In the long-standing juxtaposition of bees and wasps, the roles of these insects may be changing. Awareness of wasps' ecological importance has increased in light of research. Unlike peaceful and vegetarian bees, wasps are carnivores in larval stage. However, the "aggressive" foraging behavior of wasps towards other insects is one of wasps' main benefits from a human perspective. Wasps hunt insect species that are commonly considered pests, and a nest of wasps may eat as much insects as a pair of birds in a garden. The importance of pollinators for human food systems has received a lot of attention during the last decade, with a focus on bees. Managed honeybees (*Apis mellifera*) are crucial for crop pollination worldwide, but they may also be harmful non-native species in some places. Honeybees may transmit pathogens to native pollinator species and compete for food sources. Wasps participate also in pollination and some plant species even specialize in wasp pollination. For instance, in the case of blueberry – an important plant for humans and numerous nonhumans in the Boreal region – it has been estimated that wasps are responsible for 20% of blueberry pollination.[11]

The understanding of wasps' ecological importance has perhaps not yet reached the public, at least to the extent to change wasp perceptions broadly. What is more, the anthropocentric narrative of wasps providing essential "ecosystem services" may not be enough to disrupt the status of wasps as unloved creatures, and to provide a basis for ethical human-wasp relations. It is very general and theoretical. Human-wasp encounters are always situated and contextual and such must be our responses as well. As Ginn and colleagues

10 Noona Bäckgren, "Ärhäkät ampiaiset piinaavat nyt rannoilla ja terasseilla," *Helsingin Sanomat*, July 28, 2016, https://www.hs.fi/kotimaa/art-2000002913049.html. (Finnish quotes translated by the author.)

11 Janne Heliölä, Mikko Kuussaari and Juha Pöyry, *Pölyttäjien tila Suomessa. Kansallista pölyttäjästrategiaa tukeva taustaselvitys*, Suomen ympäristökeskuksen raportteja 34 (Helsinki: Suomen ympäristökeskus, 2021), http://urn.fi/http://urn.fi/URN:ISBN:978-952-11-5418-8.

write, the encounters are often awkward, involving violence and even death.[12] We live by the stories that are often rehearsed. Language is performative – words have an impact on the world and on human-nonhuman relations. In the precarious situation we find ourselves in the Anthropocene, we need better stories for multispecies conviviality and flourishing. For that, affective narratives and encounters are also needed, as pointed out also by Coldwell in *The Guardian*:

> When we think of bees, we think fondly of a creature that is entwined in our lives. Bees are a crucial cog in the mechanics of global agriculture. They are indispensable to the future of humanity. Without them, we would crumble. What we need is an equivalent story about wasps.[13]

Narrative forms may be ideal for thinking multispecies conviviality, as stories are revisable and able to incorporate complexity.[14] There is plenty to draw from in the shared lifeworlds of humans and wasps. Human and wasp habitats and lifestyles overlap ubiquitously – wasps are in many ways like us.[15] Wasps are flexible in choosing their nest sites and many species dwell in urban areas.[16] Wasps are active from dawn till dusk and even their preferred foods – sugars and meat – overlap with common, if not so recommendable, human food choices. Environmental changes are affecting wasp phenology and are likely to increase human-wasp encounters, highlighting the need to learn to live together.

In this chapter I aim to deconstruct the narrative of "aggressive wasps" which I find counterproductive for multispecies ethics and cohabitance. I proceed by discussing selected entomological texts that address wasp aggression. My aim is not to single out these studies as particularly problematic. They are methodologically sound within their discipline and draw extensively from previous

12 Franklin Ginn, Uli Beisel and Maan Barua, "Flourishing with Awkward Creatures: Togetherness, Vulnerability, Killing," *Environmental Humanities* 4, no. 1 (2014): 113, doi:10.1215/22011919-3614953.

13 Will Coldwell, "Why We Should All Love Wasps," *The Guardian*, May 22, 2022, https://www.theguardian.com/environment/2022/may/22/revenge-of-the-wasp-woman-dr-seirian-sumner-loves-wasps-and-advocates-for-them.

14 Brian McCormack, "Narrative, Meaning, and Multispecies Ethical Ontologies," *Humanimalia* 11, no. 1 (2019): 64, doi:10.52537/humanimalia.9478.

15 Santaoja, Torniainen and Komonen, "Human-Wasp Encounters".

16 Atte Komonen, Aleksi Nirhamo, and Jyrki Torniainen, "Social Wasps (Vespinae) in Urban Gardens and Woods," *Annales Zoologici Fennici* 57 (2020): 41, http://urn.fi/URN:NBN:fi:jyu-202001301875.

research. Rather, I want to point to the use of language in research, media, and public discourse more generally in constructing multispecies realities. Researchers are not immune to confirmation bias – cultural categories may become "smuggled" into science and shape knowledge production, making science confirm the existing cultural constructions. According to my reading, it is often not clear what is meant by wasp aggression – oftentimes it seems that the presence of wasps is enough to label them aggressive.

The talk of "aggressive wasps" is an anthropomorphic convention, projection of human characteristics and fears onto wasps. Anthropomorphism – assigning human qualities such as behaviors, emotions or appearance to nonhuman nature or objects – is a commonly used technique for instance in marketing or in children's stories. In case of insects, anthropomorphism may be problematic as insects are commonly perceived as monstrous or "the ultimate other". Furthermore, in multispecies research aiming to understand the other by engaging with their lifeworlds, anthropomorphism is avoided. I discuss the problems and opportunities with critical anthropomorphism in multispecies ethics. Finally, based on an analysis of Finnish media articles on wasps I discuss alternative, already emerging narratives that build on acknowledging shared multispecies vulnerability and care.[17]

2 Deconstructing Wasp Aggression

In a study from 1984, Parrish analyzed aggression between foraging wasps.[18] After observing yellowjackets at standardized carbohydrate baits, they distinguished four levels of wasp aggression with increasing intensity: fly-by, close-hovering, mutual-hovering, and wrestling. While not precisely defining what aggression is, Parrish gives an idea of what can be understood as aggressive wasp behavior. From a human perspective, when the main concern is wasp sting, the described "aggressive behaviors" seem rather harmless. According to the study, fly-by and close-hovering were directed to a feeding wasp, apparently with the aim of making it leave the bait. The situation could escalate to mutual-hovering, with two wasps chasing each other near the bait, and finally

17 Geographical context is important; my perspective is from the northern boreal region of Europe – in other parts of the world, vespine wasps are invasive species causing significant harm and thus the discussion of human-wasp relations is quite different.

18 Mark D. Parrish, "Factors Influencing Aggression between Foraging Yellowjacket Wasps, Vespula spp. (Hymenoptera: Vespidae)," *Annals of the Entomological Society of America* 77, no. 3 (1984): 306, doi:10.1093/aesa/77.3.306.

to wrestling: "one of the wasps would grab the other with its legs, causing the two to fall struggling to the ground." Despite apparent attempts of stinging, Parrish did not observe any deaths resulting from the wrestling. Wasps were more likely to behave "aggressively" towards other species, but they were as aggressive against nestmates as same species wasps from another colony. Parrish concluded that aggression depended on the size of the bait and the time of its availability as well as the number of wasps. Aggression between wasps could thus be explained by competition over food sources and calling it by that name could already alleviate some concerns and misunderstandings about "aggressive wasps."

In *Biology of the Vespine Wasps,* Matsuura and Yamane characterize wasp behavior casually as weak/formidable/extraordinary/violent/extreme aggression.[19] While they do not define aggression either, the book adds such behaviors as biting (between worker wasps after the loss of colony foundress) and fights between wasp queens when one attempts to colonize the other's nest. The book includes a chapter titled "The Vespines and Man," discussing ways wasps can be both beneficial and harmful to humans from an economic perspective. The authors mention how in some parts of the world people use wasp larvae as protein food, suggesting "that humans have long been predators of vespine colonies." They propose it could be "interesting to survey the relationships between the predation by primates, including humans, and the evolution of the extreme aggressiveness of some vespine species," thus suggesting the aggression of wasps might be, at least partly, outcome of human predation. At the end of the book the authors introduce 16 Vespine species living in Japan. In concluding each species description, Matsuura and Yamane give an estimate of the species' aggressiveness towards humans, such as "workers are quite aggressive" or "workers are relatively non-aggressive towards humans." What is noteworthy is that wasps' aggression was estimated when humans approached, disturbed, or even excavated the nests. In case of *Vespula flaviceps*, for instance, the authors write "people will be stung only when they approach the nest too close." It seems, then, that in many cases wasp "aggression" against humans could be substituted with "defense."

In a recent study, Jandt and colleagues wrote how "social wasps are well-known for aggressively defending their nests when these are approached, threatened, or directly attacked."[20] In their study, the authors aimed to quantify

19 Matsuura and Yamane, *Biology of the Vespine Wasps*.

20 Jennifer Jandt, Mateus Detoni, Kevin J. Loope, and Davide Santoro, "Vespula Wasps Show Consistent Differences in Colony-level Aggression Over Time and Across Contexts," *Insectes Sociaux* 67 (2020): 367, doi:10.1007/s00040-020-00768-3.

colony-level differences in wasp aggression. However, they do not define what is meant by "aggression" or "aggressive behavioral types" either. The authors' understanding of aggression is revealed implicitly in the methods section, as they explain how the study was carried out. Wasp aggression was studied in provoked and unprovoked settings. To conduct the experiments, the researchers took targets made of black cloth or black paper plates to the vicinity of wasp nests. This was the unprovoked setting – although one might ask, is it not already a provocation to place a foreign object next to the wasp nest.

In the study, wasp response was recorded by sight and sound, installing a camera and a microphone next to the target. In measuring the provoked aggression of the wasp colony, carbon dioxide was exhaled to the entrance of the wasp nest, and the entrance was scratched with a fork. The provocations aimed to mimic a potential predator, a curious mammal, or an unsuspecting gardener approaching the nest. To calculate aggression, the number of wasp "strikes" to the target was recorded, i.e., aggression equals a "strike." The "strikes could be quick, strong (wasps drove themselves into the target like a battering ram), or prolonged (wasps held onto the target, buzzed, and tried to sting the plate)."[21] The study found that some wasp colonies are consistently aggressive, while others "exhibit little to no aggressive response to disturbance". At times in the paper, "defense" seems to be used interchangeably to "aggression."

Referring to previous studies, Jandt and colleagues discuss how wasp colony aggression is likely heritable. Within-colony individual differences in wasps' aggression have been linked to differences in gene expression, which can be influenced by age as well as individual experience with the local and social environment. The authors discuss how the differences in colony aggression – both "aggressive" and non-aggressive behavioral types – can have fitness benefits in a population. While the "aggressive" colonies might defend their nests against predators more effectively, the less aggressive colonies "could go unnoticed by humans and develop undisturbed." Like Matsuura and Yamane, the study posits humans as a main wasp predator and suggests that the colony behavior may evolve in relation to that. In the Anthropocene, humans have become the main evolutionary force, driving the change of observable traits in many other species.[22]

Research on colony-level wasp aggression is valuable for increasing understanding of wasps. It may be that within discipline, there is a shared understanding of what is referred to with "aggression," so much so that in any of

21 Jandt, Detoni, Loope and Santoro, "Vespula Wasps," 372.

22 Chris Darimont et al., "Human Predators Outpace Other Agents of Trait Change in the Wild," *PNAS* 106, no. 3 (2009), 952, doi:10.1073/pnas.0809235106.

these studies it is not seen as necessary to define the key concept. The other possibility is that there is only an assumed understanding of the concept, as it is commonly used. Multispecies research draws from social sciences and humanities as well as natural sciences to construct narratives reflecting complex reality. Researchers from different disciplines become "critical friends" and interdisciplinary work highlights the importance of defining concepts and using language in such a way that it is understandable across disciplines. This calls for entomological scholarship to define concepts such as "aggression." The aim of the study discussed above seemed to be to understand wasp colony behavior better to anticipate colony response to human presence, and ultimately, to avoid wasp-human conflicts. For this aim, the used language plays an important role. The use of "aggression" as a descriptor should be carefully considered, and it should be substituted by more accurate terms such as "defensive behavior" when that is in fact the case. Wasp aggression – and even sting – may be easier to accept in the case of (purposeful) human disturbance to their colony, but as wasps enter domesticated spaces, even with the purpose of finding food sources, human defense may turn into aggression.

Researchers are not immune to common uses of language, and there are diverse motivations behind the choice of research topics and vocabulary. Sumner and colleagues discuss how unloved wasps have received less research attention than bees, but some entomologists recognize the "feral charisma"[23] of wasps, and behind the practice of nonchalantly calling wasps aggressive may be a zoomorphic wish that some of that charisma would transfer to people studying them. In the times of eco-social crises of existential dimensions, however, we should better seek recognition in becoming ambassadors of multispecies conviviality in ways that do not reflect badly on our nonhuman fellows.

After my limited review of entomological literature, the definition of wasp aggression remains in the air. The talk of wasps as aggressive is anthropocentric, highlighting the potential harm to humans. It may also be an anthropomorphic convention, a projection of human fears. If we are assigning human qualities to wasps, it is useful to have a look at what aggression is in humans. Anderson and Bushman define human aggression in short as "any behaviour directed toward another individual that is carried out with the proximate (immediate) intent to cause harm."[24] In the case of wasps, the jury is probably still out on what terms can their behavior be called intentional. In any case,

23 Jamie Lorimer, "Nonhuman Charisma," *Environment and Planning D: Society and Space* 25, no. 5 (2007): 911, doi:10.1068/d71j.

24 Craig Anderson and Brad Bushman, "Human Aggression," *Annual Review of Psychology* 53, no. 1 (2002): 27, doi:10.1146/annurev.psych.53.100901.135231.

the motivation behind the "aggressive" behaviors of wasps seems to be scaring the other off in defense of the colony or a food source, rather than primarily causing harm. Thus, wasp aggression does not match the definition of human aggression.

Human aggression has been explained with several domain-specific theories, including cognitive neoassociation (an unpleasant experience stimulates thoughts, memories, behavioral tendencies etc. related to previous negative experiences), social learning (forms of social behavior are acquired by direct experience or observing others), social interaction (aggression as seeking social influence, motivated by other primary goal), script (set of rehearsed associated concepts in memory involving action plans), and excitation transfer (arousal from a previous event may be misattributed to a following event) theories. Anderson and Bushman have integrated these into a so-called general aggression model (GAM), that better explains aggressive acts that are based on multiple motives. The model considers factors related to both the person (e.g., genotype) and the situation (e.g., alcohol). According to the model, various inputs affect the individual's internal state (their affect, arousal, and cognition), which may provide bottom-up motivation for aggressive acts. Human aggression may be more complex than aggression in wasps, but understanding human aggression may help to understand the various individual and environmental factors affecting wasps' aggressive behavior. We should consider which parts of the complex "general wasp aggression model" apply in each context, rather than labeling wasps "aggressive."

Even though the news headlines – and scholarly studies – emphasize wasp aggression, there are other voices. In August 2018, a headline in Helsingin Sanomat, the major Finnish daily paper, read "Wasps' despair manifests now as feistiness."[25] 2018 was a wasp-rich year in Finland. The article came out in the late summer, when wasp colonies reach peak size and start to run out of food, new queens fly out and the workers become unemployed, as the colony withers towards the autumn. This is the peak moment for conflicted wasp-human encounters. An entomologist was interviewed to explain "aggressive" wasp behavior. They responded:

> Wasps are not aggressive to start with. But in August they may get stressed, as the nest starts to run out of food. The workers need to find food for the larvae at any cost, and they follow the smell of food to human tables. The hungry brood waiting in the nest makes the wasps persistent. Looking for

---

25 Pihla Loula, "Ampiaisten epätoivo näkyy nyt ärhäkkyytenä," *Helsingin Sanomat*, August 2, 2018, https://www.hs.fi/kotimaa/art-2000005776859.html.

> food on plates and buzzing in front of the face may feel aggressive. But wasps cannot be described as aggressive by nature, unlike some people. Of course, wasps' innate defense reaction is strong, and many interpret it as aggression.

The interviewed entomologist continued explaining that wasps have a keen sense of smell, unlike humans. They described wasps as misunderstood species: wasps only come to "bother" people when looking for food for their offspring, just as people provide for their families. Wasps approaching humans does not mean attack; there might just be an interesting smell in the human breath. In such case, the entomologist suggested holding the breath and taking a step back, and the wasp might lose interest. The entomologist emphasized that wasps do not attack humans to start with. They sting only when attacked or provoked first. Sometimes a wasp might get trapped under human clothing and sting before its presence was even detected. "It may sting in panic." For peaceful cohabitance, the entomologist proposed providing wasps with a "bar" – placing overripe fruit or scraps of meat a few meters away. The entomologist was deliberately pointing to parallels in wasp and human lives, and in conflict situation named humans as culpable, turning the conventional narrative upside down. For redemption, they offered the narrative of care.

3 Critical Anthropomorphism and Precarious Vulnerability

Anthropomorphism refers to the tendency of assigning human qualities – such as behaviors or emotions – to nature, nonhuman animals and other species, or objects. Anthropomorphism is a commonly used strategy for instance in children's books that treat nonhuman life. Anthropomorphism – for instance, talk of "Mother Nature" or referring to climate change as "Earth having a fever" – may make people feel more connected to nature, and the connectedness may foster conservation behavior.[26] Anthropomorphism is therefore not only semiotic and symbolic but may have effects to the well-being of nonhuman others. Geographer Jamie Lorimer has discussed how anthropomorphic nonhuman charisma affects even conservation policies. We tend to favor (mammalian) species that have a recognizable face and that communicate in ways compatible with us humans as visual animals. Species that are perceived as ugly or

---

26 Kim-Pong Tam, Sau-Lai Lee, and Melody Manchi Chao, "Saving Mr. Nature: Anthropomorphism Enhances Connectedness to and Protectiveness Toward Nature," *Journal of Experimental Social Psychology* 49, no. 3 (2013): 514, doi:10.1016/j.jesp.2013.02.001.

annoying receive less attention.[27] Most directly anthropomorphic practices probably affect the well-being of companion animals. While some practices may be beneficial to them, treating them like humans may be detrimental to their health.[28]

In human perspective, insects have remained the ultimate unintelligible other.[29] The multiplicity, "monstrosity," autonomy, and parasitism of insects are often experienced to threaten the idea of a bounded, individual subjectivity.[30] Due to this otherness, insects are hard to anthropomorphize,[31] or may be anthropomorphized negatively, portraying wasps as human-like characters with questionable morals, like in the fable *The bee and the wasp,* or labeling wasps aggressive when in fact they are defending themselves against provocation. To deal with insect otherness we may need to anthropomorphize them differently, but it is not necessarily beneficial to the insects. For instance, in marketing insects for human food, anthropomorphism has been used as a strategy: Western consumers are not generally willing to consume insects perceived as disgusting but anthropomorphized "cute" insect pictures on food packaging may make entomophagy more acceptable.[32]

In multispecies research attempting to speak responsibly about other species as beings with their own lifeworlds and intentions, anthropomorphism is almost a curse word.[33] Anthropomorphism may be understood as inappropriately imposing human perceptions onto nonhuman beings, denying their ways of experiencing the world. In addressing the methodological challenges in attending to nonhuman others, human-animal studies scholars have been calling for a methodological "critical anthropomorphism" which endorses qualitative interpretations of nonhuman behavior but is complemented with species-specific knowledge from natural sciences or phenomenological

27 Lorimer, "Nonhuman Charisma," 911.

28 Daniel Mota-Rojas et al., "Anthropomorphism and Its Adverse Effects on the Distress and Welfare of Companion Animals," *Animals* 11, no. 11 (2021): 3263, doi:10.3390/ani11113263.

29 Hugh Raffles, *Insectopedia* (New York: Vintage, 2011).

30 James Hillman, "The Satya Interview: Going Bugs with James Hillman," *Satya*, January, 1997, http://www.satyamag.com/jan97/going.html.

31 Known popular culture exceptions include Disney's Jiminy Cricket, or the movie *A Bug's Life*.

32 Zining Wang and Jaewoo Park, "'Human-like' is Powerful: The Effect of Anthropomorphism on Psychological Closeness and Purchase Intention in Insect Food Marketing," *Food Quality and Preference* 109 (2023): 104901, doi:10.1016/j.foodqual.2023.104901.

33 Mollie Holmberg, "Beyond Anthropomorphism: Attending to and Thinking with Other Species in Multispecies Research," *ACME: An International Journal for Critical Geographies* 21, no. 2 (2022): 172, https://acme-journal.org/index.php/acme/article/view/2033.

practices of attention.[34] Critical anthropomorphism recognizes the phenomenal alterity of different species' experiences in the world, rather than endorsing a sentimental anthropomorphism that treats other animals as "people in disguise." At the same time, it does not assume we would not share anything with other species' experiences either. Burghardt has claimed that critical anthropomorphism as an empathetic attempt to understand the other is among the only means to escape anthropocentrism.[35]

Reading beyond the scandalous headlines of aggressive wasps in Finnish news media, there are already more nuanced, critically anthropomorphic narratives. In the news article titled "Feisty wasps are now tormenting people on beaches and terraces"[36] an entomologist was interviewed on wasps' "misbehavior." In response, they explained the phases of a wasp colony cycle. Only the wasp queens overwinter. When the foundress queen has produced a new generation of queens and males, she dies in the autumn. The new queens fly out to mate and find a place to spend the winter. At this point the worker wasps, whose tasks have been to maintain and defend the nest and find food for the larvae, are left with nothing to do. Sumner has described the situation as wasps being furloughed,[37] Matsuura and Yamane write of colonies or nests being orphaned.[38]As the worker wasps have lost the meaning of their lives, they may approach humans in their search for carbohydrates and proteins.

In the interview, the entomologist called the predicament of wasps an existential crisis. The term was probably introduced half-sarcastically, as in the same article they explained how to make a wasp trap to get rid of unwanted wasps. However, the idea of wasps in an existential crisis seemed to resonate with journalists and readers of Finnish media, and it has been rehearsed several times since in writing about wasps' autumnal mischiefs. Referring to wasps as being furloughed, in an existential crisis, or orphaned, may be interpreted as anthropomorphism – as appointing human characteristics to the insect. If read sarcastically, the narrative may alienate humans even further

34 Maisie Tomlinson, *"Critical Anthropomorphism" and Multi-Species Ethnography: An Investigation of Animal Behaviour Expertise*, Ph.D. thesis (Manchester: University of Manchester, 2021), https://research.manchester.ac.uk/en/studentTheses/critical-anthropomorphism-and-multi-species-ethnography-an-invest.

35 Gordon Burghardt, "Critical Anthropomorphism, Uncritical Anthropocentrism, and Naïve Nominalism," *Comparative Cognition & Behavior Reviews* 2 (2007): 136, https://psycnet.apa.org/record/2008-07027-009.

36 Bäckgren, "Ärhäkät ampiaiset".

37 Seirian Sumner, "Why wasps become so annoying at the end of summer," *The Conversation*, August 27, 2020, https://theconversation.com/why-wasps-become-so-annoying-at-the-end-of-summer-145053.

38 Matsuura and Yamane, *Biology of the Vespine Wasps*, 85.

from wasps. It may be taken to emphasize wasps' otherness and the impossibility of insects experiencing something like an existential crisis – understood fundamentally as a human condition requiring high-level cognitive and affective capabilities and, e.g., awareness of one's mortality. However, describing wasp behavior as an existential crisis, resulting from them being orphaned and furloughed, has basis in wasp ecology. Taking the narrative of existential crisis seriously and accepting that the same vocabulary can be used to accurately describe human and insect lives may offer opportunities for identifying shared multispecies vulnerabilities, developing empathy and understanding, and providing information on wasp lives behind their "aggressive" behavior.

Critically anthropomorphic narratives invite ecological and entomological knowledge that supports the qualitative interpretation of the situation. The narrative allows understanding better the lifeworlds of the nonhuman others and developing multispecies care via the notions of precarity and vulnerability. The narrative may become a pedagogical moment, revealing our entanglements and vulnerabilities with other species, helping to rethink our place in the world.[39] Precarity has become a global, generalized condition.[40] It does not refer only to human unstable incomes and material welfare; it is "embedded in unprecedented encounters between species, and in existential instability in the face of environmental disturbances on a planetary scale."[41] The existential crisis brought upon by anthropogenic environmental changes touches wasps and humans alike, and the intimacy and complexity of human-wasp relations render our shared condition precarious. We are together in this, but still, human and wasp vulnerability is asymmetrical – being entangled with us is not often beneficial for the other creatures.

Deborah Bird Rose and Thom van Dooren have asked what hope there is for unloved, less visible, less beautiful creatures that are less entangled in our culture, when even creatures who are vividly present in our imaginative lives are nonetheless on the edge of loss.[42] Unloved creatures that bite or sting or seem monstruous or aggressive are awkward creatures that do not fit off-the-shelf

---

39 Affrica Taylor and Veronica Pacini-Ketchabaw, "Learning with Children, Ants, and Worms in the Anthropocene: Towards a Common World Pedagogy of Multispecies Vulnerability," *Pedagogy, Culture & Society* 23, no. 4 (2015): 507, doi:10.1080/14681366.2015.1039050.

40 Anna Tsing, *The Mushroom at the End of the World: On the Possibility of Life in Capitalist Ruins* (Princeton University Press, 2015).

41 Pieta Hyvärinen, "Beekeeping Expertise as Situated Knowing in Precarious Multispecies Livelihoods," *Sosiologia* 56, no. 4 (2019): 365, https://journal.fi/sosiologia/article/view/124543.

42 Deborah Bird Rose and Thom van Dooren, "Unloved Others: Death of the Disregarded in the Time of Extinctions. Introduction," *Australian Humanities Review* 50 (2011): 1, https://australianhumanitiesreview.org/2011/05/01/issue-50-may-2011/.

ethics.[43] Franklin Ginn and colleagues write how being together in multispecies entanglements is often awkward. With Donna Haraway they argue "for an ethic of multispecies co-flourishing in which the outcomes are never certain, ethical judgments stick close to the action of worlding rather than abstract principles, and in which emotion and reason both play their parts." Vulnerability, violence, and death are inevitable part of our ongoing engagements with nonhuman others – not only negative elements to be repressed or solved. Becoming less uncomfortable with vulnerability and putting ourselves at risk may be productive ethical practices. Empathizing on a general level with the existential crisis narrative is not enough – we need to persistently pay attention to the worlds we construct in language and prepare to respond in the unexpected situated multispecies encounters. Nonhuman animal narratives developed in a multispecies ethics of conviviality are about making, rather than representing, worlds.[44]

4 Words for Sharing Worlds

> It matters what matters we use to think other matters with; it matters what stories we tell to tell other stories with; it matters what knots knot knots, what thoughts think thoughts, what descriptions describe descriptions, what ties tie ties. It matters what stories make worlds, what worlds make stories.[45]

The quote from Donna Haraway urges to pay attention to the words, language, and stories we use to narrate multispecies entanglements. Potawatomi scholar Robin Wall Kimmerer has called for "grammars of animacy" for describing ways of thinking and speaking that recognize other beings having intentions and lifeworlds. According to them, the language of animacy is close to extinction as research reduces nonhuman life to e.g., abstract species.[46] Geographer Mollie Holmberg has engaged indigenous, feminist, and queer scholarship with more-than-human geographies to go beyond anthropomorphism and

43 Ginn, Beisel and Barua, "Flourishing with Awkward Creatures," 113.

44 McCormack, "Multispecies Ethical Ontologies," 20.

45 Donna Haraway, *Staying with the Trouble. Making Kin in the Chthulucene* (Durham: Duke University Press, 2016), 12.

46 Robin Wall Kimmerer, *Braiding Sweetgrass. Indigenous Wisdom, Scientific Knowledge and the Teachings of Plants* (Minneapolis: Milkweed Editions, 2013).

think how to attend to other species in multispecies research.[47] What can be learned from indigenous scholarship and grammars of animacy, for one, is to not use metaphors to describe human-nonhuman relations. Nonhuman life-worlds and animacy must be taken seriously, as factual descriptions of multi-species realities instead of, for instance, myths. Wasp aggression starts to look like a myth, whereas the narrative of wasps in an existential crisis is an accurate description of their precarious experience.

Animal ethics seems to be in trouble with insects. Monsó and Osuna-Mascaró have listed several obstacles for including insects into moral considerations: an outdated view of insects as "lower beings" or less than animal, empirical uncertainties considering insect sentience, the cognitive and affective biases that make us view insects as annoying or disgusting, and a tendency to give more weight to scientific uncertainty that to moral risk (for example: killing wasps is not problematic since we do not have certainty of their capability to suffer). There are further obstacles stemming from insect – or more generally arthropod – biology and ecology to include insects in moral consideration and ethical treatment. From the sheer number of insects, it results that animal ethics based on individual welfare do not quite fit. While insects are often perceived as less than individuals, in some sense they are also more than individuals. Going through metamorphosis, in different stages of their lives the insects' needs and interests are completely different. For instance, wasp larvae need to be fed on animal protein food, whereas adult wasps feed on nectar. Hence, one individual may be considered different individuals throughout the course of its life. Also, the colonies of eusocial, "less than individual" insects such as wasps are superorganisms that have adaptiveness and problem-solving capacity beyond an individual. As eusocial insects such as wasps put the colony welfare before their own, forcing them into an individualistic welfare-based ethics would amount to unwarranted anthropocentrism.[48]

French philosopher Jacques Derrida proposed to get rid of the category "animal" as it is overtly generalizing and hiding a vast variety of differences between species and individuals.[49] Derrida proposed instead the term "animot," highlighting that animal is just a word, not an essentialist counterpart to human. Insects are very different from mammals, but the category of

---

47 Holmberg, "Beyond Anthropomorphism: Attending to and Thinking with Other Species in Multispecies Research" (2022): 172.

48 Susana Monsó and Antonio J. Osuna-Mascaró, "Problems with Basing Insect Ethics on Individuals' Welfare," Animal Sentience 29, no. 8 (2020).

49 Jacques Derrida, *The Animal That Therefore I Am* (New York: Fordham University Press, 2008).

"insect" is also problematic as has become apparent in preparing this book. For instance, ticks and spiders are not insects, even though intuitively we categorize them all as "bugs." Talking of "wasps" is still overtly generalizing as different wasp species have different lifestyles. Learning about species-specific wasp ecology could make us humans better prepared for encounters, but applying the knowledge in situated practice would require keen species identification skills. This has proved difficult, as humans commonly confuse wasps with other black-and-yellow insects, let alone identifying the species. It would be also useful to be able to identify the sex of encountered wasp, as only females may sting. As this level of preparedness in human-wasp encounters is demanding, I believe rehearsing multispecies narratives of animacy is the best way to cultivate our response-ability. Critically anthropomorphic narratives allow us to reconfigure default responses in the encounters. In paraphrasing Haraway, narratives of animacy may allow developing intimacy without proximity to the awkward other.

In this chapter I aimed to deconstruct the story of wasps as aggressive and proposed narratives for wasp-human conviviality instead. There is a significant difference in perceiving wasps as *being* aggressive, or alternatively wasps *behaving* aggressively. The first one is essentialist; the latter invites to ask questions about the reasons behind seemingly aggressive behavior. The questions led to narratives where wasp "aggression" seems to be rather defense against provocation, or persistence in providing food for the offspring. "Aggressive wasps" is an anthropocentric construction, and it is time to let it go. There are narratives of mutual human-wasp vulnerability at the times of the Anthropocene and destruction, but we are unequal in vulnerability, which calls for multispecies care. It may be easier to extend care to wasps being aware of their ecological importance – of the "ecosystem service" narrative. But human-wasp encounters are always situated, and such must be our responses as well. However difficult insect ethics are it is individual wasps that we encounter. Those encounters are often awkward, they may involve violence, and even death. In the encounters wasps are more vulnerable, but human vulnerability and health concerns cannot be dismissed. Furthermore, humans are differently vulnerable to wasps, depending on the living environment or profession. For instance, a berry farmer has a higher probability of encountering wasps and, thus, a higher risk of being stung, than for instance an office worker. For the berry farmer, however, the awkwardness of living with wasps could be a familiar situation and they may be better equipped with contextual responses.

To end with, I want to propose one more set of words for more nuanced wasp-human encounters. Symbiotic ontology, building on e.g., Haraway's relational ontology, is gaining increasing interest in multispecies studies. Symbiotic ontology understands life on all levels as a generally cooperative process.[50] Here again, we should be careful with our words. Symbiosis is sometimes equated with "mutual flourishing," but that is just one type of symbiotic relationships. Symbiosis means living together and describes a close relationship between two species, but the nature of the relationship needs to be specified. Commensalism is a situation where one species benefits while the other is unaffected. In mutualism both species benefit. Parasitism means one species benefits while the other is harmed. In competition, neither benefit, whereas in predation one species benefits while the other dies.

Finally, neutralism leaves both species unaffected. The symbiotic relations should not be understood as essential either – they are contextual, and human-wasp symbiosis can take any of those forms. Maybe next time when a wasp approaches us, we can take a step back and turn the question back to ourselves: "What am I in this situation?" Questioning the nature of the symbiotic situation, making space for the other and exercising curious observation are good strategies for awkward multispecies conviviality, to begin with.

Acknowledgments

I am thankful to the anonymous reviewers for their insightful observations and to the book editors for all their efforts and care. I wish to thank Dr Atte Komonen from the University of Jyväskylä, Finland, who invited me to the "Ampase" project on human-wasp cohabitance, funded by the Kone foundation, and thus got me into thinking about wasps.

Bibliography

Anderson, Craig and Brad Bushman. "Human Aggression." *Annual Review of Psychology* 53, no. 1 (2002): 27–51. doi:10.1146/annurev.psych.53.100901.135231.

Aristotle. *On the Generation of Animals*. Translated by Arthur Platt. Infomotions Inc., 2001 (original 305 BC).

Bäckgren, Noona. "Ärhäkät ampiaiset piinaavat nyt rannoilla ja terasseilla – 'Ne ovat menettäneet elämänsä tarkoituksen.'" *Helsingin Sanomat*. July 28, 2016. https://www.hs.fi/kotimaa/art-2000002913049.html.

50 Haraway, *Staying with the Trouble. Making Kin in the Chthulucene* (2016).

Burghardt, Gordon. "Critical Anthropomorphism, Uncritical Anthropocentrism, and Naïve Nominalism." *Comparative Cognition & Behavior Reviews* 2 (2007): 136–138. https://psycnet.apa.org/record/2008-07027-009.

Coldwell, Will. "Why We Should All Love Wasps." *The Guardian*. May 22, 2022. https://www.theguardian.com/environment/2022/may/22/revenge-of-the-wasp-woman-dr-seirian-sumner-loves-wasps-and-advocates-for-them.

Darimont, Chris, Stephanie Carlson, Michael Kinnison, Paul Paquet, Thomas Reimchen and Christopher Wilmers. "Human Predators Outpace Other Agents of Trait Change in the Wild." *Proceedings of the National Academy of Sciences* 106, no. 3 (2009): 952–954. doi:10.1073/pnas.0809235106.

Derrida, Jacques. *The Animal That Therefore I Am*, edited by Marie-Louise Mallet, translated by David Willis. New York: Fordham University Press, 2008.

Frankum, Richard and George Cruikshank. *The Bee and the Wasp. A Fable in Verse*. London: Basil Montagu Pickering, 1859. https://books.google.fi/books?id=tqJbAAAAMAAJ&hl=fi&pg=PP1#v=onepage&q&f=false

Ginn, Franklin, Uli Beisel and Maan Barua. "Flourishing with Awkward Creatures: Togetherness, Vulnerability, Killing." *Environmental Humanities* 4, no.1 (2014): 113–123. doi:10.1215/22011919-3614953.

Haraway, Donna. *Staying with the Trouble: Making Kin in the Chthulucene*. Durham: Duke University Press, 2016.

Heliölä, Janne, Mikko Kuussaari and Juha Pöyry. *Pölyttäjien tila Suomessa. Kansallista pölyttäjästrategiaa tukeva taustaselvitys*. Suomen ympäristökeskuksen raportteja 34. Helsinki: Suomen ympäristökeskus, 2021. http://urn.fi/http://urn.fi/URN:ISBN:978-952-11-5418-8.

Hillman, James. "The Satya Interview: Going Bugs with James Hillman." *Satya*, January 1997. http://www.satyamag.com/jan97/going.html.

Holmberg, Mollie. "Beyond Anthropomorphism: Attending to and Thinking with Other Species in Multispecies Research." *ACME: An International Journal for Critical Geographies* 21, no. 2 (2022): 172–187. https://acme-journal.org/index.php/acme/article/view/2033.

Hyvärinen, Pieta. "Beekeeping Expertise as Situated Knowing in Precarious Multispecies Livelihoods." *Sosiologia* 56, no. 4 (2019): 365–381. https://journal.fi/sosiologia/article/view/124543.

Jandt, Jennifer, Mateus Detoni, Kevin J. Loope and Davide Santoro. "Vespula Wasps Show Consistent Differences in Colony-level Aggression Over Time and Across Contexts." *Insectes Sociaux* 67 (2020): 367–381. doi:10.1007/s00040-020-00768-3.

Kimmerer, Robin Wall. *Braiding Sweetgrass. Indigenous Wisdom, Scientific Knowledge and the Teachings of Plants*. Minneapolis: Milkweed Editions, 2013.

Komonen, Atte, Aleksi Nirhamo and Jyrki Torniainen. "Social Wasps (Vespinae) in Urban Gardens and Woods." *Annales Zoologici Fennici* 57 (2020): 41–46. http://urn.fi/URN:NBN:fi:jyu-202001301875.

Lorimer, Jamie. "Nonhuman Charisma." *Environment and Planning D: Society and Space* 25, no. 5 (2007): 911–932. doi:10.1068/d71j.

Loula, Pihla. "Ampiaisten epätoivo näkyy nyt ärhäkkyytenä." *Helsingin Sanomat.* August 2, 2018. https://www.hs.fi/kotimaa/art-2000005776859.html.

McCormack, Brian. "Narrative, Meaning, and Multispecies Ethical Ontologies." *Humanimalia* 11, no. 1 (2019): 64–88. doi:10.52537/humanimalia.9478.

Matsuura, Makoto and Seiki Yamane. *Biology of the Vespine Wasps.* Berlin: Springer Verlag, 1990.

Monsó, Susana and Antonio J. Osuna-Mascaró. "Problems with Basing Insect Ethics on Individuals' Welfare." *Animal Sentience* 29, no. 8 (2020). doi: 10.51291/2377-7478.1589

Mota-Rojas, Daniel, Chiara Mariti, Andrea Zdeinert, Giacomo Riggio, Patricia Mora-Medina, Alondra Del Mar Reyes, Angelo Gazzano et al. "Anthropomorphism and Its Adverse Effects on the Distress and Welfare of Companion Animals." *Animals* 11, no. 11 (2021): 3263. doi:10.3390/ani11113263.

Parrish, Mark D. "Factors Influencing Aggression between Foraging Yellowjacket Wasps, Vespula spp. (Hymenoptera: Vespidae)." *Annals of the Entomological Society of America* 77, no. 3 (1984): 306–311. doi:10.1093/aesa/77.3.306.

Raffles, Hugh. *Insectopedia.* New York: Vintage, 2011.

Rose, Deborah Bird and Thom van Dooren. "Unloved Others: Death of the Disregarded in the Time of Extinctions. Introduction." *Australian Humanities Review* 50 (2011): 1–4. https://australianhumanitiesreview.org/2011/05/01/issue-50-may-2011/.

Santaoja, Minna, Jyrki Torniainen and Atte Komonen. "Developing Response-ability in Human-Wasp Encounters." *Trace: Journal for Human-Animal Studies* 9 (2023): 120–146. doi:10.23984/fjhas.120353.

Sumner, Seirian. *Endless Forms. The Secret World of Wasps.* London: William Collins, 2022.

Sumner, Seirian. "Why wasps become so annoying at the end of summer." *The Conversation,* August 27, 2020. https://theconversation.com/why-wasps-become-so-annoying-at-the-end-of-summer-145053.

Sumner, Seirian, Georgia Law, and Alessandro Cini. "Why We Love Bees and Hate Wasps." *Ecological Entomology* 43 (2018): 836–45. doi:10.1111/een.12676.

Tam, Kim-Pong, Sau-Lai Lee and Melody Manchi Chao. "Saving Mr. Nature: Anthropomorphism Enhances Connectedness to and Protectiveness Toward Nature." *Journal of Experimental Social Psychology* 49, no. 3 (2013): 514–521. doi:10.1016/j.jesp.2013.02.001.

Taylor, Affrica and Veronica Pacini-Ketchabaw. "Learning with Children, Ants, and Worms in the Anthropocene: Towards a Common World Pedagogy of Multispecies Vulnerability." *Pedagogy, Culture & Society* 23, no. 4 (2015): 507–529. doi:10.1080/14681366.2015.1039050.

Tomlinson, Maisie. *"Critical Anthropomorphism" and Multi-Species Ethnography: An Investigation of Animal Behaviour Expertise.* Ph.D. thesis. Manchester: University of

Manchester, 2021. https://research.manchester.ac.uk/en/studentTheses/critical-anthropomorphism-and-multi-species-ethnography-an-invest.

Tsing, Anna. *The Mushroom at the End of the World: On the Possibility of Life in Capitalist Ruins.* Princeton: Princeton University Press, 2015.

Wang, Zining and Jaewoo Park. "'Human-like' Is Powerful: The Effect of Anthropomorphism on Psychological Closeness and Purchase Intention in Insect Food Marketing." *Food Quality and Preference* 109 (2023): 104901. doi:10.1016/j.foodqual.2023.104901.

PART 2

Insects and Human Gaze

CHAPTER 3

Fly Eyes and Insect Vision at the Turn of the 20th Century: from Scientific Curiosity to Compound Menace

Concepción Cortés Zulueta

1 Introduction

In a series of workshops on flies' eyes and vision that I carried out at the Museo Nacional de Ciencias Naturales (MNCN, Madrid), most children expressed their rejection and negative emotions towards flies, arguing that they were a nuisance or that they landed in filth and then in food, thereby ruining it. A few even included swatters in their pictures, on their own account, when they were asked to draw a fly.[1] Exposing and contextualizing the roots and sources of these human prejudices opens the possibility of dispelling some of the everyday, deeply ingrained, rejections associated with insects and with insect representations that still persist.

In this chapter, my intention is to unravel the curious encounters that took place around the 1900s regarding the transmission of some visual conventions and of the accompanying visual knowledge.[2] They were associated with certain scientific and optic experiences, developed in previous centuries, related to magnification, visuality, and the compound eyes, and vision, of insects in general, and of flies in particular. In the context of the anti-fly campaigns, the turn of the century witnessed the initial stages of a change of attitude and a process whereby these experiences became tinged with specific negative emotions and misconceptions, that hindered and obscured insect-human interactions. The study is based on various media such as natural historical illustrations, solar microscopes, magic lantern slides, and scientific microphotographs.

1 I collaborated with the MNCN conducting four children's workshops in July–August 2023. I am indebted to Luis Barrera, Pilar López, Elena Gazo, Soraya Peña de Camus, Carolina Martín, Mercedes París and María Valladolid for their help. See: Concepción Cortés Zulueta, "¿Cómo ven las moscas? Un taller sobre los ojos compuestos de las moscas en el Museo de Verano," *NaturalMente* 41, March 2024, https://revista.mncn.csic.es/nm41/72-73/.

2 Written while a Juan de la Cierva-Incorporación fellow (Min. of Science, Spain), and as part of the project Entomornitofilias (y fobias): Impresiones y encuentros de aves e insectos (B1-2022_11), II Plan Propio Investigación, UMA.

 | DOI:10.1163/9789004715448_005

The sources employed describe or discuss these various instruments, dissect the anti-fly campaigns and their emotional impact, or refer to the dissemination of visual knowledge and popular science, including articles and lectures from the period.

Emotions have a history, encompassing both cultural and biological elements whose entanglements and manifestations vary over time.[3] In the case of insects and considering the nuances and complications of human-insect interactions – also on a biological level – it might be relatively straightforward to arouse an array of diverse negative emotions towards them. Even, to establish those emotions and to channel them into actions such as brandishing a swatter to kill every fly you see.[4] Nevertheless, this can be softened or reversed, to reach a better balance. We may no longer believe that common houseflies are killers, dangerous enemies, but the rejection felt towards them is widespread, and we usually overestimate and exaggerate the risk that they pose for human health and human houses.[5] It seems that we never recovered the tolerant attitude, even the sympathy towards those flies, that could be found in other times.[6]

2 From Solar Microscopes to Magic Lantern Slides: Optical Experiences in Front of an Audience

The company of scientific photographers and microscopists Flatters, Milborne and McKechnie – a partnership that was founded in 1909 in Manchester, England – offered, within its catalogue of educational and scientific magic lantern slides, one whose handwritten title read: "Queen Victoria Photographed through the Eye of Beetle."[7] An object materialized after her death in 1901, it

3 Rob Boddice, "The History of Emotions," *Revista de Estudios Sociales*, no. 62 (2017).

4 An impulse that could be passed on from grandmother to mother to daughter: Sally Edelstein, "Flies- The Fear Factor," *Envisioning The American Dream* (blog), October 8, 2020, https://envisioningtheamericandream.com/2020/10/08/flies-the-fear-factor/.

5 The short answer to the question of what fly species I am primarily addressing is *Musca domestica*. However, either houseflies, domestic flies or even flies are porous and ample terms when used to refer to a kind of roughly similar flies that you can find inside houses.

6 Naomi Rogers, "Germs with Legs: Flies, Disease, and the New Public Health," *Bulletin of the History of Medicine* 63, no. 4 (1989): 601–604; Theodore Tilton, *The Fly* (New York: Sheldon and Co., 1865).

7 See: "Flatters, Milborne and McKechnie" [item 1001386] in *Lucerna Magic Lantern Web Resource*, accessed August 30, 2023, https://lucerna.exeter.ac.uk/organisation/index.php?language=EN&id=1001386; Brian Stevenson, "A Very Unusual Lantern Slide of 'Queen Victoria Photographed through the Eye of Beetle,'" *Historical Makers of Microscopes and Microscope Slides* (blog), February 2014, http://microscopist.net/VictoriaBeetle.html.

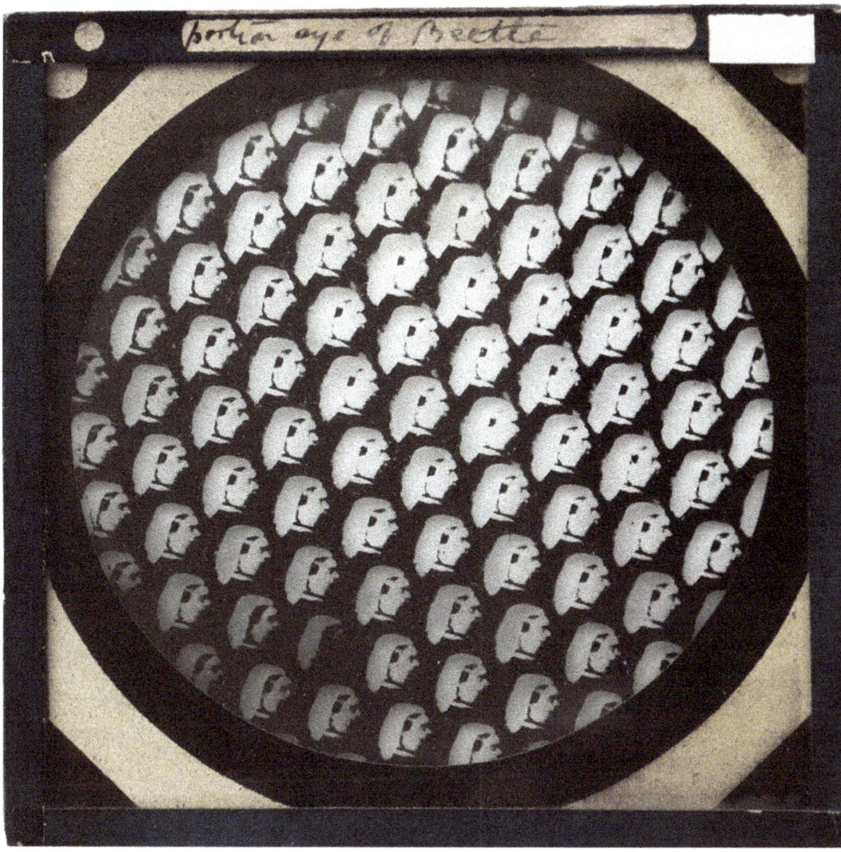

FIGURE 3.1 Magic lantern slide of "Queen Victoria Photographed through the Eye of Beetle", acquired on-line, maybe a copy of a slide by the partnership Flatters, Milborne and McKechnie
PHOTOGRAPH BY ÁNGEL VILLÉN

showed a three-quarter profile image of a mature Queen Victoria wearing a black dress and a white veil as seen, and multiplied, through the minute lenses of the compound eye of an insect – in this case, a beetle [Figure 3.1]. This object adopts and adapts certain aforementioned scientific and optic experiences, part of a lineage coming from previous centuries, and linked to the exploration of the compound eyes of insects.

These experiences involved the use of a microscope to magnify and observe very tiny things. For example, insect parts such as the mouths, feet or heads of flies, readily available for handling and dissection thanks to their ubiquity. One paradigmatic illustration of this is the head of a grey drone fly depicted by Robert Hooke (1635–1703) for *Micrographia*, where the lens of every single ommatidium making up each of the two compound eyes is individually rendered.[8] Besides amplification, another step implied further manipulating some of these insect parts, as explained and executed by Antoni van Leeuwenhoek (1632–1723). He combined a dissected fraction of the cornea of a dragonfly's eye with one of his microscopes to configure a Frankensteinian assemblage, with both organic and inorganic parts, that became animated by sources of light like the flame of a candle or a ray of sunlight. When properly attuned and focused, it was possible to employ the contraption as a telescope and look through the single glass lens of the microscope and the scrubbed chitin lenses of the compound eye. Subsequently, Leeuwenhoek watched how the flame of a candle, his neighbor's house or Delft's New Church tower were multiplied and inverted by the lenses of the insect's eye.[9] This assemblage and experiment was then replicated or evoked along the following centuries, typically framed as an optical curiosity and without necessarily assuming that flies or insects saw such iterated images.[10]

Accordingly, the "Queen Victoria Photographed through the Eye of Beetle" lantern slide adopted and adapted these previous micro-telescopic experiences, simultaneously looking through the small (micro-) and towards the distant (tele-), by using photography – or more precisely, microphotography. The photographic camera and film registered the resulting image. Thereafter, the slide was projected through light, and was thus diffused and communicated to an audience during a magic lantern lecture, in which elements of knowledge dissemination and entertainment converged. However, other prior events also mixed science and spectacle. Long before the existence of the Queen Victoria lantern slide, certain audiences were already used to the projection of microscopic images in darkened rooms. They were either static or moving images of phenomena like mineral crystals forming or living animalcules, as projected by a variation of the microscope: the solar microscope.[11] Along the 18th century,

8 Robert Hooke, *Micrographia: Or Some Physiological Descriptions of Minute Bodies Made by Magnifying Glasses with Observations and Inquiries Thereupon* (London: Jo. Martyn and Ja. Allestry, printers to the Royal Society, 1665), 174.

9 Antoni van Leeuwenhoek, *Alle de brieven van Antoni van Leeuwenhoek – The Collected Letters of Antoni van Leeuwenhoek*, ed. L.C. Palm, vol. x (Amsterdam: Swets and Zeitlinger, 1939), 125–129.

10 As Henry Baker underscored, it was reasonable to think that flies didn't see these mosaics of multiplied images, just as humans do not see double despite having two eyes: Henry Baker, *Of Microscopes and the Discoveries Made Thereby: Illustrated with Many Copper-Plates*, vol. I (London: J. Dodsley, 1785), 227.

11 Peter Heering, "The Enlightened Microscope: Re-Enactment and Analysis of Projections with Eighteenth-Century Solar Microscopes," *The British Journal for the History of Science* 41, no. 3 (2008), doi:10.1017/S0007087408000836; Elizabeth Cavicchi, "A Witness Account of Solar Microscope Projections: Collective Acts Integrating across Personal

solar microscopes made available to educated audiences augmented preparations of natural history like bits of wood, seeds, fleas, beetles, sections of flies and of other insects, that could fill a wall, and cause the corresponding impression.

The 19th century brought the standardization of science and scientists and did not favor solar microscopes. Instead, these devices became part of more spectacular enterprises, like shows and exhibitions such as the optician Philip Carpenter's Microcosm. Established on Regent Street, London in 1827, it displayed the "wonders of nature, by means of 14 lucernal and diurnal microscopes."[12] Two lithographs advertising Microcosm feature a frame of 14 rounded microscope views "on a reduced Scale," since the actual views reached from "5 inches to 3 Feet in Diameter."[13] The frame is exactly the same in both images, with views of crystals, vegetal matter, microorganisms, eels, cheese mites, water fleas, as well as insects such as a weevil, larva of the common gnat, a peacock louse, a flea, a horse fly, or the section of a fly's compound eye, with its rows of hexagonal facets. The central oval of one of the lithographs includes the text of the announcement, but the other one highlights the view of a giant fly, as if on a wall, manifesting the insistence on and the prominence of flies (and their eyes) when dealing with microscopes.[14]

In theory, when entering the dark space of the exhibition, the preparations would be seen enlarged and projected on white screens or surfaces, thanks to sun beams if the day and the weather allowed, or to artificial light. A few of these projections were of live, moving specimens, like organisms in a drop of water or cheese mites; maybe also flies, as in the lithograph. Contemplating them would have been similar to watching an early film, ever changing and never recorded. The reactions and emotions of the visitors may have oscillated

and Historical Memory," *The British Journal for the History of Science* 41, no. 3 (2008), doi:10.1017/S0007087408000885.

12 Phillip Roberts, "Philip Carpenter and the Convergence of Science and Entertainment in the Early-Nineteenth Century Instrument Trade," *Science Museum Group Journal* 7, no. 7 (2017). The quote is from the lithograph advertising Microscosm: Philip Carpenter, Charles Joseph Hullmandel, and George Johann Scharf, "Microcosm, A Grand Display of the Wonders of Nature," *Google Arts & Culture, ca.* 1827, accessed September 1, 2023, https://artsandculture.google.com/asset/microcosm-a-grand-display-of-the-wonders-of-nature-carpenter-philip-author-hullmandel-charles-joseph-printer-y-scharf-george-johann-lithographer/ygFr5K2axtYe9w.

13 Carpenter, Hullmandel and Scharf, "Microcosm".

14 Philip Carpenter, Charles Joseph Hullmandel, and George Johann Scharf, "Microcosm, A Grand Display of the Wonders of Nature," *Science & Society Picture Library, ca.* 1827, https://www.scienceandsociety.co.uk/results.asp?image=10422753. The fly seems inspired by another drawing by Hooke: Hooke, *Micrographia*, 182.

between fascination and disgust towards the enlarged creatures. In fact, the latter is predominant in a satirical print by William Heath titled "Monster soup commonly called Thames water" issued *ca.* 1828 in connection with a controversy regarding the contaminated supply of London water companies.[15] It shows a woman with her face contorted by repugnance towards what appears to be the circular projection of a solar microscope displaying the monstrous inhabitants of a drop of river water.[16] Perhaps a response intensified because in a solar microscope the organisms were alive and present, their movements instantaneously broadcasted. A similar reaction could be expected towards the magnified presence of cheese mites, and to the thought that they were there, roaming in food. This is the plot of the 1903 cinematic short *Cheese Mites*, with an initial sequence of a gentleman who becomes horrified when he directs his magnifying glass to the cheese that he was eating, then followed by a shot of the mites.[17] It is also a sample of how certain topics and tropes persisted across the decades, leaping from media like solar microscope or magic lantern shows to cinema.

Whether animated drops of water, cheese mites or flies, they all seemed tinged by a certain estrangement of the everyday. They shared human spaces but were not seen, and in the darkened but enlightened room their projections became unavoidable, weird, even fearsome. In this context, the insistence on flies and on their eyes could induce the slow and unsettling realization that the fly crossing the room was not just being looked at, but also doing the looking.

Solar microscopes opened to an audience the experience of getting closer and closer to the augmented surfaces of compound insect eyes. In order to explore further those eyes, and their gaze, an additional step pierced and peeked through their lenses. It was made possible via the blending of the preexisting magic lantern with the advances in photography and microphotography. This enabled a mass media representation, standardization and diffusion of the experiment explained by Leeuwenhoek. On the one hand, it produced slides like "Queen Victoria Photographed through the Eye of Beetle," a microphotography of the microscope preparation of a section of a beetle's eye functioning like the lens (or lenses) at the distant end of any other optical apparatus. On the other hand, photography could be used to incessantly reproduce this and

15 Daniel E. Lipschutz, "The Water Question in London, 1827–1831," *Bulletin of the History of Medicine* 42, no. 6 (1968).

16 William Heath, "Monster Soup Commonly Called Thames Water," *Philadelphia Museum of Art, ca.* 1828, https://philamuseum.org/collection/object/54821. Actually, a secondary title of the print reads "Microcosm, dedicated to the water companies."

17 Francis Martin Duncan, *Cheese Mites*, 1903, https://www.youtube.com/watch?v=wR2DystgByQ.

other similar microphotographs across different media and, combined with other transfer processes, the associated lantern slides could be multiplied, replicated, massively fabricated, and offered as part of a catalogue.[18] Afterwards, these slides featuring microphotographs could be projected in front of an audience during lectures that could be repeated tens, hundreds, thousands of times. Instead of requiring an actual preparation, or the complexities of managing and focusing a solar microscope, a lecturer could explain numerous natural history topics just with the help of a magic lantern and a commercially available set of glass slides. For instance, insect families, the various parts of their bodies, or what a fly's eye looked like, while showing an illuminated slide with its hexagonal facets.

3 Looking through Compound Eyes in Early Mass Media

The complex procedure of both looking and photographing through the lenses of an insect eye was also disseminated through mass media in journals and magazines, with the accompanying images. For instance, the article "How to Photograph Through a Fly's Eye" by Fred. W. Saxby appeared in 1898 in the British popular science periodical *Knowledge: An Illustrated Magazine of Science*.[19] The text provides, in a comprehensive manner, all the technical details necessary to reproduce the process, and to get a photograph of something seen, and focused, through the minute lenses of a fly. As in Leeuwenhoek's case, this fly was not a proper *fly* in a scientific sense, that is, an insect from the order Diptera, but a common dragonfly, from the order Odonata, considered better for the purpose because "parts of the eye may be found nearly flat."[20] Nonetheless, it was still announced as a fly in the article's title, perhaps owing to the standing and fame of flies among microscopists and microphotographers.

Saxby advises to photograph a subject "either self-luminous or white," like a window with an open landscape and plenty of natural light. Or, if artificial light was to be used, "opaque figures pasted on a ground-glass screen illuminated from behind;" even better, "white subjects by reflected light, such as a

18 The number of "Queen Victoria Photographed through the Eye of Beetle" lantern slides made by Flatters, Milborne and McKechnie is unknown. I found and bought another one on-line, of a lower quality, that may have been made and copied by another firm [Figure 3.1].

19 Fred. W. Saxby, "How to Photograph Through a Fly's Eye," *Knowledge: An Illustrated Magazine of Science* 21 (1898).

20 Saxby "How to Photograph," 188.

bust or a statuette." For his experiment, he continues, he has chosen "a small chalk bust of Her Majesty the Queen."[21] Undoubtedly, this would have been Queen Victoria, and as an explanation of this very specific choice Saxby poses a laudatory and rhetoric question to the reader: "is it not appropriate that the noblest and greatest monarch the world has ever seen should be the subject of a photograph through the most infinitesimal lens known to science?"[22] This image does not accompany the article, which is illustrated just by one photograph, that of a "Group of 'Eye-lenses' ready for the camera," magnified two hundred times.

However, the article titled "Photographing through Animals' Eyes", signed by William Franklin Watson and appeared in April 1904 in *Collier's: The National Weekly*, a United States magazine, did include three photographic illustrations.[23] The second one, a section of the "Corneal lenses from the eye of a house-fly," is featured in between a "Photograph of a house-fly made with the crystalline lens from the eye of an ox" and the "Multiple-image picture made with beetle's-eye lenses." The latter shows the iterated profile of what seems to be an anonymous man – maybe the author – looking to the left and, like Flatters, Milborne and McKechnie's Queen Victoria lantern slide, is linked to the compound eye of a beetle, instead of a fly or of a dragonfly. Flies kept figuring as protagonists, probably due to the constant insistence in their magnified eyes or to their availability, since the two other figures were photographs of a house-fly and of the lenses of the compound eye of a house-fly.

Interestingly enough, in the 1909 book *Nature through Microscope and Camera*, Richard Kerr refers to the eye of *Dytiscus marginalis* – a water beetle – and underscores it as "a beautiful object for the microscope". According to Kerr:

> The face of a watch and portraits of the late Queen have frequently been photographed through the facets or ocelli of this beetle's eye. A small portion of the eye is best for this purpose, because it is easier to flatten out a tiny area than the whole cup-shaped exterior.[24]

As with dragonflies, beetle eyes were considered convenient to photograph watches, queens, or other people through – perhaps, the Victorian beetle craze

21 Saxby "How to Photograph," 189.

22 Saxby "How to Photograph," 189.

23 William Franklin Watson, "Photographing through Animals' Eyes," *Collier's: The National Weekly*, April 30, 1904.

24 Richard Kerr, *Nature through Microscope and Camera* (London: The Religious Tract Soc., 1909), 84.

also had something to do with this.[25] But flies, and their eyes, were still very present because of the association of ideas that had been generated, as shown by Saxby's title or by two of the photographs by Watson, and through their availability. This handiness, along with the flies' disposability, may have turned these insects into easy targets, and feasible enemies, as we will see.

The "Queen Victoria Photographed through the Eye of Beetle" lantern slide succeeds a lineage of experiences and representations involving people and the compound eyes of insects. This can be seen as part of an incipient dissemination of those contents and knowledge through early mass media, such as the magic lantern and its light projections. According to the range of options sketched by Joe Kember, this specific object could be situated at a crossroads of magic lantern history, since its composite nature integrates three of the moments or trends that have been identified and studied in its evolution.[26] First, there is a scientific element: the enlargement of a section of a compound insect eye. This was a widespread motif in natural history and microscopist circles, and it can be connected to the trend that overtook the magic lantern from the mid-nineteenth century onwards, as it became a favored and extensively used medium for the popularization of science, educational discourses, and didactic lecturers. The second element, Queen Victoria's portrait, is seen and multiplied through the insect eye, and points to the political propaganda and other forms of persuasion that became common in the medium around the turn of the century. It belongs to an imperial backdrop that approached scientific and technological developments, as well as mass production, as something accomplished at the material expense of those deemed disposable, whether other animals or other humans.

Finally, the composite nature of the slide in itself, an imperial queen seen through the minute but magnified lenses of one of the eyes of a beetle, reveals a fractured superposition and a strangeness that brings to mind previous uses of the magic lantern, present until the second half of the 19th century, like phantasmagoria, gothic entertainment and its spectral amusements.[27] Despite their scientific and didactical aim, many magic lantern sessions and lectures were intended as shows, and they tried to keep a balance between education

25 Deborah Denenholz Morse and Martin A. Danahay, "Victorian Beetlemania," in *Victorian Animal Dreams: Representations of Animals in Victorian Literature and Culture*, ed. Deborah Denenholz Morse and Martin A. Danahay (Aldershot: Ashgate, 2007).

26 Joe Kember, "The Magic Lantern: Open Medium," *Early Popular Visual Culture* 17, no. 1 (2019): 1, doi:10.1080/17460654.2019.1640605.

27 For a condensed history of the magic lantern with a careful consideration of its trends, see: Deac Rossell, "The Magic Lantern," in *Ich sehe was, was Du nicht siehst! – Sehmaschinen und Bilderwelten*, ed. Bodo von Dewitz and Werner Nekes (Göttingen: Steidl, 2002).

and entertainment. As with solar microscopes, watching the projections of giant insects while sharing the same darkened room with them could easily be experienced as uncanny. Even more so if the assumption was that you were seeing as if through the compound eyes of one of those insects, either flies, dragonflies, or beetles.

4 "Flies and Their Foes, through Microscope and Camera", a Magic Lantern Lecture

These composite slides showing an iterated, multiplied image as seen through the lenses of the compound eye of an insect were projected, presented, and explained in front of an audience during certain magic lantern lectures. There is a particular and relevant example of this. On November 10th, 1908, at the ordinary meeting of The Royal Photographic Society of Great Britain, held at 66, Russel Square in London, Frank P. Smith – the naturalist and documentalist better known as Percy Smith (1880–1945) – delivered a magic lantern educational lecture featuring one of these slides. The title of this lecture was "Flies and Their Foes, through Microscope and Camera" and it was "illustrated by [both] the lantern and cinematograph," as it was subsequently chronicled in the December 1908 issue of *The Photographic Journal*.[28]

From the outset, Smith stated that he was going to deal with the topic more like an entomologist and less as a photographer, because he had produced his studies – the photographic lantern slides and the films – "simply as scientific records, using ordinary plates and blindly following given formulae."[29] Thus, Smith downplays his creative virtues in order to highlight his scientific merits. This also confirms that he made the lantern slides, manipulating the devices and various insects and other animals that were available to him.

Smith proceeded to systematically explain the characteristics of flies and address several of their groups: blow-flies, gnats, fleas – which, he explained, at that point some scientists classified as flies. Next, the anatomic parts of flies: the "beautiful and complex structure of the fly's foot," "in great favor among microscopists;" its hairy pads; the spiracles in the thorax and abdomen, and

28 Frank Percy Smith, "Flies and Their Foes, through Microscope and Camera," *The Photographic Journal* 48–XLVIII, no. 2–II (1908). I became aware of this lecture through Oliver Gaycken's book.

29 Smith, "Flies and Their Foes," 367.

the eyes of the fly.[30] Although the record of the lecture does not specify it, surely each of these steps – flies' anatomical details, enlarged specimens of the various groups – was illustrated with lantern slides. But when we get to the eyes of the fly, the chronicle does register a specific slide, the only one mentioned and singled out:

> The eyes of the fly consisted of a multitude of little lenses, and Mr. Smith here showed a remarkable photomicrograph, in the making of which a small black cross had been photographed through such an eye, and an image produced in each particular lens. It would have been possible, he said, to have covered a whole plate with these well-defined crosses.[31]

Here, the motif that was microphotographed was the small black cross instead of Queen Victoria. However, the lantern slide registered a similar process of multiplying an object through the lenses of the compound eye of an insect.

Smith continued his lecture providing examples of harmful flies, like the mosquito that transmitted the malaria, or the tsetse fly, "causing sleeping sickness and working great havoc on the African continent."[32] Right afterwards, he issued a stark warning against domestic flies – named in the unsigned minutes as "the fly" or "house fly" – whose useful scavenging didn't compensate their other filthy actions, like favoring decomposed matter and then proceeding with the unsavory "habit of sampling everything on the breakfast table."[33] According to the review of the lecture delivered by the photographer and filmmaker:

> In time, doubtless, people would realise what a disease-producing agent the fly was, and, moreover, it was probable that there were a great many injurious species of flies which at present were not known to be injurious, and when the causes of some obscure diseases were at length discovered, it might be found that the fly had a good deal to do with them.[34]

In connection with this, Smith pointed out the helpful efforts against flies of several fly foes, underscored by the title of the lecture: spiders, lizards, or frogs,

30 Smith, "Flies and Their Foes," 367.
31 Smith, "Flies and Their Foes," 367.
32 Smith, "Flies and Their Foes," 368.
33 Smith, "Flies and Their Foes," 368.
34 Smith, "Flies and Their Foes," 368. As seen in this passage, neither the terms themselves nor the limits between "the fly" as a species or group and other species of flies are clearly defined.

among others. He concluded with a few filmic sequences linked to some of his short movies, made for the Charles Urban Trading Company. These fragments:

> [...] illustrated the wonderful strength and endurance of the blue-bottle [fly], and also the results of an attempt to "train" [sic] the insects to use miniature dumb-bells and perform acrobatic feats. Another film brought home the objectionable habit of the fly of feeding on putrid matter, while others, again, illustrated the characteristic action of the fly's trunk [...].[35]

The fact that the lantern slide of the small black cross through the eyes of a fly was salient among the slides projected during this lecture is relevant for two reasons. On the one hand, the lecturer was Frank Percy Smith, the film-maker and wildlife documentary pioneer that, at the time, was working for the producer Charles Urban (1867–1942), and who would become known for the seminal film series *Secrets of Nature*.[36] This situates the said slide in a highly influential context. As part of Urban's quest to establish cinema as an educational force without renouncing its entertainment appeal, Smith made films like *The Acrobatic Fly* (1910?) or *The Balancing Blue-bottle* (1908) – according to its description, footage of the latter was shown during the lecture. On the other hand, a war against flies had been declared since the end of the 19th century, in the shape of an anti-fly campaign that started in the United States, but later spread globally to many other countries. Both Charles Urban and Percy Smith ended up at the center of this anti-fly campaign.

5 Fly Wars: the Turn of the 20th Century Anti-fly Campaigns

The description of Percy Smith's lecture suggests that he projected sequences from *The Fly Pest*, a short filmed by him and produced by Urban that was released in 1910 as part of the anti-fly campaign.[37] Among them, the cited "habit of the fly of feeding on putrid matter," and afterwards "sampling everything on the breakfast table." Likewise, the chronicle refers in passing to the distribution of pamphlets and other actions by relevant institutions, or to potential future

35 Smith, "Flies and Their Foes," 368.

36 Mary Field and Frank Percy Smith, *Secrets of Nature* (London: Faber and Faber, 1934).

37 Frank Percy Smith, *The Fly Pest*, 1910; Oliver Gaycken, *Devices of Curiosity: Early Cinema and Popular Science* (Oxford and New York: Oxford University Press, 2015), 143–156.

findings that would probably establish flies as the cause of many illnesses, contributing some actions and arguments in the realm of the campaign.[38]

According to Naomi Rogers, this anti-fly campaign was a way of advancing and consolidating medical entomology, and it contributed to establishing entomologists as institutional, public health figures. It was also impelled by the necessity, and convenience, of providing a simple, visible and cheap solution to public health concerns linked to the prevalence of illnesses such as typhoid fever, enteric diseases, or in some cases, even poliomyelitis.[39] The presence or absence of household insects could be easily seen and checked in ways not possible with germs and bacteria, and "one insect in particular was familiar to every housekeeper and health official: the common housefly."[40] The connection between flies and germs was made tangible in laboratories, through experiments that implied photographing the bacteria cultivated on gelatine plates after a fly had walked across them, and dramatically exposing the fly's tiny steps.[41] And thus claiming they were infectious, and dangerous:

> The early decades of the twentieth century saw the publication of dozens of books and pamphlets warning the lay community and the medical profession of the habits of houseflies. In these works, the fly was transformed from a friendly domestic insect into a threat to health and hearth. It was portrayed as horrific, in fact physically repulsive, and its dangers were exaggerated so that at times it became as certain a killer as the mosquito that spread yellow fever and malaria. Health officials sought to develop the idea of germs into a practical and comprehensible weapon against disease. To do this they portrayed flies as germs with legs.[42]

Indeed, at the time there were studies that pointed out common flies as the cause of certain illnesses, a theory to be refuted or disputed years later. One example was Alice Hamilton's work and publication regarding a 1902 typhoid epidemic in Chicago, where she suggested that the flies were responsible for the spread of the disease.[43] However, later on she discovered that the actual

38 On the turn of the century anti-fly campaign, see: Rogers, "Germs with Legs"; Bill Marsh, "Visual Education in the United States and the 'Fly Pest' Campaign of 1910," *Historical Journal of Film, Radio and Television* 30, no. 1 (2010), doi:10.1080/01439680903577235.

39 Rogers, Naomi. "Dirt, Flies, and Immigrants: Explaining the Epidemiology of Poliomyelitis, 1900–1916." Journal of the History of Medicine and Allied Sciences 44, no. 4 (1989): 486–505. https://doi.org/10.1093/jhmas/44.4.486.

40 Rogers, "Germs with Legs," 601.

41 Rogers, "Germs with Legs," 607–608.

42 Rogers, "Germs with Legs," 601.

43 Rogers, "Germs with Legs," 606–607.

cause behind the epidemic was a leak of sewage into the pipes of drinking water, but few listened:

> The truth was more shocking than my ingenious theory, and it never came to light, so far as the public was concerned. For years, although I did my best to lay the ghosts of those flies, they haunted me and mortified me, compelling me again and again to explain to deeply impressed audiences that the dramatic story their chairman had just rehearsed had little foundation in fact.[44]

It was generally easier and more clear-cut to blame domestic flies, or the lazy people who tolerated the filth that attracted flies, than to fund and to launch costly reforms of buildings, streets, sewages, etc., whose benefits were not immediately and visibly apparent. Moreover, everyone – even children – could contribute to the piles of dead flies that became visible signs of the fly wars and of their success. Therefore, the responsibility of fighting illness and filth was placed on people instead of on public institutions.

As part of the anti-fly campaigns, there were public appeals in newspapers, poetry competitions, exhibitions, posters, pamphlets as well as other publications, lectures, and the already mentioned Urban and Smith's film *The Fly Pest*. Two thirds of the film (*ca.* twelve minutes in total) portrayed and described flies as another natural history subject, as former Urban films had done with other animals.[45] But the last third was edited to show, step by step, how flies lived among filth and brought filth and disease into human food. One fly was even shown landing on a pacifier, then put into a baby's mouth. Consequently, the film was considered revolting and disgusting – it even experienced censorship episodes[46] –, but also necessary, and was a huge success, both commercially and in the context of anti-fly campaigns.[47] A book on film advertising even provided tips on how to profit from the fly films and the anti-fly campaigns by organizing "Swat the Fly!" events and screenings, creating a club for children, or a prize for whoever killed the most flies and brought them to the theatre. This way, a pile of dead flies – from time to time discreetly increased

44 Alice Hamilton, *Exploring the Dangerous Trades: The Autobiography of Alice Hamilton, M.D.* (Boston: Little, Brown and Company, 1943), 99–100.

45 Gaycken, *Devices of Curiosity*, 149.

46 Gaycken, *Devices of Curiosity*, 146.

47 Marsh, "Visual Education," 29–30.

with a hidden empty cigar box to make it look larger – could be shown at the entrance, inside a glass case, as a publicity stunt.[48]

Due to these campaigns and efforts, houseflies went from being considered as harmless and only mildly annoying inhabitants of the house, often tolerated, or even regarded with a certain amount of sympathy, to being treated as lethal threats that had to be actively exterminated. As presented by Rogers,[49] in the 1860s it was still possible to find a children's book that portrayed flies in an amiable manner, "[t]icling Baby's nose;" that instructed not to kill or hurt these insects even if "you see his wings of silk / Drabbled in the Baby's milk," or that highly praised their eyes and vision:

> Flies can see
> More than we –
> So how bright their eyes must be!
> Little Fly
> Mind your eye –
> Spiders are near by![50]

However, decades later children were taught about the dangers of flies and contributed to "Kill the fly!" essay and poetry contests as part of the campaigns.[51] Traces of this change of attitude persist even today, as evidenced in the workshops I delivered at the MNCN, where swatters popped in some of the drawings when the children were asked to draw a fly.[52]

---

48 Epes Winthrop Sargent and George P. Johnson, *Picture Theatre Advertising* (New York: The Moving picture world, 1915), 213–233.

49 Rogers, "Germs with Legs," 601–604.

50 Tilton, *The Fly*, VIII.

51 A drawing and a winning entry for an essay on the "Kill that Fly" topic, included in Britain's 1917 *Report of the National Baby Week Council* can be found in: Linda Bryder, "Mobilising Mothers: The 1917 National Baby Week," *Medical History* 63, no. 1 (2019), doi:10.1017/mdh.2018.60.

52 Ideally, it would be great to be able to deliver similar workshops in many other countries to assess beliefs and attitudes on flies. Aside from the English-speaking world, I have also found references to similar campaigns and contents across several different countries (campaigns in Argentina, posters from Portugal, German anti-fly films, anti-fly films exhibited in India or China very early on, etc.). Considering also that these attitudes and beliefs have found their way into contents across countries and cultures, my expectation is that the results of these workshops would probably be fairly similar regarding the enmity felt towards flies, and the erroneous belief in their mosaic vision.

In the following years, the popular success of *The Fly Pest* was translated at least into a sequel, called *The Menace of the Fly* (1915).[53] It included a novel sequence that was occasionally highlighted in reviews. As a brief published in *The Daily Telegraph* in 1915 stated:

> [...] by the far the most extraordinary part of this film is a section of photographs taken through the lenses of a fly's eye, thus giving an idea of how the fly sees his mortal enemy – man.[54]

This short sentence recovers the Queen Victoria lantern slide motif of looking through the lenses of a compound eye, although adapted to a new medium, cinema. At this point, and as a contrast with the microscope phase, it erroneously assumes that the optical trick is exposing how a fly really sees. It is also telling that this misconception accompanies what distinctly recalls the plot of many horror films starred by giant or invasive insect monsters.

I am not aware of a surviving copy of *The Menace of the Fly*. However, years ago I found about a British Pathé film from 1930 called *Eyes – by the Thousand!* which does include a similar sequence.[55] This two-minute film provides some facts about flies, and then displays some magnified anatomical views and "micro-cine pictures" of a fly's head, cornea and its lenses. Finally, it compares the human eye with an ordinary "cine-lens" and proceeds to show a mechanical toy through a section of lenses of the compound eye of a fly, so we see tens of tiny clowns spinning wheels.[56] This additional step regarding the microscopic-telescopic experience of looking through the facets of an insect compound eye involved movement and film.

---

53 *The Fly Pest*'s success also translated into similar films being made in other countries. I recently identified, in the Wellcome collection, excerpts that appear to belong to the German film *Gefährliche Plagegeister* [Dangerous pests, 1929], that are clearly inspired by Percy Smith's work, and also include microscopic views of fly parts, and of their compound eyes.

54 "[The Menace of the Fly]," *The Daily Telegraph*, May 27, 1915. I first saw this reference in: Jonathan Burt, *Animals in Film* (London: Reaktion Books, 2002), 124. Then, I found another review highlighting the same sequence: "The Menace of the Fly," *The Daily Telegraph*, November 10, 1916.

55 *Eyes –By the Thousand!* (British Pathé, 1930).

56 I analysed this film, as well as how flies don't see, in Concepción Cortés Zulueta, "¿Cómo no ven las moscas? Sobre mosaicos de iteraciones hexagonales y otras convenciones visuales," in Arte Ecosocial: Otras maneras de pensar, hacer y sentir, ed. Tonia Raquejo and Verónica Perales Blanco (Madrid: Plaza y Valdés, 2022), 119–142.

FIGURE 3.2 POV shot of the giant hybrid fly showing the multiplied scream of Hélène Delambre, played by Patricia Owens. *The Fly* (Kurt Neumann, 1958)

In his book, *Devices of curiosity: Early Cinema and Popular Science*, Oliver Gaycken has explained how early documentary short films such as Urban's or Percy Smith's, among others, as well as the sessions in which they were included, featured certain structures and characteristics that were then inherited by certain movie genres.[57] Gaycken focuses mostly on detective films and crime melodrama, but I believe that the same can be said of horror films in general, and horror films with giant or invasive insect monsters in particular. There is an iconic sequence in the horror film *The Fly* (Kurt Neumann, 1958)[58] that is a direct successor of what was illustrated by the "Queen Victoria Photographed through the Eye of Beetle" lantern slide, Saxby's and Watson's articles, Percy Smith's lecture, or *The Menace of the Fly* or *Eyes – by the Thousand!* films, and that acts as a transmitter of that motif together with certain knowledge assumptions and emotions. This particular scene is so paradigmatic that many have seen it even if they have not watched the entire film. It shows the scream of the terrified scientist's wife from the perspective of the hybrid creature that he has turned into because of his misdoings: half human and half fly with a giant fly head and huge compound fly eyes [Figure 3.2]. Therefore, in what constitutes the climax of the movie, the camera adopts the giant fly's point of view, and the screaming face of the woman multiplies, in a similar fashion as the bust of Queen Victoria did in the lantern slide.

57 Gaycken, *Devices of Curiosity*, 158–188.
58 Kurt Neumann, *The Fly* (Regal Films, Twentieth Century Fox, 1958).

Queen Victoria or wife of mad scientist, there are similitudes between the multiplied images of both women, but also differences. First, the educational lantern slide reproduces an optical experience, while the fictional filmic scene is just evocating and representing a convention linked with it – there are no actual fly lenses involved, only a kaleidoscope-like effect. Moreover, a preceding massive diffusion of knowledge, of magnified images of fly's compound eyes, and of the way their lenses iterated the objects was necessary in order for the scene to be understood and assimilated by the audience. Finally, where the slide offered wonder, curiosity and significance – even though it also insinuated a dash of strangeness and unease – the film, through a multiplication that intensified the impact of the woman's scream, massively disseminated fear, disgust and rejection.

6 An Insect Planet, a Shared Fate

Drawing from a preceding succession of conventions on insect vision and compound eyes, but also departing from them, *The Fly* and other similar contents mix different kinds of misconceptions about common houseflies. These are simultaneously related to how they see and to the alleged dangers they pose for humans, appearing intimately entangled with negative emotions. Since they are usually conveyed through means that are decoupled and distanced from an explicit and declared context of transmission of knowledge – like audiovisual fiction – these fallacies are more difficult to dispute or rebuke. This is because people's internalized assumptions of fly or insect vision often bound with a type of disgust whose explanations and particularities relate to a specific historical period and to the lineage of optical experiences I have delineated and exposed, jumping and adapting from one media to the next.

All of the above, I believe, has favored a conception of insects in general as monstruous others with alien perceptions, as menaces that it is best to kill, to exterminate, to keep away from humans and from human homes. Because, since flies are so handy and accessible, so ubiquitous, the ways they are perceived and how we perceive our relationship with them presumably tinge and encourage misconceptions and negative views about insects in general, as well as the aversion or indifference that has surely contributed to the current decay of insects. Now that insect populations are in constant decline around the globe, and the phenomenon has been labelled as an insect collapse or an insect apocalypse, it is more relevant than ever to look for, identify and expose the sources of human rejection and disgust towards insects. So we care more about the fate of the insects and about the fate of the planet they inhabit, sustain, and make possible. And therefore, we can mend our shared fate.

Bibliography

Baker, Henry. *Of Microscopes and the Discoveries Made Thereby: Illustrated with Many Copper-Plates*. Vol. I. London: J. Dodsley, 1785.

Boddice, Rob. "The History of Emotions." *Revista de Estudios Sociales*, no. 62 (2017): 10–15.

Bryder, Linda. "Mobilising Mothers: The 1917 National Baby Week." *Medical History* 63, no. 1 (2019): 2–23. doi:10.1017/mdh.2018.60.

Burt, Jonathan. *Animals in Film*. London: Reaktion Books, 2002.

Carpenter, Philip, Charles Joseph Hullmandel, and George Johann Scharf. "Microcosm, A Grand Display of the Wonders of Nature." *Science & Society Picture Library, ca.* 1827. https://www.scienceandsociety.co.uk/results.asp?image=10422753.

Carpenter, Philip, Charles Joseph Hullmandel and George Johann Scharf. "Microcosm, A Grand Display of the Wonders of Nature." *Google Arts & Culture, ca.* 1827. Accessed September 1, 2023. https://artsandculture.google.com/asset/microcosm-a-grand-display-of-the-wonders-of-nature-carpenter-philip-author-hullmandel-charles-joseph-printer-y-scharf-george-johann-lithographer/ygFr5K2axtYe9w.

Cavicchi, Elizabeth. "A Witness Account of Solar Microscope Projections: Collective Acts Integrating across Personal and Historical Memory." *The British Journal for the History of Science* 41, no. 3 (2008): 369–383. doi:10.1017/S0007087408000885.

Cortés Zulueta, Concepción. "¿Cómo no ven las moscas? Sobre mosaicos de iteraciones hexagonales y otras convenciones visuales." In *Arte Ecosocial: Otras maneras de pensar, hacer y sentir*, edited by Tonia Raquejo and Verónica Perales Blanco, 119–142. Madrid: Plaza y Valdés, 2022.

Cortés Zulueta, Concepción. "¿Cómo ven las moscas? Un taller sobre los ojos compuestos de las moscas en el Museo de Verano," *NaturalMente* 41, March 2024. https://revista.mncn.csic.es/nm41/72-73/.

Cortés Zulueta, Concepción. "A Horror Multiplied by the Eyes of Every House Fly: Compound Misconceptions and Prejudices on Filmic Insects", in press.

Edelstein, Sally. *Flies- The Fear Factor. Envisioning The American Dream* (blog). October 8, 2020. https://envisioningtheamericandream.com/2020/10/08/flies-the-fear-factor/.

Eyes – By the Thousand! British Pathé, 1930.

Field, Mary and Frank Percy Smith. *Secrets of Nature*. London: Faber and Faber, 1934.

"Flatters, Milborne and McKechnie." In *Lucerna Magic Lantern Web Resource*. Accessed August 30, 2023. https://lucerna.exeter.ac.uk/organisation/index.php?language=EN&id=1001386.

Gaycken, Oliver. *Devices of Curiosity: Early Cinema and Popular Science*. Oxford and New York: Oxford University Press, 2015.

Hamilton, Alice. *Exploring the Dangerous Trades: The Autobiography of Alice Hamilton, M.D.* Boston: Little, Brown and Company, 1943.

Heath, William. "Monster Soup Commonly Called Thames Water." *Philadelphia Museum of Art, ca.* 1828. https://philamuseum.org/collection/object/54821.

Heering, Peter. "The Enlightened Microscope: Re-Enactment and Analysis of Projections with Eighteenth-Century Solar Microscopes." *The British Journal for the History of Science* 41, no. 3 (2008): 345–367. doi:10.1017/S0007087408000836.

Hooke, Robert. *Micrographia: Or Some Physiological Descriptions of Minute Bodies Made by Magnifying Glasses with Observations and Inquiries Thereupon.* London: Jo. Martyn and Ja. Allestry, printers to the Royal Society, 1665.

Kember, Joe. "The Magic Lantern: Open Medium." *Early Popular Visual Culture* 17, no. 1 (2019): 1–8. doi:10.1080/17460654.2019.1640605.

Kerr, Richard. *Nature through Microscope and Camera.* London: The Religious Tract Soc., 1909.

van Leeuwenhoek, Antoni. *Alle de brieven van Antoni van Leeuwenhoek – The Collected Letters of Antoni van Leeuwenhoek*, edited by L.C. Palm, vol. x. Amsterdam: Swets and Zeitlinger, 1939.

Lipschutz, Daniel E. "The Water Question in London, 1827–1831." *Bulletin of the History of Medicine* 42, no. 6 (1968): 510–526.

Marsh, Bill. "Visual Education in the United States and the 'Fly Pest' Campaign of 1910." *Historical Journal of Film, Radio and Television* 30, no. 1 (2010): 21–36. doi:10.1080/01439680903577235.

Martin Duncan, Francis. *Cheese Mites.* 1903. https://www.youtube.com/watch?v=wR2DystgByQ.

"[The Menace of the Fly]." *The Daily Telegraph.* May 27, 1915.

"The Menace of the Fly." *The Daily Telegraph.* November 10, 1916.

Morse, Deborah Denenholz and Martin A. Danahay. "Victorian Beetlemania." In *Victorian Animal Dreams: Representations of Animals in Victorian Literature and Culture*, edited by Deborah Denenholz Morse and Martin A. Danahay, 35–51. Aldershot: Ashgate, 2007.

Neumann, Kurt. *The Fly*. Regal Films, Twentieth Century Fox, 1958.

Smith, Frank Percy. "Flies and Their Foes, through Microscope and Camera." *The Photographic Journal* 48–XLVIII, no. 2–II (1908): 367–370.

Smith, Frank Percy. *The Fly Pest.* 1910.

Roberts, Phillip. "Philip Carpenter and the Convergence of Science and Entertainment in the Early-Nineteenth Century Instrument Trade." *Science Museum Group Journal* 7, no. 7 (2017).

Rogers, Naomi. "Germs with Legs: Flies, Disease, and the New Public Health." *Bulletin of the History of Medicine* 63, no. 4 (1989): 599–617.

Rogers, Naomi. "Dirt, Flies, and Immigrants: Explaining the Epidemiology of Poliomyelitis, 1900–1916." Journal of the History of Medicine and Allied Sciences 44, no. 4 (1989): 486–505. https://doi.org/10.1093/jhmas/44.4.486.

Rossell, Deac. "The Magic Lantern." In *Ich sehe was, was Du nicht siehst! – Sehmaschinen und Bilderwelten*, edited by Bodo von Dewitz and Werner Nekes. Göttingen: Steidl, 2002.

Sargent, Epes Winthrop, and George P. Johnson. *Picture Theatre Advertising*. New York: The Moving picture world, 1915.

Saxby, Fred. W. "How to Photograph Through a Fly's Eye.'" *Knowledge: An Illustrated Magazine of Science* 21 (1898): 187–189.

Stevenson, Brian. "A Very Unusual Lantern Slide of 'Queen Victoria Photographed through the Eye of Beetle.'" *Historical Makers of Microscopes and Microscope Slides* (blog). February 2014. http://microscopist.net/VictoriaBeetle.html.

Tilton, Theodore. *The Fly*. New York: Sheldon and co., 1865.

Watson, William Franklin. "Photographing through Animals' Eyes." *Collier's: The National Weekly*. April 30, 1904: 23–25.

CHAPTER 4

Encounters with the Insect World: Care and Human – Insect Relationships in Wildlife Documentaries

Heidi Mikkola

1 Introduction

The narratives about insects have traditionally been situated within the realm of the "insect world," a concept referring to entomological "world-making." This notion encompasses the scales, perspectives, and cognitive frameworks associated with insects.[1] The notion of the insect world functions as a spatial metaphor for our understanding of insects, highlighting that it differs from "our world" in terms of temporal and spatial parameters, rendering humans incongruous within it; as Adam Dodd has argued "[w]ithin the insect world, humans enjoy no ontological propriety."[2] To bridge this scalar gap and make the insect world perceptible within human epistemologies, technology is essential. In wildlife documentaries about insects, the significance of technology is underscored in the visualization of the insect world, while micro and macro cinematography are widely used to visualize minute creatures and their habitats. Wildlife documentaries serve as illustrations of scientific facts and popularizing natural history, balancing science and storytelling. Broadcaster and naturalist David Attenborough[3] has noted that, when they started making *Life in the Undergrowth* (2005) in the early 2000s, invertebrates were "largely unexposed" in wildlife documentaries, while bigger animals tended to receive more visual presence in films. Insects seem to be marginalized when it comes to other animals we find more approachable, animals that we can look in the eye and have "the encounter value"[4] with. Jamie Lorimer[5] notes that insects are

1 Adam Dodd, "Minding Insects: Scale, Value, World," in *The Management of Insects in Recreation and Tourism*, ed. Raynald Harvey Lemelin (Cambridge: Cambridge University Press, 2012), 31, 33.

2 Dodd, "Minding Insects," 33.

3 David Attenborough, *Life on Air. Memoirs of the Broadcaster* (London: BBC Books, 2010), 386.

4 Donna Haraway, *When Species Meet* (Minneapolis: University of Minnesota Press, 2007), 46.

5 Jamie Lorimer, "Nonhuman Charisma," *Environment and Planning D: Society and Space* 25, no. 5 (2007): 920.

 | DOI:10.1163/9789004715448_006

"radically different to anthropocentric norms," and their charisma differs from anthropomorphic cuddly charisma, which is that of mammals and other vertebrates. However, as Lorimer argues, insects have charisma, and they attract human attention, whether natural historical curiosity or aesthetics on film. In this chapter, I examine three cases of insect documentaries: *Life in the Undergrowth* (2005), *Alien Empire. A Journey to the World of Insects* (1996), and *Microcosmos* (1996). The first two documentaries are BBC television series: *Life in the Undergrowth* is a five-part, presenter-led documentary series, and *Alien Empire* has six episodes with voice-over narration. *Microcosmos* is a French full-length documentary made for a bigger screen, released in film theaters, and has no voice-over narration, except in the very beginning and end of the film. Even if the two latter ones are from the 1990s, their visuality and approach to insects and their environment are still very relevant today. I have chosen these documentaries because they solely focus on insects and invertebrates, and each takes a different approach in representing insects, varying from one another in terms of narrative and visuality while offering different kinds of entry points to the insect world. These documentaries offer insights into the portrayal of insects, human – insect interactions, and the depiction of insects within the context of the documentary moving images and the wildlife mode.

I approach these documentaries through viewing, understanding it as a form of encounter.[6] I understand moving images not only as representations, but also as affective assemblages that involve the agencies of insects, humans, technology, aesthetics, and the viewer. I question what kind of relatings are brought forth while encountering the insect world in these documentaries. I explore this by employing two different but also intertwined analytical frameworks within documentary films. On the one hand, I delve into the visuality and aesthetics of the films, while on the other hand, I examine their narrative and scientific approach to insects. Both frameworks bring forth engagement and knowledge that encourage the viewer to engage with the insect world. I explore this in the context of wildlife mode and documentary films, which bring forth the relations between visuality, knowledge, and technology. Donna Haraway[7] notes that technologies are not just mediators but rather organic partners in world-making encounters. As Belinda Smaill[8] argues, the documentary generates an anticipation that the viewer will be "a subject of knowledge, offered access to a world that is knowable and testable, fulfilling a

6 Belinda Smaill, "Encountering Animals: Re-Viewing the Cinema of Jean Painlevé," *Antennae: The Journal of Nature in Visual Culture* 42 (2017).

7 Haraway, *When Species Meet*, 249.

8 Smaill, "Encountering Animals," 97.

desire for knowledge and an epistephilic pleasure." This brings up the question of anthropocentrism and how to undo it within the perceptual paradigm of documentary knowledge. I contest this anthropocentrism by the concept of care. Bellacasa[9] argues that "relations of thinking and knowing require care," and this resonates within the documentary mode, which is closely intertwined with viewers' desire for expanded understanding. In the context of documentary knowledge, especially in wildlife mode and its take on natural history knowledge, this can offer a different approach to insect – human relations in the film. Bellacasa emphasizes the concept of "naturecultural" visions as a means to contest the divisions between knowledge systems regarding nature and align with sociotechnical imaginaries, hence redirecting the focus toward nonhuman life and acknowledging the interconnectedness of multiple entities and agencies. By shifting the attention to the vitality and agency of nonhuman worlds and by considering humans as one ontological agency among many, these visions contest the objectification of nonhuman and subjectification of the human. In this way, they encourage an attentive approach that avoids automatic adherence to a strictly anthropocentric perspective.[10]

Care, as understood by Joan Tronto and Berenice Fisher,[11] is "everything that we do to maintain, continue and repair 'our world' so that we can live in it as well as possible. That world includes our bodies, our selves, and our environment, all of which we seek to interweave in a complex, life sustaining web." Bellacasa also emphasizes the material world and interdependencies between humans and nonhumans as a "matter of care." She understands "care as a concrete work of maintenance, with ethical and affective implications, and as a vital politics in interdependent worlds."[12] The notion of "matters of care" serves as a concept to think with, as Bellacasa argues. Instead of merely revealing facts, it proposes that we engage with them and foster more caring relationships. Therefore, it is not primarily about explaining how things are constructed; rather, it focuses on our involvement in their potential becomings.[13]

I examine care and the insect worlds in the context of wildlife mode, which is more diverse than the tradition of "blue chip," which is known to exclude the animal world from any human or cultural influence. These traditional wildlife modes of "blue chip" documentaries are characterized by their high production

9 Maria Puig de la Bellacasa, *Matters of Care: Speculative Ethics in More than Human Worlds* (Minneapolis: University of Minnesota Press, 2017), 69.

10 Puig de la Bellacasa, *Matters of Care*, 141.

11 Berenice Fisher and Joan Tronto, "Towards a Feminist Theory of Caring," in *Circles of Care: Work and Identity in Women's Lives*, ed. E.K. Abel and M.K. Nelson (Albany: SUNY Press, 1990), 40.

12 Puig de la Bellacasa, *Matters of Care*, 5.

13 Puig de la Bellacasa, *Matters of Care*, 66, 69.

value, exclusion of human presence and surroundings from the visuals, and a tendency to avoid environmental issues. Typically, these documentaries adopt dramatic narratives, which are often presented through voice-over narration, centered around large animals, especially megafauna, without delving into the historical context.[14] However, documentaries that focus on insects are often excluded from the broader conceptualization of wildlife documentaries, as pointed out by MacDonald.[15]

Nevertheless, this traditional wildlife mode potentially oversimplifies the intricate and nuanced relationship between humans and nonhumans. It overlooks the fact that these two entities share a continuum while acknowledging that the worlds of animals are intertwined with an ecological context that also encompasses human worlds.[16]

In the first section, I examine natural historical knowledge, the figure of David Attenborough in the narrative of the insect world, and nonhuman charisma. In the following section, I explore macro cinematography and insects as mediators between caring ecological connections, knowledge, and the viewer. In the final section, I examine care in narratives and visuality of *Alien Empire*, along with how the human and insect worlds merge in the documentary series.

2 *Life in the Undergrowth* and the Figure of Natural History

BBC's series *Life in the Undergrowth* starts with David Attenborough's voice-over introducing the insect world for viewers: "An eye from a different world. We don't often see a snail that way. And that's because we've only recently had the tiny lenses and electronic cameras that we need to explore this miniature world. But when we meet its inhabitants face to face, we suddenly realize that their behavior can be just as meaningful to us as the behavior of many animals more our own size." Here, the insect world is depicted as a "different world," one in which we gain perceptual access through technology that enables us to have meaningful encounters with miniature creatures. *Life in the Undergrowth* is a presenter-led series narrated by David Attenborough. He is not only an authoritative voice-over, taking the viewer into the insect world with him, but he is very much involved in the images as a presenter while depicting scales next to insects. *Life in the Undergrowth* has a strong natural history and evolutionist

14 Derek Bousé, *Wildlife Films* (Philadelphia: University of Pennsylvania Press, 2000), 14–15.

15 Scott MacDonald, "Up Close and Political: Three Short Ruminations on Ideology on the Nature Film," *Film Quarterly* 59, no. 3 (2006): 19.

16 Smaill, "Encountering Animals," 93.

discourse seeking to explain insects' anatomy and behavior to viewers. However, this evolutionary mode involves the authority to classify other life forms, but it also places humans within the same continuum as other species. *Life in the Undergrowth* brings forth different visual registers for the viewer to engage with. The first is the documentary's micro and macro cinematography, which alters perceptions by enabling the observation of creatures that are too tiny to engage with without technological aids. This is not just an observation in the sense that the viewer would not be involved in the movement of images and bodies of insects, but rather, it is an encounter that allows to stay with the proximity of insects. The other visual register is that of Attenborough's presence on the screen and his encounters with insects as sort of a messenger between the human and insect worlds. These scenes bring together human and insect bodies, as well as the question of scale. Natural history and evolutionist discourse can be seen in both visual registers; however, there are sequences that are more poetic in their depictions and where the voice-over narration is silent. Here, I focus on sequences involving Attenborough's presence and where he is in contact with specific insects. Unlike animals that show reciprocity, insects are generally perceived as more autonomous, displaying minimal reactions to human presence,[17] which is embodied by the sometimes awkward presence of Attenborough while he tries to fit into the same framing with insects by crouching and lying on the ground.

In the scene with a springtail, Attenborough lies on the ground, pointing at soil with a needle. In the next shot, there is a springtail next to a point of a needle appearing bigger than the minute insect. Attenborough's voice tells that there are numerous species living on land, but they are so tiny – a half a millimeter long – that we seldom see them. In the framing, the springtail seems to wash itself, while Attenborough describes its body as "the size of a full stop," and that "drying out is a very real problem for them." The springtail jumps while the pin of the needle is next to it, and again, the voice-over describes that "it is like human jumping over the Eiffel tower." This kind of macro cinematography takes the viewer's perception to a creature that is magnified that enables observation, making it visible, while the voice-over is giving familiar analogy to human "world" like a full stop or jumping over the Eiffel tower to give scales. With these analogies, the insect's abilities are compared with those of the bigger animals, here humans, to not only make observations but also new understandings between familiar and difference, while the ecological problem "drying out" is brought up. However, this kind of descriptive and observational

17 Lorimer, "Nonhuman Charisma," 920.

encounter embodies a specific type of science that deals only with established and unquestionable knowledge – the natural history paradigm.[18] In the context of nature documentaries compared with other scientific television documentaries, Michael Jeffries has argued that the natural history paradigm portrays the natural world by blending the traditional principles of ecological balance and adaptation with a sense of romantic awe and wonder that avoids representing scientific topics that are "changeable, challenging, contingent."[19] However, Lorimer combines the practices of natural history and nonhuman charisma. He understands practices of natural history – that Attenborough and the viewer engages here –as "one of tuning in, or 'learning to be affected' by the target organism." Practices of natural history are here understood not only as observation and collecting, but also as negotiations to achieve "ecological proximity and corporeal understanding by a host of technological apparatuses."[20] In this way, natural history knowledge is seen as an affective force between insects and Attenborough and the viewers, while insects are not just objects of study: they have agency that lures the viewer's attentiveness.

The material environment and animals adjust and affect the technology used, as Attenborough has noted in his memoirs. While filming *Life in the Undergrowth*, the film crew had to use technical refinement and slow-motion to shoot, for example, ants' movements that would otherwise be too fast to perceive. Attenborough explains that, because of technology, they began to see ants as individuals with their own personalities and behaviors.[21] He was affected by ants and their charisma. It seems like something changed in the way he perceived them, and he formed a caring relationship with the ants. In this context, perceiving ants as individuals is against natural history specimen logic that depicts individuals as specimens of the species. As Bellacasa argues, caring is fundamentally relational, involving an engagement that assumes responsibility for our interactions and their impact on others. Caring approaches encompass the process of establishing relationships that exhibit attentiveness to the needs of others.[22] If we pay attention to the *practice* of how Attenborough is present in frames with insects, we get a kind of encounter other than observational natural history as argued by Jeffries. In these sequences where Attenborough is visible, the human body serves as a sort of reference point

18 Morgan Richards, "Greening Wildlife Documentary," in *Environmental Conflict and the Media*, ed. Libby Lester and Brett Hutchins (New York: Peter Lang, 2013).

19 Michael Jeffries, "BBC Natural History versus Science Paradigms," *Science as Culture* 12, no. 4 (2003): 543.

20 Lorimer, "Nonhuman Charisma," 917.

21 Attenborough, *Life on Air*, 392.

22 Puig de la Bellacasa, *Matters of Care*.

for scales, but the scales are not predetermined in an anthropocentric way. In some scenes, Attenborough is standing and providing explanations about insects, but a challenge arises when he needs to be next to an insect that is being shown in a close-up as he squeezes into the shot, resulting in a distortion of the scale between the insects and Attenborough's body. For example, there is a sequence where an ant is examining another insect's egg and the ant is captured in front of the camera while Attenborough is in the background, placed in a perspective where the ant and Attenborough appear almost the same size. The human body appears disproportionately small, but at the same time, it is too large for the frame that is adjusted to portions of the ant. The human body no longer fits into the image. Zachary Horton has argued that the production of human knowledge has consistently revolved around the process of taming of scale. It involves confining scalar difference within established domains and structuring those domains into a spatially and conceptually interconnected framework, with the unmarked scale of the human serving as a pivotal reference point, as a "scale-stable human subject."[23] However, in the sequences where Attenborough is next to insects, trying to squeeze in the framing that is adjusted for minute creatures, he becomes like Alice in Wonderland, a character without stable scales. This aligns with Dodd's[24] argument about insect worlds that are not accustomed to accommodating humans. When an image is attuned to the scales of insects, it is the human who appears too large. This awkward scale and perspective that is framed in terms of the insect makes the viewer question the pregiven relationship between the human presenter, the ant, and the traditional anthropocentric visual framing.

Although in the beginning of *Life in the Undergrowth*, the world of insects is portrayed as "a different world," throughout the series, the difference lies in the scales depicted between Attenborough and the insects present in the same scenes. Attenborough's role as a naturalist figure can be perceived as too large in relation to the scales of the insects, which themselves are understood within the context of natural history knowledge and observation. The voice-over reminds the viewers that if insects would disappear, "[t]he land's ecosystems would collapse. The soil would lose its fertility. Many of the plants would no longer be pollinated. Lots of animals, amphibians, reptiles, birds, mammals would have nothing to eat. [...] These small creatures are within a few inches of our feet, wherever we go on land – but often, they're disregarded. We would do very well to remember them." However, even though focusing on the "safe

23 Zachary Horton, *The Cosmic Zoom: Scale, Knowledge, and Mediation* (Chicago: University of Chicago Press, 2021), 7.

24 Dodd, "Minding Insects".

science" of natural history and the "awe and wonder" of the insect world, *Life in the Undergrowth* does not address interdependencies between human and insect worlds, except briefly in a few sequences, primarily through voice-over narration. Nonetheless, the documentary manages to acquaint the viewer with insect life and behavior, positioning insects as matters of care, one fueled by the curiosity of natural history. According to Lorimer, such practices of natural history, as exemplified by the viewer's intimate engagement with insect imagery, serve the process of "learning to be affected" by the documentary's subjects.

3 *Microcosmos* and Aesthetics of the Microscopic World

Microcosmos is a full-length French documentary made for a big screen and screened at the Cannes Film Festivals. It does not have a voice-over narration, except just at the beginning and end of the film, which is atypical for a wildlife documentary, and the film does not have the same kind of natural historical discourse as *Life in the Undergrowth*. The only voice-over narration in the beginning and very end of the documentary takes the viewer into an insect world: "A meadow in early morning somewhere on earth. Hidden here is the world as vast as our own where weeds are like impenetrable jungles, stones are mountains, and even the smallest pond becomes an ocean. Time passes differently here, an hour is like a day, a day is like a season, and the passing of a season is a lifetime. But to observe this world we must fall silent now and listen its murmurs." This emphasizes the abundant diversity and complexity of life that exists within this hidden realm, one only accessible to the viewer by being quiet. After the introduction, there is no voice-over narration, and the viewer engages with the insect world through images and sounds on and off screen.

Claude Nuridsany and Marie Pérennou, the directors of *Microcosmos*, express their dissatisfaction with the conventional portrayal of the animal world in wildlife films, believing that such portrayals often exhibit a bias toward natural science.[25] In their interview, the directors explained that the emotional connection they felt when observing animals in real life was lacking in many wildlife documentaries. Their intention was to introduce viewers to an "unknown world." To achieve this, they had to develop new filming techniques and sought the assistance of robot engineers to create specialized equipment

25 Scott MacDonald, "Interview with Claude Nuridsany and Marie Pérennou," in *Adventures of Perception: Cinema as Exploration: Esseys/Interviews* (Berkeley: University of California Press, 2009), 188–189.

capable of capturing images on such a small scale.[26] The relationship with insects is intimately connected with technology and visual perception in Western culture.[27] This relationship that involves technology as an essential part of perception, as Nuridsany and Pérennou also note, has its history in the invention of the microscope that brought about a significant shift in the way insects were observed. However, the early microscopic illustrations, as well as early cinematography about insects, which were shot in studios or laboratories, isolated their subjects from their natural habitat, following specimen logic by representing insects against blank backgrounds as isolated objects.[28] Giraud and Hollin[29] have studied relationships of care in laboratory animals, but their theorization of relations of care can be applied to the technology used in filming insects, as well as the viewer's engagement with the images shaped by insects. Caring involves the space to be affected, and Giraud and Hollin argue that, through this caring relationship and affective engagement, where animals are seen as agents, not just under observation, animals are enabled to "speak back" in ways that reshape their environment.[30]

Unlike a traditional laboratory setting for filming that isolates insects, *Microcosmos* presents insects in relation to their habitats, highlighting their agency and the interconnectedness of their life worlds. *Microcosmos* brings forth affective engagement with specific insects through macro cinematography and close-ups, as well as with an ecological connection in which the insects intertwine as agents in their habitat. The moving image has the capacity to affect through its tactile nature, promoting a sense of closeness between viewers and nonhuman life, rather than reaffirming a hierarchical relation.[31] For example, in a scene of a snail drinking from a puddle, we see the snail and its movement across soil covered in moss. When the snail's mouth touches water, surface tension expands the water, which appears like a different substance while the snail touches it. This depiction of water defies our usual perception of it on a human scale. The snail is not portrayed as a separate or isolated object as a specimen, but rather, the framing emphasizes the snail's

26 MacDonald, "Interview," 190.

27 Dodd, "Minding Insects"; Oliver Gaycken, *Devices of Curiosity: Early Cinema and Popular Science* (New York: Oxford University Press, 2015); Janice Neri, *The Insect and the Image: Visualizing Nature in Early Modern Europe, 1500–1700* (Minneapolis: University of Minnesota Press, 2011).

28 Neri, *The Insect and the Image*, XII–XIII.

29 Eva Giraud and Gregory Hollin, "Care, Laboratory Beagles and Affective Utopia," *Theory, Culture & Society* 33, no. 4 (2016).

30 Giraud and Hollin, "Laboratory Beagles," 30.

31 Smaill, "Encountering Animals," 84.

connection to it and the tactile nature of the environment, moss, and soil. According to film scholar Janet Harbord[32], the technologies used to observe insects and capture their rapid movements, both in the laboratory and on film, need to be designed with the capabilities and capacities of insects in mind. This resonates with caring, keeping insects' needs in mind and their capacity of "speaking back." Harbord views the insect as a messenger or translator in the laboratory of cinema, bridging the human world, the world of machines, and the world of insects.[33] The medium of the moving image functions as a laboratory, allowing for adjustments to human perception. Here, the snail can be seen as a messenger or translator for creating understandings of different aesthetics and ecologies.

In *Microcosmos*, the use of macro cinematography takes the viewer's perception of the world of insects and small creatures. The technique bridges the distance between human perception and the minute details of insects, allowing viewers to engage with their behaviors and interactions that would otherwise go unnoticed. The camera brings the viewer into a sense of proximity to the microcosmic realm, fostering an observational intimacy with insects. Gaycken argues that in the micro- and macroscopic images in cinema, the camera, through its observational capabilities, provides viewers with a feeling of being intimately connected to the world on film. This closeness goes beyond magnification: it encompasses observational proximity that promises deeper comprehension.[34] In a scene about a mosquito's metamorphose, an insect emerges from a still surface of water. The visuals do not rely on scientific bias, but the focus is on the poetic connection between the insect's body and its environment, while the water takes on a tactile quality. Human perception is decentered, while the insect rising from water is not recognizable at first, and without voice-over explanation, the viewer is left without a pregiven relationship, although at the end of the scene, when the mosquito spreads its wings, it takes a shape that is familiar, giving the viewer the pleasure of recognition. Haraway emphasizes curiosity and knowing more as part of caring about: "[c]aring means becoming subject to the unsettling obligation of curiosity, which requires knowing more at the end of the day than at the beginning".[35] The scene brings forth an event that often goes unnoted, and the film builds an encounter that does not carry previous preconceptions of the viewer

32 Janet Harbord, *Ex-Centric Cinema: Giorgio Agamben and Film Archaeology* (London: Bloomsbury Publishing, 2016), 138.

33 Harbord, *Ex-Centric Cinema*, 145.

34 Gaycken, *Devices of Curiosity*, 39–40.

35 Haraway, *When Species Meet*, 36.

toward mosquitos but rather emphasizes the importance of attentiveness and the viewer's engagement with the image. This visual proximity, which does not get objectified by voice-over narration but leaves space for the viewers to make their own connections, brings forth insects as subjects that blur the boundaries between familiarity and difference; that of anthropomorphism, sensibilities of insects and the possibilities to examine different encounters that are not necessarily pregiven by natural history bias. Even if *Microcosmos* does not bring forth extinction or ecological problems *per se*, it engages the viewer in different kinds of ecological aesthetics that emphasize the caring relationship with the small and fragile, adjusting and sensitizing the viewer's perception to different kinds of relations to familiar environments.

4 *Alien Empire* and Staying with the Trouble on the Shared Planet

Alien Empire differs from *Microcosmos* and *Life in the Undergrowth* in its depictions of human – insect relations. The two later documentaries introduced an "insect world" that is based on scalar difference, like Dodd[36] has noted, using it as a metaphor to understand insects and highlighting its distinction from "human world," rendering humans incongruous within it. However, *Alien Empire* does not visually differentiate insects and human worlds but instead focuses on the interaction and conflicts that invertebrates and human primates may cause to each other. *Alien Empire* is a wildlife documentary produced by BBC natural history unit, and it has voice-over narration given by actor John Shrapnel. Here, I focus on insect – human relationships depicted in *Alien Empire*. Even though there are traditional entomological and anatomical descriptions of insects and micro and macro cinematography, there are also sequences that have a broader approach to insect – human relationships like the use of pesticides, health and economic issues, and cultural approaches like eating insects, focusing on insects' ambiguous relationship with humans.

In the beginning of the series, the narration introduces the "insect world" quite differently than in the other two series discussed earlier: "Somewhere on the outer reaches of the galaxy. There's as small planet dominated by an alien life form, creatures too numerous to count, they can survive conditions no other creature can tolerate. Other life forms try to destroy them, but their resilience has been their strength for hundreds of millions of years. They lurk in

36 Dodd, "Minding Insects," 33.

the darkest of corners or control entire landscape. They seem to come from the world of science fiction, but they belong to the world of science fact. Welcome to the world of insects." Here, the insect world is not a place of awe and wonder of minute creatures, or traditional principles of ecological balance as argued by Jeffries,[37] but conflicted with other lifeforms that are humans. Although the series' name *Alien Empire* is a metaphor for the insect world and the voice-over compares insects to aliens and machines, echoing Cartesian biases, the visuals of the series can be viewed as contesting the visual tradition between human and insect worlds. The voice-over has a god-like position to describe insects, but humans are also seen as species while their doings are described as that of any other animals.

Conflict between humans and insects is demonstrated in a montage sequence about cockroaches and people trying to get rid of them. The scene begins with a voice-over narration: "One of the longest running feuds has been with cockroaches. They contaminate food, carry diseases and generally offend the human sensibilities." The camera then zooms in on a close-up shot of a cockroach's head. A montage follows, showing an exterminator walking into a hotel basement, a cockroach scurrying through pipes, and a young boy riding a tricycle in a hallway. These shots reference various film genres, such as horror, thrillers of the 1990s, and film noir. The narration continues, mentioning that humans spend millions of pounds each year trying to combat the cockroach problem with chemicals, and in the pursuit of an insect-free life, humans saturate their homes, fields, rivers, and the world with toxins. However, the narration suggests that chemical engineering may never be the final solution. The scene concludes with the statement that there is still only one guaranteed method of killing a cockroach as the young boy on the tricycle runs over the cockroach in the hallway. The camera captures this moment from a low angle, with a shaky camera following the cockroach. Visually, the sequence connects insects in Western popular culture and broadens the typical visualization of insects in wildlife mode. In its visual suspense, the cockroach sequence makes the viewer take the insect's side in its escape from the terminator that represents destructive toxins. However, this is a kind of situated caring that is not necessarily applied in encounters outside the filmic event. Lorimer[38] emphasizes that different organisms "can be both awe-some and awe-full." It is easy to care about insects, even in their strangeness and awkwardness, when they are in "insect world," "out" in nature like usually visually depicted in wildlife documentaries. However, when they come into

37 Jeffries, "BBC Natural History," 543.

38 Lorimer, "Nonhuman Charisma," 918.

our homes, eat people's crops, or carry deceases, it is harder to care about them. However, it is this uneasy and diverse relationship between humans and insects that is brought up in the series. Care is providing a way of "staying with the trouble."[39] As Bellacasa reminds us, care in "naturecultures" is not innocent activity. The obligations tied to caring in "naturecultures" extend beyond mere "stewardship" or a pastoral form of care where humans exert control over the natural world.[40]

John Berger[41] argues that animals are constantly being watched and observed by humans to the point where their ability to observe humans has become insignificant. However, Burt critiques this stand and emphasizes that the act of looking involves not only the act of being observed, but also the reciprocal nature of being looked at. Burt suggests that this dynamic goes beyond straightforward objectification and instead plays an active role in establishing relationships between humans and animals.[42] This is connected to Giraud and Hollin's concept of "speaking back,"[43] in which the caring relationship includes animal agency, and they are not just ones observed – or objects to be cared for. In documentary film, witnessing is one of the key concepts.[44] In *Alien Empire,* this mode of witnessing is given to insects, and they have agency while humans are depicted as one species among any other. It is not only that people observe insects but insects observe people as well, witnessing evolution, like one of the scene's voice-over narrates: "[o]nly a few million years ago, they [insects] witnessed a large brained ape climb down from the tree, stand up and invent technology." Also, the composition and framing in some of the scenes emphasize this shared observation: while insects are in the front of the camera, in the background, there are people out of focus, like children playing or street view from the city. Insects are not depicted as isolated to an "insect world" that does not habituate humans, but instead, both species share habitats and environments. This kind of depiction is very much against the tradition where insects are represented against a blank background where their only purpose is to become observed; indeed, Neri[45] argues that this "[s]pecimen logic turns nature into object by decontextualizing select creatures and items – that is,

39 Donna Haraway, *Staying with the Trouble: Making Kin in the Chthulucene* (Durham: Duke University Press, 2016).

40 Puig de la Bellacasa, *Matters of Care*, 164.

41 John Berger, *Why Look at Animals?* (London: Penguin Books, 2009), 27.

42 Jonathan Burt, "John Berger's 'Why Look at Animals?' A Close Reading," *Worldviews: Global Religions, Culture, and Ecology* 9, no. 2 (2005): 207.

43 Giraud and Hollin, "Laboratory Beagles".

44 Bill Nichols, *Introduction to Documentary* (Bloomington, IN: Indiana University Press, 2001).

45 Neri, *The Insect*, XIII.

by removing them from their habitats, environments, and settings." This same logic can be traced to traditional wildlife documentary mode as well because it had a tradition of depicting nature as separated from culture and human influences. Beside the voice-over of John Shrapnel, another kind of narrator can be found on the screen: a computer animated, real-looking fly whose point of view takes the viewer to different scenes of human – insect encounters. Although voice-over explains how humans classify insects, the camera follows a fly that enters a door "Entomology" written on it. The fly flies past folders on a bookshelf with insect etymological names written on their covers. Later, the fly takes the viewer to a natural history museum and laboratory where scientists conduct research about mosquitos and dengue fever. The fly is witnessing and observing the conduction of scientific research while the camera takes the fly's point of view, like the first-person perspective in a video game. Natural history and scientific knowledge become objects of insect's – and the viewer's – observation, but also a matter of care. The visuals of the point-of-view shots blur the object – subject boundaries while the fly becomes an inadvertent observer and participant in the scientific process, acting as an unexpected mediator between humans and the insect world.

The last episode, "War of the Worlds," introduces conflicts and codependences between insects and people from different cultures, like short sequences about the African tradition of eating insects and organized beetle battles in Thailand. Insects are connected to cultural traditions but also with economic systems, such as silk factories and farming. In a sequence about the US alfalfa industry that heavily relies on leafcutter bees for pollination, the narration stresses that "[l]eafcutter bees are recent partners, humans have destroyed so many wild populations of insects with their chemical sprays, they are now having to forge new relationships." Once the bees complete their tasks in one field, they are transported to the next. These scenes highlight the intricate and mutually beneficial relationships between humans and various insect species while stressing the troubles of human – insect relationships. Besides the economic benefits insects produce, the sequences shed light on how human – insect care is reciprocity: insects can perform care for us and with us. Bellacasa argues that even if human – nonhuman relationships inherently involve care, our caring actions can also lead to disconnections. It is impossible to care for everything because not everything holds the same importance in the world. Just as life and death are inseparable, caring and disconnecting are intertwined.[46] Here, the use of pesticides to take care of crops has led to the death of pollinators and disconnection in caring about them; however, this has

46 Puig de la Bellacasa, *Matters of Care*, 78–79.

led to a new kind of relationship with leafcutter bees that benefits humans and these particular bees as well.

The ways in which *Alien Empire* makes cultural and economic factors related to care visible are noteworthy. In the documentary, insects are not solely depicted within the confines of a natural history bias or as inhabitants of a world detached from human culture and environmental issues. Although the way the traditional wildlife mode depicts natural history is criticized as "safe science" or "old ecology" of awe and wonder, *Alien Empire* takes a different paradigm characterized by change, crisis, and challenge while bringing forth caring about insects and ecologies. *Alien Empire* "stays with the trouble" and engages the viewer with a world shared with insects and humans and not always in a pleasant way.

5 Pollination

The exploration of insect – human relationships within the realm of wildlife documentaries has provided understandings of caring aesthetics and meaningful encounters. The analysis of *Life in the Undergrowth*, *Microcosmos*, and *Alien Empire* has brought forth the diverse strategies employed by these documentaries to engage viewers with the insect world. The documentary mode generates anticipation for a comprehensible, verifiable world of knowledge,[47] while this pursuit of knowledge can be hindered by anthropocentrism. However, Bellacasa's notion that "thinking and knowing require care" deeply resonates in documentary knowledge. Care involves maintenance, ethics, and recognition of the connections between humans and nonhumans.

The documentaries serve as more than just representations of insects: they can bring forth a speculative and caring mode of engagement. The importance of technology in mediating these encounters is evident because it enables us to traverse the scale differences between human and insect worlds. Although some narratives and visuality of the documentaries I have explored can contest anthropocentrism between the human and insect worlds and encourage thinking with insects, they also remind us that caring is not a one-sided endeavor, but a reciprocal relationship that calls for attentiveness, curiosity, and understanding. Moreover, this analysis has demonstrated that caring about the insect world extends beyond a mere discourse – it manifests as ecological connection. The documentaries compel us to consider the complex interplay

47 Smaill, "Encountering Animals," 97.

between humans and insects, acknowledging that caring is not always straightforward. Although we may emphasize the importance of individual insects on screen, the challenges of caring for creatures like cockroaches or mosquitos in our daily lives reveal the nuances of situated caring.

In a world characterized by environmental challenges and biodiversity loss, no longer can we view insects as mere background characters in the narrative of our planet; instead, they emerge as essential cohabitants with whom we forge interwoven destinies. Framing insects and humans within separate worlds is unsustainable. Although there might be a discourse aimed at discussing and visualizing insects within distinct "worlds," it is important to recognize that these worlds are inherently interconnected at the material level and cannot be examined in isolation: insects, humans, and technologies make worlds together.

Bibliography

Attenborough, David. *Life on Air. Memoirs of the Broadcaster*. London: BBC Books, 2010.

Berger, John. *Why Look at Animals?* London: Penguin Books, 2009.

Burt, Jonathan. "John Berger's 'Why Look at Animals?' A Close Reading." *Worldviews: Global Religions, Culture, and Ecology* 9, no. 2 (2005): 203–218.

Bousé, Derek. *Wildlife Films*. Philadelphia: University of Pennsylvania Press, 2000.

Dodd, Adam. "Minding Insects: Scale, Value, World." In *The Management of Insects in Recreation and Tourism*, edited by Raynald Harvey Lemelin, 23–36. Cambridge: Cambridge University Press, 2012.

Fisher, Berenice and Joan Tronto. "Towards a Feminist Theory of Caring." In *Circles of Care: Work and Identity in Women's Lives*, edited by E.K. Abel and M.K. Nelson, 35–62. Albany: SUNY Press, 1990.

Gaycken, Oliver. *Devices of Curiosity: Early Cinema and Popular Science*. New York: Oxford University Press, 2015.

Giraud, Eva, and Gregory Hollin. "Care, Laboratory Beagles and Affective Utopia." *Theory, Culture & Society* 33, no. 4 (2016): 27–49.

Haraway, Donna. *When Species Meet*. Minneapolis: University of Minnesota Press, 2008.

Haraway, Donna. *Staying with the Trouble: Making Kin in the Chthulucene*. Durham: Duke University Press, 2016.

Harbord, Janet. *Ex-Centric Cinema: Giogio Agamben and Film Archaeology*. London: Bloomsbury Publishing, 2016.

Horton, Zachary. *The Cosmic Zoom: Scale, Knowledge, and Mediation*. Chicago: University of Chicago Press, 2021.

Jeffries, Michael. "BBC Natural History versus Science Paradigms." *Science as Culture* 12, no. 4 (2003): 27–545.

Life in the Undergrowth. Producer: BBC, 2005.

Lorimer, Jamie. "Nonhuman Charisma." *Environment and Planning D: Society and Space* 25, no. 5 (2007): 911–932.

MacDonald, Scott. "Up Close and Political: Three Short Ruminations on Ideology on the Nature Film." *Film Quarterly* 59, no. 3 (2006): 4–21.

MacDonald, Scott. "Interview with Claude Nuridsany and Marie Pérennou." In *Adventures of Perception: Cinema as Exploration: Esseys/Interviews,* 184–197. Berkeley: University of California Press, 2009.

Microcosmos (*Microcosmos: Le peuple de l'herbe*). Producer: Jacques Perrin, 1996.

Neri, Janice. *The Insect and the Image: Visualizing Nature in Early Modern Europe, 1500–1700*. Minneapolis: University of Minnesota Press, 2011.

Nichols, Bill. *Introduction to Documentary*. Bloomington, IN: Indiana University Press, 2001.

Puig de la Bellacasa, Maria. *Matters of Care: Speculative Ethics in More than Human Worlds*. Minneapolis: University of Minnesota Press, 2017.

Richards, Morgan. "Greening Wildlife Documentary." In *Environmental Conflict and the Media,* edited by Libby Lester and Brett Hutchins, 171–186. New York: Peter Lang, 2013.

Smaill, Belinda. "Encountering Animals: Re-Viewing the Cinema of Jean Painlevé." *Antennae: The Journal of Nature in Visual Culture* 42 (2017): 82–99.

PART 3

Science and Knowledge

CHAPTER 5

Wild Spiders in Fragile Knowledge Networks: Spiders in Medicine, Natural History, and Silk Production in Eighteenth-Century Europe

Laura Hollsten

1 Introduction

"It is certain a great Advantage may be made of this Insect, which the Publick has always look'd on as troublesome and dangerous, on account of its Venom," wrote the French president of the Court of Auditors of Montpellier, François Xavier Bon de Saint Hilaire in 1710.[1] Bon himself was enthusiastic about spiders as potentially useful animals but his statement shows that a negative view of spiders was prevalent. Another early modern observer to think favorably of spiders was Carl von Linné, who in a speech in 1739 admired "the skillful web of the spider, his residence in the middle of it, so that he can feel even the smallest fly touching its smallest string, just as you would imagine the soul, in the brain of a body, feels where the nerves converge."[2] While Bon focused on the usefulness of the spider and Linné admired it without instrumental motives, both ventured that spiders merited more appreciation than they received.[3]

1 François Xavier Bon de Saint Hilaire, "A Discourse Upon the Usefulness of the Silk of Spiders," *Philosophical Transactions of the Royal Society* 27, 325–36 (1710): 14, https://www.jstor.org/stable/103101.

2 Linné marvels at the spider's sensitivity and its web in a speech on insects that he gave at the Royal Academy of Science in Stockholm in 1739. See Carl von Linné, *Märkwärdigheter uti insecterne* (Stockholm, 1739). Linné uses the masculine pronoun for the spider: "Se på *Spinnelens* konstiga nät, dess Residence i nätets medelpunct, at han må känna den minsta fluga, som rörer dess minsta sträng, liksom man föreställer sig själen uti en krops hjärna känna där nerverne gå tilsammans," trans. L.H.

3 Both Bon's characterization and the title of Linné's speech show that spiders at the time were considered to be insects, a classification which lasted up until the early nineteenth century when Jean-Baptiste Lamarck separated the arachnids, mollusks, and crustaceans each into a class of their own. The word "insect" is used instead of arachnids in the text where sources are paraphrased.

Spiders are the largest order of arachnids and undoubtedly the most charismatic of the arthropods.[4] Spiders have eight legs and six to eight eyes but rely on touch, vibration, and taste to navigate. They have a small nervous system, and some species exhibit intelligent behavior, such as the ability to count and make plans. In addition, spiders have silk-extruding spinnerets, and they can inject venom into prey. Admired for their skills, spiders have all over the world been the subject of folklore and mythology, connected with both positive and negative characteristics such as wisdom, patience, and dangerousness.[5] People have been intrigued by spiders' ability to spin and weave, and spiders have played an important role in folk medicine and traditional knowledge. Silk spinning, medical use, and venom have been the most important components in popular as well as scholarly knowledge about spiders before the development of early modern natural history and taxonomy. These characteristics persisted in eighteenth-century works of natural history and popularized knowledge.

This chapter investigates knowledge about spiders in early modern Europe, with some focus on Sweden. The study is based on natural-historical works on insects and spiders from the first part of the eighteenth century. The theoretical inspiration derives from the history of knowledge and the understanding that natural-historical knowledge is created in multispecies knowledge networks, and that these networks can be vulnerable.[6] Knowledge is understood to be formed in interactive processes where human and non-human actors are seen as co-constructors of knowledge in multispecies networks. A distinction is made between various forms of knowledge: intellectual, traditional, and useful or practical knowledge. Intellectual knowledge was made by and circulated among scholars, natural historians, and learned societies while traditional knowledge is here understood as "a cumulative body of knowledge and beliefs handed down through the generations by cultural transmission."[7] Useful knowledge is knowledge that can be applied in everyday life. The forms

4 On charismatic animals, see Jamie Lorimer, "Nonhuman Charisma," *Environment and Planning D: Society and Space* 25, no. 5 (2007).

5 Katarzyna Michalski and Sergiusz Michalski, *Spider* (London: Reaktion Books, 2010), chapters 3 and 7.

6 Lorraine Daston, "The History of Science and the History of Knowledge," *Know. A Journal on the Formation of Knowledge* 1 (2017); Johan Östling et al., "The History of Knowledge and the Circulation of Knowledge," in *Circulation of Knowledge: Explorations in the History of Knowledge*, ed. Johan Östling et al. (Lund: Nordic Academic Press, 2018).

7 Fikret Berkes, "Biodiversity, Religion Traditions", in *Encyclopedia of Biodiversity*, ed. Samuel M. Scheiner (Elsevier, Academic Press, 2024); Viktor Ulicsni, Ingvar Svanberg, and Zsolt Molnár, "Folk Knowledge of Invertebrates in Central Europe - Folk Taxonomy, Nomenclature, Medicinal and Other Uses, Folklore, and Nature Conservation," *Journal of Ethnobiology and Ethnomedicine* 12 (2016).

of knowledge overlap as natural historians often were involved in scholarly pursuits as well as in the making of useful knowledge. Intellectual knowledge, in turn, often carried elements of traditional knowledge. On the whole traditional knowledge was closer to everyday mainstream knowledge than intellectual knowledge.

The networks operate on two levels: both as wider scholarly networks in Europe, and as local knowledge networks consisting of humans and spiders, and, in some cases, other species and devices.[8] Rather than traveling from center to periphery, from "professional" natural historians to amateurs, knowledge is understood to be constructed at various points of networks consisting of natural historians, animals, assistants, correspondents, and texts, as well as instruments and tools. While enabling the formation and circulation of knowledge, knowledge networks were, as suggested by Dániel Margócsy, fragile, in the sense that each part of the network was vulnerable to breakdowns.[9] As the chapter shows, the vulnerable part in the network was often the spider, or rather the relationship between humans and spiders which was marked by human ignorance of how to cooperate with the spider in a successful way. Hence, while the objective is to examine the formation of knowledge about spiders, the aim is also to bring the spiders into the study, not simply as objects of knowledge, but as crucial parts of knowledge networks.

The first section of the chapter considers how collecting, scientific observation, classification, and nomenclature contributed to the formation of scholarly knowledge about spiders by studying works on natural history and early modern developments in entomology and arachnology. In the second section traditional knowledge about spiders is studied through the example of

8 Knowledge about the natural world circulated through networks in various parts of the "empires of knowledge," as described by Paula Findlen in *Empires of Knowledge. Scientific Networks in the Early Modern World* (Oxford: Routledge, 2018). See also Emma Spary, "Botanical Networks Revisited," in *Wissen im Netz: Botanik und Pflanzentransfer in europäischen Korrespondenznetzen des 18. Jahrhunderts*, ed. Regina Dauser et al. (Berlin: Akademie Verlag, 2008). For social networks in the study of insects, see Koen Scholten, "Pierre Lyonet's (1706–1789) Study of Insects: Displaying Virtue and Gaining Social Status through Natural History," *Studium: Tijdschrift voor Wetenschapsen Universiteitsgeschiedenis* 11, no. 4 (2018). On multispecies networks, see Bruno Latour, *Reassembling the Social: An Introduction to Actor-Network-Theory* (Oxford: Oxford University Press, 2007); Ursula K. Heise, "Multispecies Futures and the Study of Culture," in *Futures of the Study of Culture: Interdisciplinary Perspectives, Global Challenges*, ed. Doris Bachmann-Medick, Jens Kugele and Ansgar Nünning (Berlin and Boston: De Gruyter, 2020).

9 Dániel Margócsy, "A Long History of Breakdowns: A Historiographical Review," *Social Studies of Science* 47, no. 3 (2017), http://www.jstor.org/stable/44652509.

physicotheology and medicine. Finally, useful, practical knowledge is investigated through the case of spider silk.

2 Intellectual Knowledge: Authorities, Experiments, and Instruments

According to Aristotle, spiders, scorpions, and centipedes belonged to the category of insects, a classification which continued up to the nineteenth century. Pre Linnaean taxonomic thinking in the early modern period was influenced by a hierarchical view of the creations, summed up in the concept of the Great Chain of Being.[10] In this hierarchical chain everything, from the lowest forms of life up to the spheres of angels and God, had its own proper place.[11] The idea of the Great Chain begins with Plato and Aristotle, continues in late antiquity, was prevalent in the Middle Ages and continued to characterize thinking in early modern times, although the tradition was challenged by new empirical knowledge which led to the development of new taxonomies.

What was the place of spiders in the Great Chain of Being? Animals were classed from the highest to the lowest. The lion was considered as the noblest of animals while eagles ranked highest among birds. Besides noble qualities, the usefulness of animals determined their position in the chain so that the snake was found at the bottom of the scale since it was both potentially poisonous, while also symbolizing evil in the Christian imagination. A further principle for the categorization of animals had to do with their reproductive ability: live birth was considered more valuable than egg-laying, while warm-blooded mammals and birds ranked higher than bloodless invertebrates.[12] Above the serpent, but in the lowest sphere of the animal world, were the insects, including spiders. Insects were ranked so that beneficial insects such as bees, or insects with favorable properties such as ants, were higher on the scale than beetles and mosquitos. Although the case of spiders in the Great Chain is not specifically addressed in the sources, it can be speculated that spiders, characterized by industriousness and remarkable skills, might be classed

10 In addition to scholarly ideas, ethnobiological classifications and popular taxonomies were prevalent.

11 Arthur O. Lovejoy, *The Great Chain of Being* (Cambridge: Harvard University Press, 1976/1936); Clarence Glacken, *Traces on the Rhodian Shore. Nature and Culture in Western Thought from Ancient Times to the End of the Eighteenth Century* (Berkeley, Los Angeles and London: University of California Press, 1967).

12 See e.g. Charles Bonnet, *Traité d'insectologie,* vol. 1 (Chez Durand, libraire, 1745).

in similar categories as ants or bees; as useful "insects" although potentially venomous.

Renaissance knowledge of nature was literary and anthropocentric – in addition to the benefits and harms of animals in relation to humans, earlier descriptions and anecdotes were included in the descriptions. A prominent representative of the ideal is Ulisse Aldrovandi (1522–1605), a professor in Bologna and author of several works of natural history, among them *De Animalibus Insectis* (1602) which formed a part of a larger work on animals. Like Aristotle, Aldrovandi maintained that insects had neither nerves nor stomach, a view later refuted by the Italian Marcello Malpighi and Dutch Jan Swammerdam.[13] Seventeenth-century natural historians continued to refer to Aristotle in their works. The history of arachnology continues with zoologists such as Robert Hooke, Antoni van Leeuwenhoek, John Ray, and Swammerdam; the latter identified four orders of insects according to their metamorphosis where the first consisted of spiders, mites, lice, fleas, woodlice, worms, scorpions, leeches, centipedes, and snails. In the eighteenth century the study of invertebrates developed into a more systematic endeavor. Thanks to the development of microscopes, it became easier to investigate insects with greater precision. The early modern period saw a variety of competing taxonomic systems, but eventually Carl von Linné introduced the binomial naming system.

The Swede Carl Clerck (1709–1765) was the first to use the binomial system for spiders in *Svenska Spindlar* (Swedish Spiders) published in 1756 and preceding Carl von Linné's classification in the second edition of *Systema Naturae* by a year.[14] Carl Clerck can be characterized as an amateur – even though he enrolled at the University of Uppsala in 1726, he could not afford the studies and had to earn his living in government service as a tax collector. Clerck was inspired to devote himself to natural history, particularly butterflies and spiders, after hearing von Linné's lecture on insects in Stockholm in 1739.[15] Eventually, Clerck became part of the learned circle around the Swedish Academy of Science. Some of the scientific names proposed by Clerck were adopted by Linné in his *Systema Naturae* in 1758 with only minor modifications. Clerck's

13 Brian W. Ogilvie, "Order of Insects: Insect Species and Metamorphosis between Renaissance and Enlightenment," in *The Life Sciences in Early Modern Philosophy*, ed. Ohad Nachtomy and Justin E.H. Smith, online edn. (Oxford Academic, 2014), accessed March 17, 2023, doi:10.1093/acprof:oso/9780199987313.003.0012.

14 Torbjörn Kronestedt, "Carl Clerck and What Became of His Spiders and Their Names," *European Arachnology 2008* 1 (2010).

15 Torbjörn Kronestedt, "Carl Clercks 'Svenska spindlar' 250 år [Carl Clerck's 'Svenska spindlar' 250 years]," *Fauna och Flora* 102 (2007).

Svenska Spindlar attracted attention abroad and was eventually translated into English by T. Martyn in 1793, combined with Eleazar Albin's (1736) earlier work on English spiders.[16] Clerck based his treatise on careful observations of spiders but also on Martin Lister's work on spiders in *Historiae Animalium Angliae* (1678) when it comes to species unknown to him. Clerck describes sixty-eight spiders, whereas Linnaeus in *Systema Naturae* 10th Ed., published the following year, describes only thirty-seven species.[17] In addition to Lister, Clerck refers to the Swedish entomologist Charles De Geer and the German theologian Christian Lesser, both of whom are mentioned in the preface of his work, showing that he was familiar with current knowledge about insects.

Fieldwork and careful observation were important elements in the formation of scholarly knowledge. As Mary Terrall has noted, the insects had to be caught, fed, and kept at a suitable temperature.[18] This involved deploying a variety of practices and tools; in Clerck's case, specially constructed tin-boxes with glass bottoms for catching the spiders so as not to damage the fragile creatures. In a paper on the catching and feeding of spiders, Clerck also includes an illustration of a terrarium for housing and feeding them, with flies caught in a similar fashion as the spiders.[19] Well-situated natural historians were able to rely on assistants who could help with the collecting of the specimens.[20] Little is known about Clerck but according to Torbjörn Kronestedt he was of modest means, and it is possible that he conducted his research without assistance.[21] In a speech at the Swedish Academy of Science in 1764, shortly before his death, Clerck told his audience that he had been "crawling on the ground and in the bushes in order to collect insects," and describes how he had held a representative of *Nivalis* captive and fed it, but how the spider managed to

16 The English translation of the work is entitled *Aranei, or a natural history of spiders: including the principal parts of the well-known work on English spiders by Eleazar Albin, as also the whole of the celebrated publication on Swedish spiders by Charles Clerck.*

17 Kronestedt, "Carl Clercks 'Svenska spindlar'".

18 Mary Terrall, *Catching Nature in the Act: Réaumur and the Practice of Natural History in the Eighteenth Century* (Chicago: University of Chicago Press, 2014), 20.

19 Carl Clerck, *Om spindlars fångande och födande* (Kgl. Vetenskapsakademiens handlingar, 1761), 243, doi:10.5962/t.173032.

20 On collecting, see Mary Terrall, "Following Insects Around: Tools and Techniques of Eighteenth-Century Natural History," *British Journal for the History of Science* 43, no. 4 (2010); Mary Terrall, *Catching Nature in the Act*, 79.

21 Kronestedt, "Carl Clercks 'Svenska spindlar'".

escape.[22] A passage in the speech shows how much time and patience the observation of the spiders demanded:

> I have been looking for a Spider for several years, one which can only be seen during a short period, only in spring, after a night of severe frost when the morning is calm and the sun shines warm, moving about when no shadow appears and scares him. I have observed him in cracks in logs and planks, but have not yet been able to catch him. He has sat outside warming himself in the sun, but at the slightest glimpse [...] ran into the crack, from where I have not been able to induce him to come out; and when forcibly taken out, only a few crushed pieces of him remained.[23]

When addressing his audience, Clerck acknowledged the significance of his learned colleagues, the network of the learned community, stating that his collection would have remained useless had it not been for the diligence of the academy in bringing his knowledge to light.[24] Hence Clerck's example aptly shows the various stages of knowledge-making and circulation: the inspiration and motivation, the practices and tools, the collecting and ordering, the importance of networks and publicity. It also shows how, in some cases, the crucial part of the network, the spider itself, could be elusive and fragile. At least some degree of cooperation from the spiders was required and this often posed a challenge.

3 Traditional Knowledge: Physicotheology, Medicine and Poison

According to a long tradition of Occidental thought, prevalent in the eighteenth century, nothing was created in vain, but all beings were useful in one manner or another. Physico-theology was a way of thinking according to which divine purpose and design lay behind all creation, while maintaining that science could be seen as compatible with Christian beliefs in the providence and benevolence of God. *Maxima in minimis*, the great in the small, is one of the underlying ideas of the physicotheologians; by studying the smallest of creatures in nature, it was thought possible to acquire knowledge about God's

22 Carl Clerck, *Tal, innehållande några anmärkningar om insecterne, hållet för Kongl. Vetensk. Academien, den 7 Martii 1764, etc.* (Stockholm: Lars Salvius, 1764), 5, 101.

23 Clerck, *Tal*, 8, trans. L.H.

24 Clerck, *Tal*, 8, trans. L.H.

creation where everything was part of God's great plan. According to Brian Ogilvie, physical theology encouraged general considerations of insect behavior and notions of an "economy of nature" that emerged in the eighteenth century. He argues that physical theology was not only dependent on natural history but that it contributed to new ways of thinking about nature through its pursuit of detailed evidence.[25]

One proponent of the connection between the usefulness of insects for humans and the omnipotence of God was the German theologist Friedrich Christian Lesser, who treats the natural history of insects and their uses in *Insecto-theologia*, published in 1738.[26] Lesser was convinced that God's power and benevolence were manifest even in the tiniest insects. A long section deals with the use of insects as medicine; both spiders themselves and their webs are mentioned among the remedies.[27] The spider *Araneus diadematus* was considered useful for curing fever: according to the instructions, the spider was to be placed in a nutshell and worn in a band around the neck or placed on the pulse. The spider amulet is an example of traditional knowledge circulating in natural-historical works from antiquity to early modern times. The amulet was thought to prevent ague, as the spider and its oil were believed to be remarkably efficacious in curing fevers.[28] Spider webs could also be used on wounds to stop bleeding, a method often mentioned in connection with folk medicine.[29]

However, the spiders were also understood to be the cause of ill health. The perception of spiders as causes of illness derives from the fact that some of the approximately 50,300 species of spiders are venomous. Although none of the European spiders are dangerous to humans, the beliefs concerning their harmfulness led to medical discussions about tarantism in southern Europe in the seventeenth century.[30] The negative view of spiders is likely to derive from

25 Brian Ogilvie, "Maxima in Miminis Animalibus. Insects in Natural Theology and Physico-theology," in *Physico-theology: Religion and Science in Europe, 1650–1750*, ed. Ann Blair and Kaspar von Greyerz (Baltimore: Johns Hopkins University Press, 2020).

26 Here, the English translation is used. Friedrich Christian Lesser, *Inseco-theology: Or a Demonstration of the Being and Perfections of God, From a Consideration of the Structure and Economy of Insects* (Edinburgh, 1799), https://archive.org/details/insectotheology00lyongoog/page/n234/mode/2up.

27 Lesser, *Inseco-theology*, 203.

28 James Newman and Catherine Newman, "Oh What a Tangled Web: The Medicinal Uses of Spider Silk," *International Journal of Dermatology* 34, no. 4 (1995); see e.g. Dioscorides, *De Materia Medica* (Johannesburg: Ibidis Press), 205.

29 Newman and Newman, "Oh What a Tangled Web".

30 Tarantism is a form of hysterical behavior with roots in Italy, believed to result from the bite of the wolf spider *Lycosa tarantula*. See Jean Fogo Russell, "Tarantism," *Medical History* 23, no. 4 (1979).

the Bible, where spiders are depicted as dangerous and deceitful. Moreover, one of the most common phobias in Western culture is arachnophobia.[31] Still, many of the naturalists who spent time in proximity with spiders found them to be harmless. François Xavier Bon de Saint Hilaire assured his readers that "Spiders are not Venomous, having been very often bit by them my self, without any ill Consequence. And as for their Silk, it is so far from having any Venom, that every body makes use of it to stop Bleeding and heal Cuts; and indeed its Natural *Gluten* is a kind of Balsam, that cures small Wounds, by defending them from the Air."[32] Carl Clerck likewise reports that he often had been bitten by spiders and felt a pinch, without further after-effects, also noting that chickens and other small birds eat them without any harmful consequences.[33] According to Lesser, spiders "may be fatal to man, while they not only prove delicate morsels to many birds, but even specific remedies against their diseases."[34] This was in line with the physicotheologist view according to which everything could be of use to somebody; what might be poison to one animal becomes salutary to others. Traditional knowledge about spiders was often anthropocentric and the view on spiders used in medicine was instrumental. However, although a pronouncedly anthropocentric philosophy, the physicotheological view of the usefulness of all creatures to other animals than humans also suggests a wider understanding of ecological relationships where spiders had their given place.

4 Useful Knowledge: Spiders as Animals of Production

While spider webs, according to traditional knowledge, were understood to be useful for healing wounds resourceful naturalists developed new ways of using spider thread. Eighteenth-century natural historians were inspired by the Enlightenment ethos which said that useful knowledge should benefit individuals and society.[35] Prominent natural historians all over Europe were interested in the practical uses of plants and animals, particularly as substitutes for imported goods. As Edward D. Melillo has noted, Europe found itself

31 Megan Cavell, "Arachnophobia and Early English Literature," *New Medieval Literatures* 18 (2018).

32 Bon de Saint Hilaire, "A Discourse," 14.

33 Clerck, *Svenska Spindlar*, 6.

34 Lesser, *Insecto-theology*.

35 Useful knowledge as both propositional and prescriptive, see Joel Mokyr, *The Gifts of Athena: Historical Origins of the Knowledge Economy* (Princeton, NJ: Princeton University Press 2002); Peter Dear, "What is the History of Science the History *Of*? Early Modern Roots of the Ideology of Modern Science," *Isis* 96, no. 3 (2005).

to be on a knowledge periphery when it came to extracting useful substances out of insects.[36] Carl von Linné was greatly concerned that up to three quarters of Sweden's export expenditures were spent on silk imports, and he was interested in replacing imported products such as cochineal, Spanish flies and silkworm with indigenous alternatives.[37]

All spider species, with the exception of *Palystes* (huntsman) spiders, produce silk. The silk passes through spinnerets and the spider extrudes it through its spigots. The spiders use their silk to weave webs for catching prey, but it also plays a role in reproduction, where the male weaves a small sperm web on which to deposit its sperm.[38] In his work on Swedish spiders, Carl Clerck remarks on the superior qualities of the spider web – the thread did not melt in water, did not burn in fire, and maintained a sticky, gluelike character.[39] The ambition to use spiders for the production of silk is an example of the entanglement of intellectual and practical knowledge.

A Swedish natural historian and entomologist who took a strong interest in the practical knowledge and usefulness of insects and arachnids was Charles De Geer, a wealthy merchant and owner of the ironworks at Löfsta manor on the Swedish east coast. De Geer's main work, *Mémoires pour servir à l'histoire des insects* (1752–78), the title of which can be seen as an homage to René Antoine Ferchault de Réaumur (1683–1756), is a comprehensive work on entomology but his views on the practical use of spiders are to be found in a speech he gave at the Academy of Science in Stockholm a decade earlier.[40] De Geer concentrated on the practical uses of insects in the speech which discusses the "use the insects and their study bestows on us." As an entomologist and a merchant, he had a keen interest in the national economy and was enthusiastic about the possibilities of finding a substitute for the silk made from the silk moth *Bombyx mori*. De Geer had learned about spider silk from Réaumur and Bon, both of whom experimented with the production of spider silk.

---

36 Edward D. Melillo, "Global Entomologies: Insects, Empires, and the 'Synthetic Age' in World History," *Past and Present* 223 (2014).

37 Lisbet Koerner, *Nature and Nation* (Cambridge: Harvard University Press, 1999), 133.

38 Ricki Lewis, "Unraveling the Weave of Spider Silk," *BioScience* 46, no. 9 (1996), doi:10.2307/1312891; Catherine E. Scott, Alissa G. Anderson and Maydianne C.B. Andrade, "A Review of the Mechanisms and Functional Roles of Male Silk Use in Spider Courtship and Mating," *The Journal of Arachnology* 46, no. 2 (2018).

39 Clerck, *Svenska Spindlar*, 9.

40 Charles De Geer, *Mémoires pour servir à l'histoire des insectes* (Stockholm: Grefing & Hesselberg, 1752–1778). The title alludes to René Antoine Ferchault Réaumur's *Mémoires pour servir à l' histoire des insects* (Paris, 1734–1742); Charles De Geer, *Tal om nyttan, som Insecterne och deras skärskådande, tilskynda oss* (Stockholm: Lars Salvius, 1744).

The first to conduct experiments with spider silk was François Xavier Bon de Saint Hilaire, who published his results in *A Discourse upon the Usefulness of the Silk of Spiders*. Bon praises the qualities of spider silk, remarking that the lack of wider knowledge about it was due to "[t]he prejudice that is entertained against so common and despicable an Insect."[41] The method used by Bon recalled the process where silk was extracted from the silkworm. Bon explains that he took 12 or 13 ounces of "bags" (sacs or cocoons) from commonly found short-legged spiders and beat them with a small stick to free them from dust. He then washed them in warm water and let them steep in a pot with soap, salpeter (potassium nitrate), and some pieces of Arabic gum, and let them boil for two or three hours. The sacs were then rinsed and dried, after which the material could be carded. The result was an ash-colored silk which could be spun, and which, according to Bon, took dye well.[42] The spiders, according to Bon, laid a greater quantity of eggs than silk flies, 6–700, and they hatched themselves without care in August or December, dying soon afterwards. The young spiders that were bred from these eggs lived for 10–11 months without eating. When the weather turned warm, they started spinning and searching for food. In his experiment, Bon ordered as many large, short-legged spiders as could be found to be brought to him in August and September. The spiders were then housed in containers, in individual paper cells with air holes, and were fed flies. According to Bon the spiders fared well in their solitary confinement.[43]

Two or three pairs of gloves and stockings were made up and presented to the *Académie Royale des Sciences* in Paris while the natural historian Hans Sloane at The *Royal Society of London* received one pair.[44] The results attracted much attention as they promised to make an important contribution to the French silk industry. Thus, in 1709 the French government asked the naturalist René-Antoine Ferschault de Réaumur to assess Bon's experiment and to find out if silk made from spiders could be lucrative on a larger scale. Réaumur was by this time a well-established natural historian and a member of the *Académie des Sciences*. In his paper *Examen de la soie des Araignées* from 1710 Réaumur describes how he collected egg sacs from spiders and had them

41 Bon de Saint Hilaire, "A Discourse," 14.

42 Bon de Saint Hilaire, "A Discourse," 11–12.

43 Bon de Saint Hilaire, "A Discourse," 13.

44 Eleanor Morgan, "Sticky Layers and Shimmering Weaves: A Study of Two Human Uses of Spider Silk," *Journal of Design History* 29, no. 1 (2016): 8, 23, doi: 10.1093/jdh/epv019.

washed, boiled, and dried, and subsequently carded with fine combs, and the thread was woven on a weaving frame.[45]

Réaumur found the process to be difficult and time consuming. One of his objections had to do with the ferocity and unsociability of the spiders: in captivity they began exhibiting cannibalistic behavior, the larger spiders eating the smaller and weaker individuals.[46] Some spiders are cannibals, consuming each other for food while for other species, such as the black widow and the redback, it's a part of the mating process where the female devours the smaller male after mating.[47] Réamur's view of spiders as unsociable concurs with early modern understandings of the relationship between humans and the rest of creation. Domestic animals were seen as links between humans and wild animals,[48] and since the reclusive and cannibalistic spiders were difficult to domesticate, they could be perceived as wild and uncivilized. Hence, the spiders had to be kept in separate cells, and to obtain the silk they had to be immobilized. Réaumur's estimate was that 55,296 spiders were necessary to produce 500 grams of silk, compared to approximately 2,500 silkworms. His conclusion was that spider silk was more expensive, laborious to produce, and of inferior quality compared to silk produced by the *Bombyx mori* silk moth. The opinion of Réaumur, a celebrated and well-connected natural historian, was likely to have been more valued than that of Bon who can be considered an amateur when it comes to natural history, and whose position in the knowledge networks was not as central as that of Réaumur.

One difficulty with spider silk had to do with the demanding and potentially expensive breeding process, since spiders had to be kept in isolation. The catching of the spiders, by contrast, appears to have been found easy. Natural historians were used to catching and feeding single spiders but breeding them as domestic animals was difficult when compared to silk moths.[49] In contrast to spiders, the silk moth was domesticated around 3,500 BCE, and

---

45 René Antoine Ferchault Réaumur, "Examen de la soie des araignées," in *Mémoire de l'Académie royale des sciences* (1710), imprimé en 1732.

46 Réaumur, "Examen," 392.

47 Pierre Lesne, Marie Trabalon, and Raphaël Jeanson, "Cannibalism in Spiderlings Is Not Only about Starvation," *Behavioral Ecology and Sociobiology* 70, no. 10 (2016), http://www.jstor.org/stable/44857119.

48 Jennifer Mason, *Civilized Creatures: Urban Animals, Sentimental Culture, and American Literature, 1850–1900* (Baltimore, MD: JHU Press, 2005), 8–10; Harriet Ritvo, *The Platypus and the Mermaid, and Other Figments of the Classifying Imagination* (Cambridge, MA: Harvard University Press, 1997), 40.

49 Thomas Lecocq, "Insects: The Disregarded Domestication Histories," in *Animal Domestication*, ed. Fabrice Teletchea (IntechOpen, 2019), doi:10.5772/intechopen.81834.

in silk production the whole life cycle of the moth was controlled by humans; as a result of selective breeding they had lost their ability to fly.[50] Taming an animal requires at least some degree of cooperation from it, if only the act of staying alive for the duration of the product to be harvested. This modest feat was achieved by feeding the silk moth. Bon and Réamur fed the spiders, but this may not have been enough for the spiders to stay alive. Further challenges were the presumed aggressivity of the spiders and the task of producing silk of a high consistency of quality on a large scale.

Some of these problems were addressed by the Spanish Jesuit priest Raimondo Maria de Termeyer (1740–1814?). Termeyer published a treatise, originally in Italian, on spider silk based on experiments he had made in America and Italy. He set out to systematically refute Réaumur's objections to the successful production of spider silk on a larger scale, although he admits that "[t]he great authority of the French entomologist is certainly of great weight."[51]

Termeyer found the species *Linnaeus diadema* particularly suitable for the enterprise. These spiders produced five or six cocoons a year, resulting in 1,200 eggs, each of which in their turn would generate 4,000 young spiders annually. Hence, twelve spiders would eventually produce 50,000 eggs. Of particular interest is Termeyer's opinion that spiders were not as voracious and cruel by nature as Réaumur maintained, but that their aggressiveness may have been due to scant feeding and "the narrowness of their prison." Instead of individual cells, Termeyer recommended rearing spiders in a room or even outside, and to provide them with lots of flies and worms, some of which they could catch themselves. But although the rearing of the spiders may have been more charitable than in Bon's and Réaumur's experiments, the method of extracting the thread was cruel. Termeyer devised a contrivance to hinder the spider from touching the extremities of its abdomen and thereby cutting the best quality thread. The device was made from cork, a sheet of iron and iron pins, and the spider was placed on it so that the sheet of iron fell between the corslet and the abdomen of the animal, holding it in place with a half moon-shaped aperture. The spider lay on its back and the thread was drawn out onto a reel as the silk was taken directly out of the spider's body. This device kept the spider firmly in place, prevented it from extending its legs, and forced it to keep on spinning strong, high-quality thread.[52] The device where the spider was kept alive and forced to

50 Hui Xiang et al., "The Evolutionary Road from Wild Moth to Domestic Silkworm," *Nature Ecology & Evolution* 2 (2018), doi:10.1038/s41559-018-0593-4.

51 Raimondo Maria de Termeyer, *Researches and Experiments Upon Silk from Spiders, and Upon Their Reproduction* (Essex Institute, 1866), 53.

52 Termeyer, *Researches and Experiments*, 66–67.

produce can be understood in terms of Jason Hribal's insights into how various solutions have been created to prevent animal resistance.[53]

The spider's position as an animal of production in the knowledge network involving the spider, Termeyer himself, possible assistants, flies, worms, and the device, was active in the sense that it produced a product. If considered as a commodity however, its role was passive.[54] In both interpretations, this view of the spider is very different from the one expressed in Carl von Linné's verse, cited at the beginning here, which describes the soul, the brain, and nerves of a spider (although Linné would probably have approved of the spider silk venture).[55] The calculations made by both Bon, Réaumur and Termeyer show that spiders were considered as mere numbers in a production apparatus.

5 Spiders in Knowledge Networks

An investigation into knowledge networks including spiders offers a case which shows how knowledge was formed, on the one hand, in networks where natural historians, assistants, correspondents, specimens, texts and illustrations circulated among the learned community in Europe, and on the other hand, in networks consisting of actors such as humans, spiders, and other species, as well as instruments and devices. The knowledge about spiders was cumulative in the sense that it was built on and grew out of previously existing works and traditions. Intellectual and traditional knowledge about spiders circulated via earlier authorities from antiquity and the Renaissance period, conveyed by earlier literary works on natural history. In early modern times, new forms of scholarly knowledge emerged, both continuing and breaking with older traditions circulating in knowledge networks.

In the seventeenth and eighteenth centuries fieldwork, experiments, and closer observation, together with new tools and technology, contributed to increasing exploration, ordering and exploitation of the natural world in Europe. The Swedish amateur natural historian Carl Clerck was part of a knowledge network in which he engaged in the formation of scholarly and empirical

---

53 Jason Hribal, "'Animals Are Part of the Working Class': a Challenge to Labor History," *Labor History* 44, no. 4 (2003): 449. On insects as workers, see Jennifer Bonnell, "Occupational Hazards: Honeybee Labour as an Interpretive Device in Animal History," in *Traces of the Animal Past: Methodological Challenges in Animal History*, ed. Jennifer Bonnell and Sean Kheraj (Calgary: University of Calgary Press, 2022).

54 On the discussion of animals as a passive commodity or active workers, see Charlotte Blattner, "Should Animals Have a Right to Work? Promises and Pitfalls," *Animal Studies Journal* 9, no. 1 (2020): 44.

55 Carl von Linné, *Märkwärdigheter uti insecterne.*

knowledge by observing, naming, and ordering spider species. The vital participant in the network, however, was the spider itself. Although natural historians such as Clerck chased, cajoled, and fed spiders to render descriptions as exact as possible, the studied spiders could be elusive and fragile. Spiders are wild animals and the benefits of consenting to cooperate with humans consisted merely of food. Successful observation of a spider demanded that it was caught and kept alive while being studied, which required tools and instruments for transporting spiders and terraria for housing and feeding them. In Clerck's case the spider can be considered an uncooperative participant in a knowledge network.

The position of spiders in networks where traditional and medical knowledge was formed was instrumental when it came to using spiders as medicine, as in the case of wearing an amulet containing pulverized spider, for which the spider had to be killed, dried, and crushed. The spider thread used for wounds, by contrast, was a product that could be collected without killing the spider. Traditional knowledge about the use of spiders in medicine can be characterized as useful knowledge in the sense that it was perceived as beneficial.

Another form of useful or practical knowledge about spiders concerned not only their usefulness to individuals but also to the nation and its economy. François Xavier Bon de Saint Hilaire, Réaumur, and Raimondo Maria de Termeyer developed their respective methods for breeding spiders with a view to the production of silk. Descriptions of the experiments and results circulated in the learned circles of Europe, and were conveyed to, among others, the Swedish natural historian Charles De Geer. The endeavor necessitated breeding spiders on a large scale as well as arriving at technical solutions, developed with a view to the problems posed by spiders' unwillingness to cooperate in knowledge networks. These difficulties prompted the development of new techniques aimed at controlling the bodies of the spiders which can be observed in the formation of scholarly knowledge and practical, or "useful" knowledge.

The vulnerable part in the network was the spider; or rather the relationship between humans and spiders, which was marked by human ignorance concerning how to keep spiders alive. The networks within which intellectual, empirical, traditional, and useful knowledge was formed were fragile because the main element, or actor, the spider, was a wild animal who had to be caught, bred, and fed – in the cases where it was not simply killed. Taming a spider so that it could be bred as a domestic animal proved to be difficult or impossible. Hence, the fragility of spider life resulted in breaks in knowledge networks. The skillful weaver could not be made great advantage of, as Bon had hoped, without a great deal of effort and high costs which may have exceeded the profit that could be made. This was not because spiders posed a danger to humans, but because of they were wild, untamed creatures who would not cooperate.

Bibliography

Berkes, Fikret. "Biodiversity, Religion Traditions." In *Encyclopedia of Biodiversity*, edited by Samuel M. Scheiner, 18–29. Elsevier: Academic Press (2024).

Blattner, Charlotte. "Should Animals Have a Right to Work? Promises and Pitfalls." *Animal Studies Journal* 9, no. 1 (2020): 32–92.

Bon de Saint Hilaire, François Xavier. "A Discourse Upon the Usefulness of the Silk of Spiders." *Philosophical Transactions of the Royal Society* 27, no. 325–336 (1710): 2–16. https://www.jstor.org/stable/103101.

Bonnell, Jennifer. "Occupational Hazards: Honeybee Labour as an Interpretive Device in Animal History." In *Traces of the Animal Past: Methodological Challenges in Animal History*, edited by Jennifer Bonnell and Sean Kheraj, 49–72. Calgary: University of Calgary Press, 2022.

Bonnet, Charles. *Traité d'insectologie*. Vol. 1. Chez Durand, libraire, 1745.

Cavell, Megan. "Arachnophobia and Early English Literature." *New Medieval Literatures* 18 (2018): 1–43.

Clerck, Carl. *Svenska Spindlar: uti sina hufvud-slägter indelte samt under några och sextio särskildte arter; beskrefne och med illuminerade figurer uplyste*. Stockholm: Lars Salvius, 1757.

Clerck, Carl. *Om spindlars fångande och födande*. Kgl. Vetenskapsakademiens handlingar, 1761. Holding Institution: University of Alberta Library. doi:10.5962/t.173032.

Clerck, Carl. *Tal, innehållande några anmärkningar om insecterne, hållet för Kongl. Vetensk. Academien, den 7 Martii 1764, etc.* Stockholm: Lars Salvius, 1764.

Daston, Lorraine. "The History of Science and the History of Knowledge." *Know. A Journal on the Formation of Knowledge* 1 (2017).

Dear, Peter. "What is the History of Science the History *Of*? Early Modern Roots of the Ideology of Modern Science." *Isis* 96, no. 3 (2005): 390–406.

Dioscorides. *De Materia Medica*. Johannesburg: Ibidis Press, 2000.

Findlen, Paula. *Empires of Knowledge. Scientific Networks in the Early Modern World*. Oxford: Routledge, 2018.

Geer, Charles De. *Tal om nyttan, som Insecterne och deras skärskådande, tilskynda oss*. Stockholm: Lars Salvius, 1744.

Geer, Charles De. *Mémoires pour servir à l'histoire des insectes*. Stockholm: Grefing & Hesselberg, 1752–1778.

Glacken, Clarence. *Traces on the Rhodian Shore. Nature and Culture in Western Thought from Ancient Times to the End of the Eighteenth Century*. Berkeley, Los Angeles and London: University of California Press, 1967.

Heise, Ursula K. "Multispecies Futures and the Study of Culture." In *Futures of the Study of Culture: Interdisciplinary Perspectives, Global Challenges*, edited by Doris Bachmann-Medick, Jens Kugele and Ansgar Nünning, 274–287. Berlin and Boston: De Gruyter, 2020.

Hribal, Jason. "'Animals Are Part of the Working Class': A Challenge to Labor History." *Labor History* 44, no. 4 (2003): 435–453.

Jorink, Eric. "Insects, Philosophy and the Microscope." In *Worlds of Natural History*, edited by H. Curry, N. Jardine, J. Secord and E. Spary, 131–148. Cambridge: Cambridge University Press, 2018.

Koerner, Lisbet. *Nature and Nation*. Cambridge: Harvard University Press, 1999.

Kralj-Fišer, Simona and Matjaž Gregorič. "Spider Welfare." In *The Welfare of Invertebrate Animals*, edited by Claudio Carere and Jennifer Mather, 105–122. Vol. 18. Cham: Springer, 2019.

Kronestedt, Torbjörn. "Carl Clercks 'Svenska spindlar' 250 år [Carl Clerck's 'Svenska spindlar' 250 years]." *Fauna och Flora* 102 (2007): 6–14.

Kronestedt, Torbjörn. "Carl Clerck and What Became of His Spiders and Their Names." *European Arachnology 2008* 1 (2010): 105–117.

Latour, Bruno. *Reassembling the Social: An Introduction to Actor-Network-Theory*. Oxford: Oxford University Press, 2007.

Lecocq, Thomas. "Insects: The Disregarded Domestication Histories." In *Animal Domestication*, edited by Fabrice Teletchea. IntechOpen, 2019. doi:10.5772/intechopen.81834.

Lesne, Pierre, Marie Trabalon and Raphaël Jeanson. "Cannibalism in Spiderlings Is Not Only about Starvation." *Behavioral Ecology and Sociobiology* 70, no. 10 (2016): 1669–1678. http://www.jstor.org/stable/44857119.

Lesser, Friedrich Christian. *Insecto-Theology: Or a Demonstration of the Being and Perfections of God, From a Consideration of the Structure and Economy of Insects*. Edinburgh, 1799. https://archive.org/details/insectotheologyoolyongoog/page/n234/mode/2up.

Lewis, Ricki. "Unraveling the Weave of Spider Silk." *BioScience* 46, no. 9 (1996): 636–638. doi:10.2307/1312891.

Linné, Carl von. *Märkwärdigheter uti insecterne*. Stockholm, 1739.

Linné, Carl von. *A General System of Nature: Through the Three Grand Kingdoms of Animals, Vegetables and Mineral …* London: printed by Voss and Morris for Lackington, Allen and Co., 1800–1804.

Lorimer, Jamie. "Nonhuman Charisma." *Environment and Planning D: Society and Space* 25, no. 5 (2007): 911–932.

Lovejoy, Arthur O. *The Great Chain of Being*. Cambridge: Harvard University Press, 1976/1936.

Mason, Jennifer, *Civilized Creatures: Urban Animals, Sentimental Culture, and American Literature, 1850–1900*. Baltimore, MD: JHU Press, 2005.

Margócsy, Dániel. "A Long History of Breakdowns: A Historiographical Review." *Social Studies of Science* 47, no. 3 (2017): 307–325. http://www.jstor.org/stable/44652509.

Melillo, Edward D. "Global Entomologies: Insects, Empires and the 'Synthetic Age' in World History." *Past and Present* 223 (2014): 233–270.

Michalski, Katarzyna and Sergiusz Michalski. *Spider*. London: Reaktion Books, 2010.

Mokyr, Joel. *The Gifts of Athena: Historical Origins of the Knowledge Economy*. Princeton, NJ: Princeton University Press, 2002.

Morgan, Eleanor. "Sticky Layers and Shimmering Weaves: A Study of Two Human Uses of Spider Silk." *Journal of Design History* 29, no. 1 (2016): 8–23. doi: 10.1093/jdh/epv019.

Newman, James and Catherine Newman. "Oh What a Tangled Web: The Medicinal Uses of Spider Silk." *International Journal of Dermatology* 34, no. 4 (1995): 290–292.

Ogilvie, Brian W. "Order of Insects: Insect Species and Metamorphosis between Renaissance and Enlightenment." In *The Life Sciences in Early Modern Philosophy*, edited by Ohad Nachtomy and Justin E.H. Smith, 222–245. Online edition, Oxford Academic, 2014. Accessed March 17, 2023. doi:10.1093/acprof:oso/9780199987313.003.0012.

Ogilvie, Brian W. "Maxima in Minimis Animalibus. Insects in Natural Theology and Physico-theology." In *Physico-theology: Religion and Science in Europe, 1650–1750,* edited by Ann Blair and Kaspar von Greyerz. Baltimore: Johns Hopkins University Press, 2020.

Réaumur, René Antoine Ferchault. "Examen de la soie des araignées." *Mémoire de l'Académie royale des sciences* (1710). Imprimé en 1732. 386–408.

Réaumur, René Antoine Ferchault. *Mémoires pour servir à l' histoire des insects*. Paris, 1734–1742.

Ritvo, Harriet. 2005), 8–10; Harriet Ritvo, *The Platypus and the Mermaid, and Other Figments of the Classifying Imagination*. Cambridge, MA: Harvard University Press, 1997.

Russell, Jean Fogo. "Tarantism." *Medical History* 23, no. 4 (1979): 404–259.

Scholten, Koen. "Pierre Lyonet's (1706–1789) Study of Insects: Displaying Virtue and Gaining Social Status through Natural History." *Studium: Tijdschrift voor Wetenschapsen Universiteitsgeschiedenis* 11, no. 4 (2018): 245–259.

Scott, Catherine E., Alissa G. Anderson and Maydianne C.B. Andrade. "A Review of the Mechanisms and Functional Roles of Male Silk Use in Spider Courtship and Mating." *The Journal of Arachnology* 46, no. 2 (2018): 173–206.

Spary, Emma. "Botanical Networks Revisited." In *Wissen im Netz: Botanik und Pflanzentransfer in europäischen Korrespondenznetzen des 18. Jahrhunderts*, edited by Regina Dauser, Stefan Hächler, Michael Kempe, Franz Mauelshagen and Martin Stuber. Berlin: Akademie Verlag, 2008.

Termeyer, Raimondo Maria de. *Researches and Experiments Upon Silk from Spiders, and Upon Their Reproduction*. Essex Institute, 1866.

Terrall, Mary. "Following Insects Around: Tools and Techniques of Eighteenth-Century Natural History." *British Journal for the History of Science* 43, no. 4 (2010): 573–588.

Terrall, Mary. *Catching Nature in the Act: Réaumur and the Practice of Natural History in the Eighteenth Century*. Chicago: University of Chicago Press, 2014.

Ulicsni, Viktor, Ingvar Svanberg, and Zsolt Molnár. "Folk Knowledge of Invertebrates in Central Europe-Folk Taxonomy, Nomenclature, Medicinal and Other Uses, Folklore, and Nature Conservation." *Journal of Ethnobiology and Ethnomedicine* 12 (2016): 1–40.

Xiang, Hui, Xiaojing Liu, Muwang Li, Ya'nan Zhu, Lizhi Wang, Yong Cui, Liyuan Liu et al. "The Evolutionary Road from Wild Moth to Domestic Silkworm." *Nature Ecology & Evolution* 2 (2018): 1268–1279. doi:10.1038/s41559-018-0593-4.

Östling, Johan, David Larsson Heidenblad, Erling Sandmo, Anna Nilsson Hammar and Kari H. Nordberg. "The History of Knowledge and the Circulation of Knowledge." In *Circulation of Knowledge: Explorations in the History of Knowledge*, edited by Johan Östling, Erling Sandmo, David Larsson Heidenblad, Anna Nilsson Hammar and Kari H. Nordberg, 9–33. Lund: Nordic Academic Press, 2018.

CHAPTER 6

Valuing Birds and Insects in America, *c.*1815–1920: a Multispecies Perspective

Sophie FitzMaurice

1 Introduction

In *An Introduction to Entomology* (1815), the parson-naturalist William Kirby defended the nascent science of entomology from two major objections. The first was that entomology was frivolous, its practitioners "mere trifler[s]." Kirby answered this criticism at length, first drawing his readers' attention to the presence of insects in the writings of ancient philosophers and in scripture. More persuasive for Kirby, a natural theologian, was the notion that studying insects was revelatory of God's design. Since God created nature, which included insects, studying nature brought the inquisitor closer to God: "Who then shall dare maintain, unless he has the hardihood to deny that God created them, that the study of insects and their ways is trifling or unprofitable?" But, for the less reverent, Kirby pointed to entomology's practical applications, claiming that "the study of insects may be productive of considerable utility." Insects, he argued, were useful or noxious, beneficial or injurious, important to humans "both as sources of good or evil."[1]

The second objection was an ethical one. In order to understand the insect world, entomologists killed and dissected the objects of their study, subjecting wings, abdomens, and antennae to careful scrutiny under microscopes. Though Kirby differed from his co-author, William Spence, on the question of insect "instinct," he conceded that entomologists' methods could cause pain to insects, and that this capacity for pain constituted legitimate grounds for objection. "That the science of Entomology cannot be properly cultivated without the death of its objects, and that this is not to be effected without putting them to some pain, must be allowed," Kirby wrote. But he justified the

1 William Kirby and William Spence, *An Introduction to Entomology or, Elements of the Natural History of Insects*, vol. 1 (London: Longman, Hurst, Rees, Orme, and Brown, 1815), 22, 27, 29. For an analysis of the ideological beliefs of Kirby and Spence see J.F.M. Clark, "History from the Ground Up: Bugs, Political Economy, and God in Kirby and Spence's *Introduction to Entomology* (1815–1856)," *Isis* 97, no. 1 (March 2006).

DOI:10.1163/9789004715448_008

potential harm caused to insects on the basis of the "hints for useful improvements in the arts and sciences" that entomology could provide.[2] In short, insects could suffer and were indeed worthy of moral consideration, but any obligation to avoid causing them harm was outweighed by the practical benefits of entomology.

The defense of entomology that Kirby articulated in *An Introduction to Entomology* calls attention to the contested place of insects in Anglo-American thought at the opening of the nineteenth century – the moment of entomology's emergence as a distinct branch of natural science.[3] The idea that insects could experience pain was not uncontroversial in 1815, when the first volume of *An Introduction to Entomology* was published. A Cartesian tradition that regarded all animals as automata – unfeeling if highly complex mechanisms that lacked an internal life – exercised a strong hold on popular beliefs about insects. For those who did afford nonhumans the capacity for pain, that capacity – or "sensibility" – was often correlated to size. "It is well known that, in proportion as we descend in the scale of being, the sensibility of the objects that constitute it diminishes," Kirby wrote, referencing insects' place near the bottom of the *scala naturae* as further justification for the methods of entomologists.[4] But nor was it the case that insects' diminutive size necessarily placed them at an "empathetic distance" from people, or "expunged their dissections of violence," as evidenced by Kirby's need to defend himself from the "charge of inhumanity and cruelty."[5]

By the second half of the nineteenth century entomologists no longer had to defend their pursuits from charges of frivolity. Entomology had firmly

2 Kirby and Spence, *Introduction to Entomology*, 56, 57. Spence had a more secular view of insect instinct than Kirby, as Clark argues in "History from the Ground Up."

3 *An Introduction to Entomology* was read widely on both sides of the Atlantic, and many of the first entomologists in America were English-born. J.F.M. Clark argues that *An Introduction to Entomology* was a "founding text" of entomology. Clark, "History from the Ground Up," 28. Frank Egerton disagrees, but acknowledges that the book was central to the popularization of the science. Frank N. Egerton, "Ecological Aspects of Entomology During the 1800s," *Ecological Society of America Bulletin* 94, no. 1 (January 2013).

4 The idea of a *scala naturae*, or chain of being, was undermined by Darwin's suggestion of common evolutionary ancestors. See Mark V. Barrow Jr., *Nature's Ghosts: Confronting Extinction from the Age of Jefferson to the Age of Ecology* (Chicago: The University of Chicago Press, 2009), 20–21.

5 Uli Beisel et al. write that "[insects'] proximity, size and number makes them easy to collect, count and cultivate; their empathetic distance, alien form and short life spans expunges their dissections of violence and their deaths of sacrifice." Uli Beisel, Ann H. Kelly and Noémi Tousignant, "Knowing Insects: Hosts, Vectors and Companions of Science," *Science as Culture* 22, no. 1 (2013), 6.

established itself as a useful science with broad applications to agriculture. In Britain, France, Germany, the Russian Empire, and the United States, learned societies, journals, international exhibitions, and government agencies produced and disseminated information about insects to a broad readership. In the United States, home to a disproportionately large share of nineteenth-century entomologists, newspapers were by the 1870s urging that entomology be taught in common schools.[6] At the same time, the second objection from which Kirby defended entomologists – that entomology was cruel – had largely evaporated. Half a century after *An Introduction to Entomology* was published, insects were no longer trivial; many were dangerous enemies from which civilization must protect itself through the coordinated application of scientific expertise.

In part, this evolution in attitudes towards insects was the result of anthropogenic environmental change. There really were more insect "pests" in the nineteenth century than the eighteenth. Changes in agricultural production, land use, and global transportation infrastructure – processes bound up in the dynamics of settler-colonialism and extractive capitalism – resulted in the introduction of insect species to areas where they were previously unknown, and to the rapid multiplication of species considered pests.[7] This process was particularly marked in the United States, where the rapid rate of change in land use caused populations of both native and non-native ("invasive") species to mushroom. Spikes in populations of Colorado potato beetles (*Leptinotarsa decemlineata*) and Rocky Mountain locusts (*Melanoplus spretus*) in the 1850s, 1860s, and 1870s – events that provoked "a near-biblical sense of crisis" among the American public – underscored the social importance of entomology, ultimately leading to the science's institutionalization within the federal government.[8]

But environmental change did not occur in a cultural vacuum. Shifts in attitudes towards insects cannot be understood outside the context of the cultural

---

6 See, for example, editorial in *The Burlington Patriot* [Burlington, KS], February 16, 1877. One study suggests that nearly twenty percent of "noteworthy entomologists" in the nineteenth century were American. Scott A. Elias, "A Brief History of the Changing Occupations and Demographics of Coleopterists from the 18th through the 20th Century," *Journal of the History of Biology* 47, no. 2 (May 2014).

7 See James E. McWilliams, *American Pests: The Losing War on Insects from Colonial Times to DDT* (New York: Columbia University Press, 2008).

8 Brian Bonhomme, "Indispensable Allies or Overly-Sentimentalized Critters?: Debating the Merits of Insectivorous Birds as Defenders of Crop, 1850–1914," *The Historian* 83, no. 1 (May 2021), 69.

and institutional evolutions that both gave meaning to, and sought to measure and manage, changes in the nonhuman world. These include the professionalization of zoological science, the application of scientific technique to agricultural production, and the cultural invention of insects as pests.

Focusing on the closely related disciplines of economic entomology and economic ornithology in the United States, where those disciplines had their greatest influence, this chapter demonstrates how American attitudes towards insects and birds evolved in conjunction with one another. Economic entomologists and ornithologists adopted a dichotomous model of "beneficial" and "injurious" species to assess the value of birds and insects. This model – which was based on dualistic understandings of the natural world with deep origins in theological thought – greatly assisted in the passage of bird protection legislation. However, by positioning birds as "farmers' friends" on the basis of their consumption of insects, economic ornithology also helped cement the perception of insects as enemies. By the turn of the twentieth century, many economic ornithologists valued *all* birds as enemies of insects, consigning insects to the role of enemies of agricultural production – a particularly loaded charge given the venerated place of agrarianism in American culture. Increasingly, insects' value was measured not by their position on the great chain of being, but by the functional role they played within a grand Manichean drama between the forces of good and evil.

Despite a fair amount of scholarship on economic entomology and economic ornithology, few studies have analyzed the relationship between these disciplines. But, as this chapter will demonstrate, economic ornithology and economic entomology developed in concert. They drew on similar underlying principles about the state of nature and used the same dichotomous framework to categorize species as either beneficial or injurious. Moreover, the two disciplines often studied the same subjects, with economic ornithologists probing the contents of birds' stomachs to identify the insect life found within. Yet the two disciplines were not mirror images of one another. Economic ornithologists and economic entomologists used different methodologies, followed different institutional trajectories, and had very different cultural and legislative impacts. While economic ornithologists had some success in translating their findings into action, economic entomologists sought to influence legislation and change public opinion with far less frequency. Drawing on scientific publications, agricultural bulletins, newspaper articles, and the private correspondence of economic ornithologists and entomologists, this chapter demonstrates how attitudes towards birds and insects evolved within a multispecies network, in which the cultural and scientific construction of insects as pests helped elevate birds as allies in need of protection. Using the case study of the northern bobwhite (*Colinus virginianus*), the chapter further argues that cultural attitudes shaped scientific inquiry as much as the obverse.

2 The Origins of Economic Entomology and Ornithology

Broadly defined, economic entomology is the study of insects in relation to human activity. As the nineteenth-century British entomologist Andrew Murray pointed out, "in a wide sense every insect may be said to affect man, for those which appear to have no direct relation to him, may still affect him by giving pleasure by their external beauty, or instruction by their anatomy and physiology."[9] However, few nineteenth-century practitioners concerned themselves with insects' aesthetic or didactic value. In this respect, economic entomology diverged considerably from the related discipline of economic ornithology, whose adherents not only insisted upon the aesthetic value of birds but frequently sought to quantify it. Economic entomology would later expand to encompass disease-carrying insects and insects in their relation to livestock and forest products, but at its inception the discipline was concerned almost exclusively with insects that either benefited or harmed crop production.

Economic entomology developed as a transnational discipline out of late eighteenth century agricultural improvement and gardening practices.[10] Numerous European entomological societies sought to disseminate "useful knowledge" about insects to a popular audience, while dispelling "prejudice" and sentimentality. For example, the French journal *L'Insectologie Agricole* declared in its inaugural edition that its objective was to dispel sentimentality and misunderstanding about insects by "populariz[ing] knowledge of insects, especially insects useful or harmful to agriculture" without the use of "langage affecté."[11] But economic entomology had its greatest influence in the United States, where it was institutionalized at multiple levels of government. Drawing from farmers' practical experience, American entomologists disseminated knowledge about "useful" and "noxious" insects in the agricultural press, with publications of economic entomology appearing in the *New England Farmer*

9 Andrew Murray, *Economic Entomology Aptera* (London: South Kensington Museum, n.d. [1870s]).
10 J.F.M. Clark, *Bugs and the Victorians* (Princeton, NJ: Princeton University Press, 2009), 162.
11 *L'Insectologie Agricole* 1, no. 1 (February 1867), 1.

as early as 1823.[12] The first state entomologist was appointed in New York in 1841, with other states soon following. However, it was the explosive growth of populations of Rocky Mountain locusts in the 1850s, 1860s, and 1870s that added momentum to economic entomology and convinced a broad public of the necessity for a sustained investigation of insect life. The Department of Agriculture appointed an official Entomologist in 1863, and in 1877 the United States Entomological Commission was established to deal specifically with the Rocky Mountain locust outbreak.[13]

Economic ornithology emerged later than economic entomology, albeit from a similar set of underlying assumptions about nature. Like economic entomologists, economic ornithologists sorted the natural world into the dichotomous categories of beneficial and injurious species. Beneficial birds were those that aided farmers by eating injurious insects (*i.e.*, insects that damaged crops). Injurious birds were those that directly harmed the economic interests of farmers by feeding on crops or orchard fruit, or that indirectly harmed farmers' interests by feeding on beneficial birds or insects.

A major tradition influencing both economic entomology and economic ornithology was the balance of nature. Appearing in different forms within many worldviews, the "balance of nature" concept holds that nature, unless disrupted by human activity, exists in a state of harmony.[14] The concept is well summarized in the following excerpt from a lecture on economic ornithology delivered by Foster Ellenborough Lascelles (F.E.L.) Beal:

> In the primitive condition of any country, before the advent of civilized man, the animal and vegetable kingdoms are so nicely balanced against each other and within themselves that only in very long periods of time

---

12 James E. McWilliams, "The Pen and the Plow: Bridging the Knowledge Gap between American Entomology and Agriculture, 1740–1870," *American Entomologist* 56, no. 1 (Spring 2010); Egerton, "Ecological Aspects of Entomology".

13 For the history of economic entomology in the United States, see Clark, *Bugs and the Victorians*, 132–153; L.O. Howard, *Progress in Economic Entomology in the United States* (Washington: Government Printing Office, 1901), 136; McWilliams, *American Pests*; W. Conner Sorensen, "The Rise of Government Sponsored Applied Entomology, 1840–1870," *Agricultural History* 62, no. 2 (Spring 1988); W. Conner Sorensen, *Brethren of the Net: American Entomology, 1840–80* (Tuscaloosa: University of Alabama Press, 1995). For the history of the Bureau of Entomology, see Gustavus A. Weber, *The Bureau of Entomology. Its History, Activities, and Organization* (Washington: The Brookings Institution, 1930).

14 For a discussion of the balance of nature concept, see Barrow, *Nature's Ghosts*, ch. 1, "Bones of Contention: The American Incognitum and the Discovery of Extinction"; Frank N. Egerton, "Changing Concepts of the Balance of Nature," *The Quarterly Review of Biology* 48, no. 2 (June 1973).

> is a species wholly exterminated or even considerably reduced, while, on the other hand, they are so held in check that not one ever increases to such an extent as to seriously conflict with the others. When civilization, so called, is introduced all of this stability is overthrown, and oscillations begin resulting in great damage to many or fewer of the elements involved, of which man himself is the most important.[15]

Closely related to the "balance of nature" concept was the concept of the "economy of nature."[16] Within the economy of nature, every species played a unique, perhaps divinely ordained role. Wilson Flagg summarized this view in his 1861 *Utility of Birds*, a foundational text of economic ornithology: "Each species of bird performs certain services in the economy of nature, which cannot be so well accomplished by any other species."[17] Both economic entomology and economic ornithology were, then, implicitly (and sometimes explicitly) teleological, positing a purpose and order to nature.[18] In this respect, the disciplines retained the theological influence of earlier forms of natural history, as expressed by Kirby in *An Introduction to Entomology*.

Yet despite their significant conceptual overlaps, the two sciences had very different institutional histories. Economic ornithology was not institutionalized until 1887, when the Division of Economic Ornithology and Mammalogy was created within the Department of Agriculture (The Division of Economic Ornithology and Mammalogy was reorganized as the Division of Biological Survey in 1896, becoming the Bureau of Biological Survey – hereafter Survey – in 1905).[19] While economic entomology was established in response to specific insect outbreaks, the Division of Economic Ornithology was established partly as a result of campaigning by the American Ornithologists' Union (AOU) for bird protection legislation. AOU members were motivated not by utilitarian

15 F.E.L. Beal, unfinished manuscript on economic ornithology, n.d. Box 47, Folder 3, Waldo Lee McAtee Papers, Library of Congress (LOC).

16 For the economy of nature, see Donald Worster, *Nature's Economy: A History of Ecological Ideas*, 1977; reprint (Cambridge: Cambridge University Press, 1994).

17 Clarence Weed and Ned Dearborn, *Birds in Their Relations to Man* (Philadelphia: J.B. Lippincott Co., 1903), 18.

18 For Americans' desire to find order in nature after Darwin, see Thomas Dunlap, *Saving America's Wildlife: Ecology and the American Mind, 1850–1990* (Princeton: Princeton University Press, 1988), 18–46.

19 For a history of the Bureau of Biological Survey, see Jenks Cameron, *The Bureau of Biological Survey: Its History, Activities and Organization* (Baltimore: Johns Hopkins University Press, 1929). See also Matthew Evenden, "The Laborers of Nature: Economic Ornithology and the Role of Birds as Agents of Biological Pest Control in North American Agriculture, ca. 1880–1930," *Forest & Conservation History* 39, no. 4 (October 1995).

concerns about agricultural production, nor by the naturalist's desire to elevate their pursuits to scientific respectability, but by a belief in the moral, aesthetic, and didactic value of birds.[20] These divergent institutional histories help explain why economic ornithologists tried and succeeded in popularizing the idea of birds as friends, while economic entomologists less frequently sought to alter attitudes towards insects.

Survey ornithologists were charged with determining the economic status of all North American birds, and they did so primarily by producing knowledge of birds' consumption of insects. To arrive at their calculations, analysts relied on a combination of field observation and stomach content analysis, with the latter method generally considered to be more reliable. Stomach analysts carefully sorted through the contents of birds' stomachs and tallied the quantity and type of insects, as well as grain, fruit pulp, and other animal matter, found therein. Survey employees processed thousands of birds' stomachs every year, hoping to determine definitively the birds' intrinsic and immutable economic status.[21]

Having made determinations about the economic status of birds, Survey employees disseminated information about useful birds to farmers through bulletins and circulars. These publications often included colored images to enable easy identification of beneficial birds. The Survey sought to dispel farmers' "prejudices" against beneficial birds, and to educate them on which species to encourage around their property. The Survey's findings were also used to advise state governments on bird protection legislation. After the passage of the Migratory Bird Treaty Act in 1918, which outlawed the killing and sale of migratory insectivorous birds, determinations about a bird's feeding habits helped establish which species would be protected under federal law.[22] Thus, while the Biological Survey has mostly been remembered for its role in predator control, it also played an important role in passing bird protection legislation.

The task of economic ornithology was to simplify the complexity of interspecies relationships into a binary, based primarily on consumption, and to produce straightforward, actionable advice for farmers. But while stomach content analysis seemed to promise an objective and definitive answer to the

20 On the didactic value of birds and Christian ornithology, see Robin Doughty, *Feather Fashions and Bird Preservation* (Berkeley: University of California Press, 1975), 32–35.

21 Matthew Evenden characterizes this as the "revealed truths of stomach data." Evenden, "Laborers of Nature," 178.

22 For an account of the Migratory Bird Treaty, see Kurkpatrick Dorsey, *The Dawn of Conservation Diplomacy: U.S.–Canadian Wildlife Protection Treaties in the Progressive Era* (Seattle: University of Washington Press, 1998).

question of a bird's function in the economy of nature, practitioners understood that the data provided by stomach analysis was inexact and open to interpretation. Most economic ornithologists also understood that birds' diets varied seasonally, and according to their stage in the life cycle. "No one can tell what any animal may eat in case of necessity," Edward Forbush pointed out. "No mammal is considered more carnivorous than the wolf, yet at times it feeds voraciously on berries."[23]

Faced with uncertainty over feeding habits, and in furtherance of the personal appreciation of birds that had led many of them to devote their lives to ornithology, some economic ornithologists began to categorize *all* birds – the entire class Aves – as beneficial. "It is evident that any agency that will reduce the insect world as a body is to be fostered and encouraged and it is exactly this function that the birds fulfil," declared Beal in a lecture on economic ornithology. As long as the balance between good and bad insects was maintained – that is, as long as different insect species diminished in proportion to one another – a decline in insects could only be positive. "It is only when a bird is shown to be continually and persistently eating a much larger proportion of useful species that it can be condemned." But this outcome was highly unlikely, since "all birds eat some insects of both groups, so that the result of their work ... is to diminish the great mass of insect life."[24]

3 Making Friends: Economic Ornithology in Practice

As much as economic ornithologists strove for scientific precision, the discipline left considerable room for interpretation. In private correspondence, if less frequently in published matter, ornithologists debated the economic status of birds, sometimes heatedly. "I feel very sure most practical entomologists after inspecting the entire lot of insects eaten by the crow would say unhesitatingly that it was doubtful if the crow did not eat enough beneficial insects to counterbalance <u>all</u> the good done by eating harmful ones," steamed Walter Barrows in a lengthy letter to Survey Chief C. Hart Merriam in 1895. "At best, the weight does not fall very heavily on either side, and that would leave all the

23 Edward Howe Forbush, *The Natural Enemies of Birds* (Boston: Wright & Potter Printing Co., 1916), 11.

24 "Two lectures on economic ornithology given at various places by Professor F.E.L. Beal," McAtee Papers, LOC.

large amount of incontestable harm done in other ways without anything to offset it. ... I am very sorry that we cannot agree as to the merits of the crow."[25]

Agreement was even more difficult when it came to calculating the economic value of charismatic or widely beloved species, such as the northern bobwhite (*Colinus virginianus*). Aesthetically interesting, with distinctive black and white facial markings and a "cherry voice [that] thrills all to whom it is familiar," bobwhites are a species of quail whose native range is predominantly east of the Rocky Mountains but which by the nineteenth century could be found throughout North America.[26] "There is, perhaps, no bird in all America more generally known than the attractive and beneficial Bobwhite," wrote one ornithologist in 1916.[27]

Many Americans valued bobwhites for aesthetic and sentimental reasons, paying little regard to the birds' consumption of so-called injurious insects. Indeed, bobwhites were known to eat grain – a behavior that rendered the birds fair game in the minds of many farmers who granted hunters access to their land, free of charge, to shoot the birds. Bobwhite hunting was alarming to economic ornithologists who considered the birds allies in their battle against insects, as well as to bird-lovers who decried the cruelty of hunting.

Sylvester D. Judd, an Assistant Biologist at the Biological Survey, set out to disprove the misconception that bobwhites were primarily granivorous, and therefore harmful. Based on an examination of the contents of 801 bobwhite stomachs, Judd concluded that the bobwhite was indeed a beneficial species. In his 1903 circular on the *Economic Value of the Bobwhite*, Judd reported a crop of 10,000 pigweed seeds in one bobwhite's stomach, alongside chinch bugs and thirty Rocky Mountain locusts. This last finding was particularly significant. The Rocky Mountain locust plague of the 1870s would still be present in the memories of many of Judd's readers.[28] Judd's circular included a full-page illustration of bobwhites eating potato bugs from a farmer's (Figure 10.2), but omitted any depictions of the birds consuming grain. Through visual representation, economic ornithologists contributed to the creation of a cultural understanding of bobwhites as farmers' friends.

25 Letter from Walter B. Barrows to C. Hart Merriam, May 20, 1895. Box 102, Folder 2, McAtee Papers, LOC. Emphasis in the original. For simplicity, I am using "Survey" to refer to what was still the Division of Economic Ornithology and Mammalogy.

26 Earle A. Brooks, *The Food of West Virginia Birds* (Charlotte, WV: West Virginia Department of Agriculture, 1916), 37.

27 Brooks, *West Virginia Birds*, 37.

28 Sylvester D. Judd, *The Economic Value of the Bobwhite* (Washington DC: Government Printing Office, 1903).

Despite the publication of the 1903 circular and Judd's embarkation on a lecture tour to disseminate the findings of his study, the popular understanding of bobwhites as granivorous persisted. One newspaper article, widely circulated throughout the Midwestern and Southern United States between 1909 and 1912, reported that "Bobwhite has been accused of robbing grain fields in the south," even though stomach examinations had concluded that grain constituted only about one fifth of the bird's diet.[29] As Judd himself acknowledged in his 1903 circular, "the bobwhite is partial to wild grapes, and so might be expected to injure cultivated varieties, especially as its California relative, sometimes in a flock of a thousand, plunders vineyards."[30] If bobwhites ate grain – thus directly damaging the products of farmers' labor – how could economic ornithologists claim they were a beneficial species?

Survey biologists were joined by other actors who hoped to rehabilitate bobwhites in public opinion. Indeed, the birds became the subject of a new direction in economic ornithology. In 1904, Margaret Morse Nice innovated a new method for assessing birds' economic status, propelled by her desire to convince farmers and landowners of the bobwhite's value. Nice began to experiment with feeding captive bobwhites raised by her colleague at Clark College, physiology professor Clifton F. Hodge. Rather than sort through the contents of dead birds' stomachs, Nice fed living captive bobwhites on a variety of foodstuffs, including insects they would not normally encounter out of captivity. Before and after feeding the bobwhites, Nice weighed the quantity of each insect and weed seed, to determine the volume of each substance the birds ate. Using this "feeding-test method," Nice calculated that a bobwhite could consume 75,000 insects – weighing seven and a half pounds – in a year, together with 5,000,000 weed seeds. Chicks ate even more insects than adults; Nice estimated that a single bobwhite chick would eat 20,000 insects over a six week period, consuming almost half of its body weight in insects every day. Hunters who believed bobwhites were granivorous and thus fair game were mistaken, Nice argued. Grain constituted only a minor part of the bobwhite's diet, while the birds had a fondness for a huge variety of injurious insects, including such renowned pests as potato beetles, grasshoppers, Rocky Mountain locusts, and chinch bugs. By basing her analysis of bobwhites' diets on birds in captivity,

29 "A Good Word for Bobwhite," *The Leavenworth Times* [Leavenworth, KS], October 30, 1909. This report was widely circulated between 1909 and 1912, appearing in numerous newspapers across the states of Kansas, Louisiana, Minnesota, Missouri, Mississippi, Nebraska, New York, New Jersey, Ohio, Oklahoma, Pennsylvania, South Dakota, Texas, and Wisconsin.

30 Judd, *Economic Value*, 199.

Nice was able to add sixty-eight insect species to the official list published by the Department of Agriculture, including mosquitos, stable flies, and coddling moths.[31] Here, as elsewhere, "injurious" insects made "beneficial" birds.

Hodge declared Nice's study "the most careful and complete investigation ever made of the food of any bird." Her findings would prove the bobwhite "is worth one hundred-fold more alive and at work than dead," Hodge claimed, attaching a dollar amount to this claim: bobwhites would save farmers $8,500,000 in weeds and $500,000,000 in "insect tax." Anticipating the objection that bobwhites ate beneficial as well as injurious insects, Hodge argued that bobwhites' consumption of insects was prolific enough to render beneficial insects redundant: "if we could have enough bobwhites, they would leave nothing for the beneficial insects to do."[32] In this, he echoed Beal's logic: fewer insects, overall, was a net positive.

FIGURE 7.1
Bobwhite eating a Colorado potato beetle. The bobwhite is presented as a beneficial bird, performing a valuable economic service by eating a pest insect. Yearbook U.S. Department of Agriculture, 1903. Museum of Comparative Zoology, Harvard University

31 Margaret Morse Nice, "Food of the Bobwhite," *Journal of Economic Entomology* 3, no. 3 (June 1910).

32 Hodge, "The Bobwhite," *Nature and Culture* 2, no. 5 (April 1911).

FIGURE 7.2 From C.F. Hodge, "The Bobwhite," *Nature and Culture* 2:5 (April 1911). Caption reads "Bobwhite Chick Three Weeks Old – Usual Occupation." The presence of the grasshopper in the bobwhite chick's mouth clearly signals its status as a useful bird. Museum of Comparative Zoology, Harvard University

As the bobwhite demonstrates, underlying economic ornithology's absolutist rhetoric of good/bad, beneficial/injurious, ally/pest, friend/enemy was a profound relativism. Bobwhites were useful to farmers when eating Rocky Mountain locusts and Colorado potato beetles, but harmful when eating grain. Whether they were beneficial or injurious *as a species* – whether their food habits ultimately "condoned" or "condemned" them – was partly a matter of interpretation. By feeding captive bobwhites on grasshoppers, mosquitos, and stable flies, Margaret Morse Nice helped cement their status as beneficial birds. Had she fed them grapes, her audience may have reached a different conclusion.[33]

4 The Aesthetic Value of Insects?

Despite Murray's suggestion that insects could benefit humans through their "external beauty," economic entomologists rarely extended the concept of insect value beyond the direct applications to agriculture. In their tome on Destructive and Useful Insects, whose third edition exceeded 1,000 pages, the entomologists Clell Lee Metcalf and Wesley

33 This is something Waldo McAtee pointed out in a critique of Nice and Hodge's publication, though he appears to have been somewhat alone in his criticism. Waldo L. McAtee, "Discussion and Correspondence. Food of the Bobwhite," *Journal of Economic Entomology* 3, no. 5 (October, 1910).

Pillsbury Flint devoted less than a page to the aesthetic value of insects. The authors attested that "insects rival birds and flowers in beauty" but acknowledged that, in most cases, this beauty could be appreciated only with the aid of a microscope.[34] In the absence of aesthetic appreciation, "beneficial" insects were limited to a handful of species that aided agricultural production – either directly, by pollination, or indirectly, through parasitism or predation of "injurious" insects. Even when they decreed a species beneficial, economic entomologists rarely elevated insects to the status of "friend," "ally," or "co-laborer" – labels frequently attached to beneficial birds. Edward Oliver Essig wrote giddily of the "mass production" of mealybug ladybirds (*Cryptolaemus montrouzieri*) – a species judged beneficial to the citrus industry – at the California State Insectary.[35] Insects were insensible products, rather than benevolent agents.

In contrast, bird conservationists and economic ornithologists successfully extended the definition of utility to encompass birds' aesthetic value. Sylvester Judd, for example, confidently claimed that "the aesthetic pleasure derived from the presence of the bobwhite has a *certain definite value*."[36] Economic ornithologists were aided in their endeavors by the contributions of conservationists, amateur birdwatchers, and Christian ornithologists who believed in the didactic value of birds. For many of these constituencies, economic ornithology's instrumentalization of birdlife was at odds with their emotional and sympathetic attachment to birds, yet the beneficial/injurious framework proved an effective tool to enact bird preservation legislation.

Economic entomologists did not, of course, have a monopoly on attitudes towards insects. Many scholars studied insects such as bees and ants for their social complexity, paying little regard to the creatures' impact on agriculture. Some, like the French scientist Louis Figuier, rejected the instrumentalist treatment of insects more comprehensively. Figuier's book *The Insect World*, published in New York, London, and Paris in 1868, celebrated insects – including species generally considered pests and parasites – on their own terms, reflecting what one scholar has identified as a "biocentric ethic" on the part of its author.[37] But Figuier did not have a major impact on attitudes towards insects in the United States, where the beneficial/injurious binary provided

34 C.L. Metcalf and W.P. Flint, *Destructive and Useful Insects: Their Habits and Control* (1928; reprint, McGraw–Hill, 1951), 73.

35 E.O. Essig, *A History of Entomology* (1931; reprint, New York: Hafner Publishing Company, 1965).

36 Judd, *Economic Value*, 194. Emphasis added.

37 Jeanette Samyn, "Cruel Consciousness: Louis Figuier, John Ruskin, and the Value of Insects," *Nineteenth-Century Literature* 71, no. 1 (June 2016).

the dominant lens for public understandings of insects. Absent an equivalent cohort championing the didactic and aesthetic value of insects, the dominant impression of insects as pests prevailed well into the twentieth century, when the development of synthetic pesticides revolutionized farmers' abilities to make war on insects.[38]

5 Conclusion

More than two centuries after Kirby and Spence published *An Introduction to Entomology*, the frameworks of economic entomology and economic ornithology continue to hold traction. Economic ornithology may no longer exist, in part having been made obsolete by the rise of synthetic pesticides, but, as Jakub Kronenberg points out, the logic that animated it endures in the contemporary concept of "ecosystems services."[39] A recent article in the *Journal of Ornithology* called for "a new economic ornithology," with the authors stressing the importance of "quantifying the services provided by birds."[40] Meanwhile, economic entomology continues to be practiced under that name; the *Journal of Economic Entomology* is in its 116th year. Present-day economic entomologists may have a more secular view of insects than their nineteenth-century counterparts, many of whom believed that insects' status as "beneficial" or "injurious" was divinely ordained. But, at least in cultural understandings of insects, the ongoing use of the beneficial/harmful framework continues to reify these categories as immutable and intrinsic properties of species, rather than constructs developed in response to specific historically contingent social and institutional needs.

The question of insect sentience has recently emerged as a research agenda in the fields of ethology and comparative psychology, and has taken on renewed

38 On the development of chemical pesticides see, for example, Edmund Russell, *War and Nature: Fighting Humans and Insects with Chemicals from World War I to Silent Spring* (Cambridge: Cambridge University Press, 2001).

39 For the demise of economic ornithology, see Evenden, "The Laborers of Nature." For economic ornithology and ecosystems services, see Jakub Kronenberg, "Betting Against Human Ingenuity: The Perils of the Economic Valuation of Nature's Services," *BioScience* 65, no. 11 (November 2015) and Jakub Kronenberg, "What Can the Current Debate on Ecosystem Services Learn from the Past? Lessons from Economic Ornithology," *Geoforum* 55 (August 2014).

40 Christopher J. Whelan, Ç.H. Şekercioğlu and Daniel G. Wenny, "Why Birds Matter: From Economic Ornithology to Ecosystem Services," *Journal of Ornithology* 156 (December 2015).

urgency in the face of mass species extinction.[41] Just as nineteenth-century environmental change prompted changes in attitudes towards insects, and the establishment of new institutions to study them, so increasing awareness of – and anxiety about – anthropogenic climate change, biodiversity loss, and mass species extinction informs the study of insects and cultural attitudes towards them in the twenty-first century.[42] But as we continue to ponder the value of insects, it is worth reflecting on how attitudes towards insects have evolved within a multispecies network. Injurious insects made beneficial birds. While the theory of economic entomology and economic ornithology provided that the opposite could be true, culture ensured that in practice it rarely was: beneficial insects never made injurious birds. The history of economic ornithology, and the case of the bobwhite in particular, reveals that these dominant scientific frameworks left space for a considerable amount of flexibility, and perhaps even guesswork. Perhaps those of us concerned about the plight of insects can take inspiration from ornithologists' success in yoking scientific discourse to a sentimental love of birds.

Bibliography

Archival Sources

Waldo Lee McAtee Papers. Library of Congress.

Published Primary Sources

Brooks, Earle A. *The Food of West Virginia Birds.* Charlotte, WV: West Virginia Department of Agriculture, 1916.

Essig, E.O. *A History of Entomology.* 1931; reprint, New York: Hafner Publishing Company, 1965.

Forbush, Edward Howe. *The Natural Enemies of Birds.* Boston: Wright & Potter Printing Co., 1916.

Hodge, Clifton Fremont. "The Bobwhite." *Nature and Culture* 2, no. 5 (April 1911): 17–32.

41 A notable example of a recent work that grapples with the question of insect sentience seriously is Lars Chittka, *The Mind of a Bee* (Princeton: Princeton University Press, 2022).

42 Entomologists disagree about the extent and rate of mass insect die-off, but it is unquestionable that we are in the midst of a catastrophic loss of life on Earth. On so-called "Insectageddon" see Francisco Sánchez-Bayo and Kris A.G. Wyckhuys, "Worldwide Decline of the Entomofauna: A Review of its Drivers," *Biological Conservation* 232 (April 2019) and Chris D. Thomas, T. Hefin Jones and Sue E. Hartley, "'Insectageddon': A Call for More Robust Data and Rigorous Analyses," Invited Letter to the Editor, *Global Change Biology*, March 1, 2019.

Howard, L.O. *Progress in Economic Entomology in the United States.* Washington: Government Printing Office, 1901.

Judd, Sylvester D. *The Economic Value of the Bobwhite.* Washington DC: Government Printing Office, 1903.

Kirby, William and William Spence. *An Introduction to Entomology or, Elements of the Natural History of Insects*, vol. 1. London: Longman, Hurst, Rees, Orme, and Brown, 1815.

McAtee, Waldo L. "Discussion and Correspondence. Food of the Bobwhite." *Journal of Economic Entomology* 3, no. 5 (October, 1910): 437–438.

Metcalf, C.L. and W.P. Flint. *Destructive and Useful Insects: Their Habits and Control.* 1928; reprint. McGraw-Hill, 1951.

Murray, Andrew. *Economic Entomology Aptera.* London: South Kensington Museum, n.d. [1870s].

Nice, Margaret Morse. "Food of the Bobwhite." *Journal of Economic Entomology* 3, no. 3 (June 1910): 295–313.

Weber, Gustavus A. *The Bureau of Entomology. Its History, Activities, and Organization.* Washington, DC: The Brookings Institution, 1930.

Weed, Clarence and Ned Dearborn. *Birds in Their Relations to Man.* Philadelphia: J.B. Lippincott Co., 1903.

Newspapers and Periodicals

The Burlington Patriot

The Davey Mirror

The Headlight

L'Insectologie Agricole

The Leavenworth Times

Lincoln County News

Literature

Barrow, Mark V. Jr. *Nature's Ghosts: Confronting Extinction from the Age of Jefferson to the Age of Ecology.* Chicago: The University of Chicago Press, 2009.

Beisel, Uli, Ann H. Kelly and Noémi Tousignant. "Knowing Insects: Hosts, Vectors and Companions of Science." *Science as Culture* 22, no. 1 (2013): 1–15.

Bonhomme, Brian. "Indispensable Allies or Overly-Sentimentalized Critters?: Debating the Merits of Insectivorous Birds as Defenders of Crop, 1850–1914." *The Historian* 83, no. 1 (May 2021): 44–69.

Cameron, Jenks. *The Bureau of Biological Survey: Its History, Activities and Organization.* Baltimore: Johns Hopkins University Press, 1929.

Chittka, Lars. *The Mind of a Bee.* Princeton: Princeton University Press, 2022.

Clark, J.F.M. *Bugs and the Victorians.* New Haven: Yale University Press, 2009.

Clark, J.F.M. "History from the Ground Up: Bugs, Political Economy, and God in Kirby and Spence's *Introduction to Entomology* (1815–1856)." *Isis* 97, no. 1 (March 2006): 28–55.

Dorsey, Kurkpatrick. *The Dawn of Conservation Diplomacy: U.S.– Canadian Wildlife Protection Treaties in the Progressive Era.* Seattle: University of Washington Press, 1998.

Doughty, Robin. *Feather Fashions and Bird Preservation.* Berkeley: University of California Press, 1975.

Dunlap, Thomas. *Saving America's Wildlife: Ecology and the American Mind, 1850–1990.* Princeton: Princeton University Press, 1988.

Egerton, Frank N. "Changing Concepts of the Balance of Nature." *The Quarterly Review of Biology* 48:2 (June 1973): 322–350.

Egerton, Frank N. "Ecological Aspects of Entomology During the 1800s." *Ecological Society of America Bulletin* 94, no. 1 (January 2013): 36–88.

Elias, Scott A. "A Brief History of the Changing Occupations and Demographics of Coleopterists from the 18th through the 20th Century." *Journal of the History of Biology* 47, no. 2 (May 2014): 213–242.

Evenden, Matthew D. "The Laborers of Nature. Economic Ornithology and the Role of Birds as Agents of Biological Pest Control in North American Agriculture, ca. 1880–1930." *Forest & Conservation History* 39, no. 4 (October 1995): 172–183.

Kronenberg, Jakub. "Betting Against Human Ingenuity: The Perils of the Economic Valuation of Nature's Services." *BioScience* 65, no. 11 (November 2015): 1096–1099.

Kronenberg, Jakub. "What Can the Current Debate on Ecosystem Services Learn from the Past? Lessons from Economic Ornithology." *Geoforum* 55 (August 2014): 164–177.

Judd, Richard W. ""A Wonderfull Order and Balance": Natural History and the Beginnings of Forest Conservation in America, 1730–1830." *Environmental History* 11 (2006): 8–36.

McWilliams, James E. *American Pests: The Losing War on Insects from Colonial Times to DDT*. New York: Columbia University Press, 2008.

McWilliams, James E. "The Pen and the Plow: Bridging the Knowledge Gap between American Entomology and Agriculture, 1740–1870." *American Entomologist* 56, no. 1 (Spring 2010): 44–53.

Russell, Edmund. *War and Nature: Fighting Humans and Insects with Chemicals from World War I to Silent Spring*. Cambridge: Cambridge University Press, 2001.

Samyn, Jeanette. "Cruel Consciousness: Louis Figuier, John Ruskin, and the Value of Insects." *Nineteenth-Century Literature* 71, no. 1 (June 2016): 89–114.

Sánchez-Bayo, Francisco and Kris A.G. Wyckhuys. "Worldwide Decline of the Entomofauna: A Review of its Drivers." *Biological Conservation* 232 (April 2019): 8–27.

Sorensen, W. Conner. *Brethren of the Net: American Entomology, 1840–80*. Tuscaloosa: University of Alabama Press, 1995.

Sorensen, W. Conner. "The Rise of Government Sponsored Applied Entomology, 1840–1870." *Agricultural History* 62, no. 2 (Spring 1988): 98–115.

Thomas, Chris D., T. Hefin Jones and Sue E. Hartley. "'Insectageddon': A Call for More Robust Data and Rigorous Analyses." Invited Letter to the Editor, *Global Change Biology*, March 1, 2019.

Whelan, Christopher J., Ç.H. Şekercioğlu and Daniel G. Wenny. "Why Birds Matter: From Economic Ornithology to Ecosystem Services." *Journal of Ornithology* 156 (December 2015): 227–238.

Worster, Donald. *Nature's Economy. A History of Ecological Ideas.* 1977; reprint. Cambridge: Cambridge University Press, 1994.

CHAPTER 7

From Harmless Nuisance to Frightening Enemy: the Perceptions of Ticks in Finland before the Beginning of the Tick Hysteria in the 1990s

Otto Latva

1 Introduction

In October 1991, a letter was published in the medical column of a newspaper in eastern Finland, in which a mother expressed concern about how her son had been bitten by a tick during the summer:

> There has been a lot of discussion recently about the tick-borne disease, which causes all sorts of harm. Our little boy was bitten by a tick in the summer, it was rubbed off and he was cured. The boy seemed to be perfectly healthy then and now, months later. Should he be tested as there has been such a scare, and how do we best get the tick off the skin? In the old days, there were plenty of ticks on people living in the countryside.[1]

The letter clearly shows how the connection between Lyme disease (borreliosis) and ticks, which had been discovered only a few years prior in 1981,[2] began to provoke worrying reactions in public discussion in the early 1990s. A doctor who replied to the above-mentioned letter in the same column had also noticed this and wrote that "there were ticks on almost every child in the old days and they seemed to survive. The tick hysteria is spreading to Finland, but only on a small scale compared to America."[3]

The question from the concerned mother and the doctor's response are among the many examples of how attitudes toward ticks in Finland became tinged with fear from the early 1990s onward. However, the conversation between the two individuals also reveals something about the period before

1 "Seuraava potilas," *Länsi-Savo*, October 7, 1991.

2 Willy Burgdorfer et al., "Lyme Disease – a Tick-Borne Spirochetosis?" *Science* 216, no. 4552 (1982), doi:10.1126/science.7043737. See also Alan G. Barbour and Jorge L. Benach, "Discovery of the Lyme Disease Agent," *mBio* 10, no. 5 (2019), doi:10.1128/mbio.02166-19.

3 "Seuraava potilas".

the so-called hysteria. They are talking about the old days, when people in rural areas had a lot of ticks, and the impact of these small arachnids onpeople was clearly not a matter of concern in the same way as it was in the early 1990s.

In this chapter, I examine the historical relationship between ticks and people in Finland before the 1990s, when fear, anger, and anxiety began to define the Finnish attitude toward ticks. I explore the shared history of humans and ticks by examining the Finnish public discussion. I will investigate how ticks have been described in different times, what perceptions were associated with these animals, and how these perceptions have changed over time.

There are about 1500 species of ticks in Finland, but this study focuses on the so-called common tick, whose Latin name is *Ixodes ricinus*. These ticks have been studied almost exclusively in the natural sciences, especially in the fields of biology and virology. In Finland, for example, studies in these fields have produced significant findings on the physiology and behavior of ticks and their distribution. These are important studies, but they do not really account for the shared history between humans and ticks and the cultural dimension of this relationship.[4] Overall, the history between ticks and humans has been previously mentioned in only a few studies and usually very briefly. These explanations mainly include only a few sentences about how Homer might have mentioned ticks as early as 800 BC or how early naturalists such as Aristotle and Pliny described ticks as hideous parasites.[5]

In the field of environmental humanities, which looks at the shared history of humans and nature and the cultural dimension of this relationship more extensively, only a few studies have been published exploring the humans – tick relationship, and most of them are included in this book.[6] However, none of the studies looks at the long-term relationship between humans and ticks, nor the changes in this relationship. Thus, my aim in this chapter is to fill this gap in our knowledge, explore the shared long-term history of humans and

---

4 See, e.g., Maija Laaksonen et al., "Crowdsourcing-Based Nationwide Tick Collection Reveals the Distribution of Ixodes Ricinus and I. Persulcatus and Associated Pathogens in Finland," *Emerging Microbes & Infections* 6, no. 1 (2017), doi:10.1038/emi.2017.17; Eeva Sajanti et al., "Lyme Borreliosis in Finland, 1995–2014," *Emerging Infectious Diseases* 23, no. 8 (2017), doi:10.3201/eid2308.161273.

5 See, e.g., John F. Anderson, "The Natural History of Ticks," *Medical Clinics of North America* 86, no. 2 (2002): 205, doi:10.1016/S0025-7125(03)00083-X.

6 See chapters by Sanna Lillbroända-Annala, Suvi Rytty, Tuomas Räsänen and Heta Lähdesmäki, and Taina Syrjämaa in this volume. See also Sanna Lillbroända-Annala and Oscar Winberg, "'Fästingen håller inget säkerhetsavstånd': Konkurrerande riskdiskurser om fästingar i media," TRACE ∴ *Journal for Human-Animal Studies* 9 (2023), doi:10.23984/fjhas.121761.

ticks in Finland, and explain how our relationship with these arachnids has changed over time and why old beliefs to understand ticks as harmless nuisance persisted for a surprisingly long time until the turn of the 1980s and 1990s, even though ticks had been known to spread diseases to both cattle and humans for decades. These questions are important because humans have coexisted with ticks in Finland for centuries, if not millennia. Our relationship with nature and species is constantly changing and has changed throughout history,[7] and our perception of the tick is no exception.

As my research data, I use the material digitized by the National Library of Finland, which contains a massive number of nearly 27 million pages of newspapers, magazines, books, and ephemera published in Finland from the early modern period to the present day.[8] This amount of data already offers a comprehensive representation of how Finnish public discourse perceived and comprehended various phenomena, as well as how it responded to them.

I have conducted a word search in the database using the word "puutiainen," which is the Finnish word for tick. The National Library's database contains a total of 1237 hits with this key word in various works published before 1991. The first mention of this small arachnid dates to 1889, which is also the year from which I begin my examination of the research data. I will end my analysis in 1991, when the relationship between humans and ticks in Finland had already become marked by fear. The time frame of my study therefore covers the period between 1889 and 1991.

Since "puutiainen" is a word with many meanings in Finnish, including the surname of people, I have had to exclude many search results that do not directly refer to this animal. In addition, some ORC errors have occurred during the digitization of the Finnish National Library's collection, and some words similar to "puutiainen" such as "punainen" (red) or "puutetta" (shortage) have been incorrectly machine-read as "puutiainen." I have also removed all such errors from the data. After filtering the data from all the irrelevant search results, the total number of relevant results for the period 1889–1991 was only 407. Although the amount of data was greatly reduced after filtering out irrelevant data, the dataset of more than 400 research sources is still a representative sample of the public discussion, given that it has been carefully collected from all of the more than 27 million pages of data. Thus, in this chapter, I will focus only on the results found with the word "puutiainen"

7 Otto Latva, *The Giant Squid in Transatlantic Culture: The Monsterization of Molluscs* (London: Routledge, 2023), 9, doi:10.4324/9781003311775.

8 National Library of Finland, https://digi.kansalliskirjasto.fi/stats?set_language=en, Accessed October 24, 2023.

and not on the sources found with the Finnish synonyms for tick such as "punkki" or "puutäi," nor with the Swedish name for tick such as "fästingar" or "skogsbässar".

Methodologically, my research contributes to digital history research as it uses digitized sources, in which printed texts have been translated into machine-encoded data.[9] My research also makes use of the search functions of the National Library's digital archive, which have enabled me to search material on ticks in a vast number of Finnish publications from different periods. In addition to the digital methods, I have analyzed all 407 sources using qualitative methods such as careful reading and content analysis.

Theoretically, my research contributes to the study of multispecies history and the tradition of human-animal studies. The starting point for such a study is to look at the past from the perspective of multispecies agency, where humans are not the only agents to have shaped the past, but where all other species, alongside humans, have by their actions and existence made the world as it is today.[10] For example, if we think that ticks had never existed on Earth, the world would certainly look very different today.

2 Changes in the Public Discussion

As you can see from Figure 8.1, people have been talking steadily about ticks in Finland from the late 1800s to the 1990s. The number of texts written about these animals has increased significantly since the 1990s, when the so-called tick hysteria started. However, ticks were by no means an uncommon topic of discussion before the 1990s, but how were they discussed then? From my data set, four major themes emerged that defined how these little arachnids were talked about in 1889–1991. They were spoken of as metaphors for human activity, unpleasant creatures that attached themselves to humans and other animals, disease vectors to cattle as well as to humans.

As Figure 8.2 shows, the debate on ticks has changed not only in the 1980s and 1990s but also earlier in the 20th century. By far the most significant change occurred in the 1950s and 1960s, when ticks' ability to spread diseases dangerous to humans became part of the public discussion. As the volume of this debate began to increase, the amount of discussion describing ticks as a vector of cattle diseases decreased. In addition to these themes, ticks were

9 Hannu Salmi, *What Is Digital History?* (Cambridge: Polity Press, 2021).

10 Margo DeMello, *Animals and Society: An Introduction to Human-Animal Studies*, (New York: Columbia University Press, 2012); Latva, *The Giant Squid.*

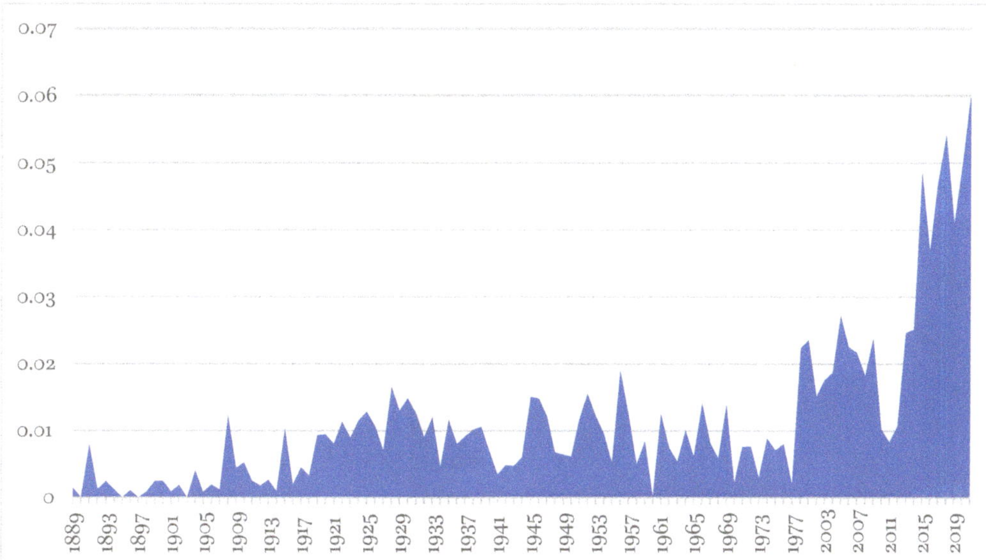

FIGURE 8.1 Occurrence of the word "puutiainen" in public discussion in Finland between 1889 and 2021
Note: The graph is based on data from which all hits that do not refer to ticks have been removed. The years are shown below the graph, and the percentage related to all digitized public discussion is shown on the left.

steadily used as a metaphor in Finnish public discourse, especially for human activity, until the 1950s and 1960s, after which they were only occasionally mentioned metaphorically. Equally, the number of texts describing ticks as merely unpleasant creatures began to decline by the 1970s at the latest.

I now turn to each of the four above-mentioned themes separately. The order in which they are dealt with is chronological, starting with the oldest ways of understanding the tick in Finland, using it as a metaphor or understanding the tick as a nuisance to humans and animals.

3 The Tick as a Metaphor in Finland

The abundance of tick-related metaphors in the dataset shows that these arachnids and their ability to burrow firmly into humans and suck blood was widely known in Finland as early as the 19th century. In these metaphors, the appearance and behavior of the ticks were most often used to describe devouring and a person who has eaten or drunk too much. For example, Aleksis Kivi, the author of *Seitsemän veljestä* (*Seven Brothers*) (1870), which became

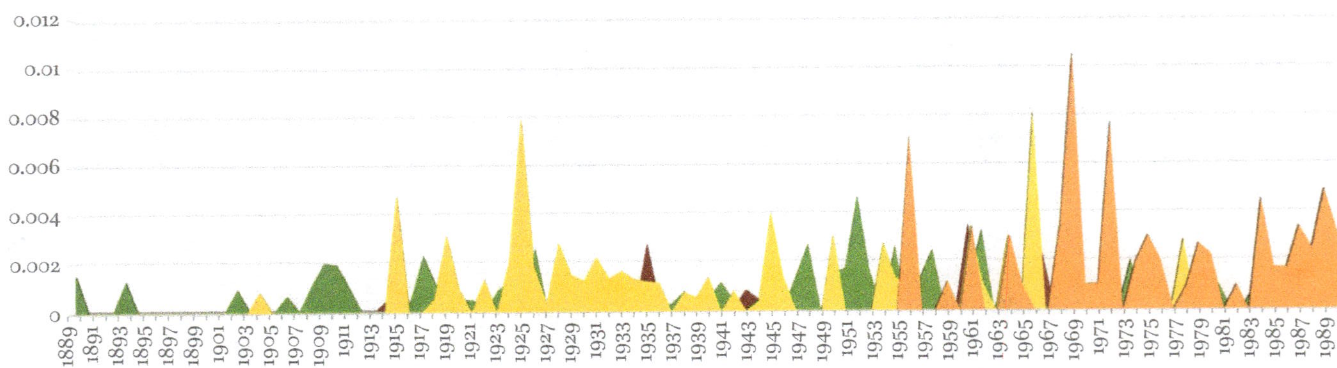

FIGURE 8.2 Graph representing discussions of the tick
Note: In the graph, orange represents the discussion of the tick as a disease vector to humans, yellow the tick as a disease vector to cattle, green the tick as a metaphor for human activity, and dark red the tick as a nuisance to human and other animals.

a Finnish national novel, compared the ticks in that book and in many other works to a human who had been overeating and drinking excessively.[11]

The first metaphorical mention of a tick in my data, dating from 1889, is similar. It describes an elderly man who had become as full as a tick from his haymaking lunch.[12] The comparison of the ticks' habit of sucking blood with overeating, especially by humans, is repeated numerous times in my dataset throughout the early 20th century. It appears, for example, in connection with the Finnish Christmas feast or, more generally, with the consumption of large quantities of food, usually in a humorous sense.[13] The last time this metaphor was mentioned in my data is in 1974.[14]

Although comparing the ticks' blood-sucking habit to the human overeating is central to tick-related metaphors, these metaphors have also been made in relation to other subjects. A metaphor closely related to devouring has been to compare large or slow-moving people and other things to ticks. For example, in 1950, it was written that after Christmas, a person is "as fat and ill-tempered as a tick."[15] During the Winter War, Soviet transport planes were also compared to ticks: "A familiar sight on the front is a heavily loaded Russian transport plane. Because of their load, they fly slowly, resembling a tick that has sucked itself full."[16]

Another notable characteristic of ticks (in these examples), which has been used as a metaphor, is that these animals bite and attach themselves to their target. Such metaphors have been written, for example, in relation to the weather, how a frost or wind bites a person like a tick.[17] What is interesting is that this metaphor has also been used in a positive sense. Such cases often involve situations in which a person falls in love with a place and decides to live there. These cases have been described in the data, for example, as "when you meet a place you like, you bite yourself into it like a tick."[18] Alongside these

11 Aleksis Kivi, *Seitsemän veljestä* (Helsinki: Gummerus, 2014 [1870]), 205–206. See also Aleksis Kivi, "Olviretki Schleusingenissä," in *Näytelmiä*, ed. E.A. Saarimaa (Helsinki: SKS, 1916 [1866]), 100.

12 "Salmelaisten heinänteko," *Savo-Karjala*, July 12, 1889.

13 See, e.g., Juho Koskimaa, "Leivän takia," *Vaasa*, June 19, 1924; "Sana aikanaan," *Etelä-Saimaa*, December 23, 1947; Olli, "Makkarakärryt," *Uusi Suomi*, January 12, 1958.

14 Anneli Ollus, "Ajattele sydäntäsi," *Helsingin Sanomat*, January 31, 1974.

15 Lippa, "Lipoittain," *Uusi Suomi*, January 6, 1950.

16 T.S., "Bolshevikkien 'erikois-hiihto-osaston' tuho Kuhmossa," *Kansan voima*, February 27, 1940.

17 See, e.g., Samuli, "Sydäntalven Somero," *Uusi Aura*, January 17, 1926; Samuli, "Salomailta sahoille," *Uusi Aura*, February 7, 1926; Samuli, "Kirvesmies," *Uusi Suomi*, June 16, 1929.

18 "Miksi, miten ja missä on retkeiltävä," *Oman kodin opas* no. 2 (1933): 3–5. See also Jalmari Sauli, "Vaiti olevat muotokuvat," *Maailma* no. 9 (1926): 390–392.

examples, tick bites have been used as a metaphor for concrete bites, especially dog bites.[19]

A much less used theme than the many metaphors of tick swelling and biting was to use the tick as a symbol for patience and perseverance. Two such cases were found in the dataset. One was a description written in 1951 of a person's perseverance in staying in business as "tenacious as a tick."[20] Another was a text written in 1944, which compared a tram standing at the terminus as a patient and tenacious tick.[21]

Metaphors related to ticks were told in Finland especially in a political context. As early as 1908, a newspaper writing appeared in which the new generation's activity in party politics was criticized by saying that they were "all as full of party politics as ticks."[22] This was the time when the first Finnish parties were founded.[23] Soon afterward, the tick metaphors began being used to criticize parties and supporters of different political tendencies.[24] According to my dataset, tick metaphors were used in the Finnish public discussion of politics until 1970.[25]

The range of tick-related metaphors mentioned above is significant because it reveals much about the past relationship of Finns with these arachnids. Above all they show that the tick was by no means an unknown or unimportant animal to Finns. The ticks' habit of sucking blood and how they attach themselves to humans and cling tenaciously to the skin was known in detail. Moreover, the above-mentioned metaphors clearly show that the tick was not a well-liked or appreciated animal, but rather a nuisance. Few of the metaphors describing its bodily functions are mentioned in a positive light, but even these cannot be interpreted as implying that the tick was somehow perceived as a pleasant animal.

There are also metaphors suggesting that the ticks were perceived as a pest not only to humans but also to other animals such as dogs and cattle. For instance, before the mid-20th century, metaphors appeared in which things

---

19 See, e.g., Samuli-Setä, "Leppo, Linta ja Lurkki," *Uusi Aura*, July 22, 1928; Tor Forslund, "Lemmikkieläimet," *Apu* no. 4 (1981): 80.

20 Jahvetti, "Elää, ken näkee," *Suomen Sosialidemokraatti*, November 20, 1951.

21 Lippa, "Lipoittain," *Uusi Suomi*, March 15, 1944.

22 Onni W.A., "Köppi," *Uusi Suometar*, November 1, 1908.

23 Juhani Mylly, "The emergence of the Finnish multi-party system System. A Comparison with Developments in Scandinavia, 1870–1920," *Scandinavian Journal of History* 5, no. 1–4 (1980), doi:10.1080/03468758008578978.

24 See, e.g., Jussi, "Puhetta ja pakinaa," *Karjala*, October 28, 1926; Sulka, "Maalaisliitto kuumehoureissa," *Sisä-Suomi*, September 25, 1930.

25 See, e.g., "Olipa kerran," *Suomen Sosialidemokraatti*, November 5, 1933; STT-Reuter, "NL:lla ohjusten torjuntaohjaus," *Uusi Suomi*, July 17, 1962; Aki, "Kiljavalta kuuluu," *Etelä-Suomen Sanomat*, September 5, 1970.

were described to be as closely related as a tick to a dog's skin or as a tick to a heifer.[26] Although tick metaphors were widely used and had many different meanings, their use began to decline in the 1950s and 1960s. Perhaps the reason for the slow decline in the use of metaphors was the fact that public discussion began to include information that ticks also spread diseases that are dangerous to humans. There may also have been other factors, such as the decline in cattle grazing in forests and increasing migration from rural to urban areas, where nature was no longer present in the same way as in the countryside. Whatever the reason, ticks were no longer used as metaphors in public discussion toward the end of the 20th century.

4 An Unpleasant Creature and a Nuisance

As the metaphors discussed above show, ticks were considered a nuisance to humans as well as to other species. This conclusion becomes even more evident when one looks at the other public discussions on ticks. Many articles describing ticks as unpleasant appeared in the early 20th century. One of the first such texts in my dataset is an account of a 1914 expedition to Sakhalin Island by the Finnish-Swedish writer Ludvig Munsterhjelm. The island is in the North Pacific Ocean on Russia's eastern border and had been used since the 1850s as a penal colony for prisoners.[27] In this connection, Munsterhjelm wrote of the island's nature that it offers an unpleasant environment for criminals, not only because of its climate but also because of its fauna. He mentions that "the island is infested with the disgusting nuisance of ticks, which are incredibly ingenious at clinging to human skin. Every night we had the arduous and painful task of cleaning ourselves of these bloodsuckers, which often left their heads under the skin, causing painful inflammations."[28]

Munsterhjelm's description of ticks as a nuisance to humans is not exceptional among my data. Many texts about ticks published in the early 20th century either directly or indirectly give a similar picture of the relationship between humans and ticks. Experiences with ticks were particularly prevalent in texts on hiking and camping.[29] For example, in 1924, an article about a traveling party's visit to a wilderness praised the beautiful nature of the area

26 See, e.g., "Olipa kerran"; Kuukunen, "Juttukirje," *Etelä-Suomen Sanomat*, March 14, 1941.

27 Andrew A. Gentes, "The Institution of Russia's Sakhalin Policy, from 1868 to 1875," *Journal of Asian History* 36, no. 1 (2002).

28 Ludvig Munsterhjelm, "Sahalinin sydänmaista," *Uusi Suometar*, December 20, 1914.

29 See, e.g., A.R., "Hawaintoja Woikaalta," *Eteenpäin*, July 27, 1924; "Leirikuvia ja selostuksia," *Partio* no. 10 (1925): 157–159.

and emphasized that, although "tick sucks blood, the traveler still wants to go deeper into the wilderness to enjoy the bounty of nature, such as berries."[30]

Ticks appeared to have been a familiar animal to Finns who spent time outdoors, but as the published metaphors and other material on ticks show, it is likely that a much larger number of Finns were familiar with these arachnids. A text describing ticks, written by biologist Heikki Väänänen in 1933, stated that "everyone has probably encountered it at least once in his or her lifetime."[31] The familiarity of ticks is also indicated by a widespread text, written in 1929 by the entomologist Niilo Vappula. His text explained in detail what was known at the time about the biology, behavior, and habitat of ticks. The text was initially published in an agricultural magazine,[32] but later that year, it was also published in *Kotiliesi*, a widely circulated and very popular so-called women's magazine.[33] The reason for the publication was mentioned in the article – several readers had asked the editorial team how to get rid of the ticks that were causing problems in their hometowns. As for the relationship between humans and ticks, Vappula said that a tick bite causes an unpleasant itching sensation but very rarely has worse consequences. He adds that a few cases of severe infections and even fever are mentioned in the research literature, but generally, the tick bite is insignificant to humans.[34]

Vappula's text and the other examples mentioned above show how ticks were understood in Finland in the first half of the 20th century as harmless, albeit irritating, pests. There is only one widespread news report of a human death involving a tick in my database. The case concerned the death of a 21-year-old man in Sääminki village in 1908, a few days after the man had been bitten by a tick and the head of the tick had remained stuck under his skin. Although the case received a lot of publicity, it did not directly affect people's attitudes toward ticks. It was also reported as a rare cause of death.[35]

Altogether, though the tick was perceived as a disgusting nuisance, people clearly did not perceive it as a frightening or dangerous animal in the late 19th century or the first part of the 20th century. There is only one exception to this, and that is the tick's effect on children. In a text written in 1943, archaeologist and writer Sakari Pälsi described children plucking ticks from the skin of calves and putting them on their own skin so that they could later pull them

---

30 A.R., "Hawaintoja Woikaalta".

31 "Eläinten maailma," *Turun Sanomat*, March 25, 1933.

32 Niilo Vappula, "Puutiainen ja sen suhde punatautiin," *Karjatalous* no. 10 (1929a): 297–298.

33 Päivi Aikasalo, *Alli Wiherheimo – Uranaisen sydän* (Helsinki: Otava, 2004).

34 Niilo Vappula, "Puuntäit eli puutiaiset kesäkiusana," *Kotiliesi* no. 11 (1929b): 430–432.

35 See, e.g., "Harwinainen kuoleman syy," *Savolainen*, August 13, 1908; "Harwinainen kuoleman syy," *Turun Sanomat*, August 16, 1908; "Harwinainen kuoleman syy," *Tampereen Sanomat*, August 18, 1908.

off at the beach.[36] This may have been connected to the children's carefree and approachable attitude to living with animals. However, I believe that adults did not do this, and such actions were probably rare among children.

Although ticks were not understood to be a threat to human life, there was still a desire to remove them from the skin of humans and animals. There are several instructions for the removal and disposal of ticks in my data. Petroleum, for example, was used before the 1900s by pouring it over a tick that had become attached to the skin. Among the substances mentioned for pouring over the ticks were spirit, grease, lamp oil, and turpentine.[37] Today, these methods of tick removal are no longer recommended, but they were used to remove ticks from the skin still in the 1990s. By that time, however, petroleum and turpentine had been replaced by Vaseline and nail varnish.[38]

However, the ticks were a nuisance not only to humans but also to other animals. My data show that dogs were particularly targeted by ticks. For example, the novel *Rai Jakkerintytär* (1927), written by the Finnish novelist Joel Lehtonen, describes a situation in which a tick attaches itself to a dog's paw, and how the dog's owner carefully tries to remove it.[39] Ticks attached to dogs were also later addressed in various texts on dog care in the 1930s and 1960s.[40] Texts in newspapers and magazines also show that people were aware that ticks were found on wild animals in the early 20th century. They were said to be an annoyance to rabbits, birds, and reptiles such as lizards.[41]

In addition to dogs and other animals, ticks were found especially in cattle. Many texts show how cattle visiting forest pastures were sometimes found to be very abundant in ticks. In his text, Vappula said that "in some areas ticks are said to be so abundant that in the evening, when the cattle return from grazing, a bucket of one liter can be filled with ticks and used as food for the chickens."[42] The text may be an exaggeration, but it repeats the same story as many other newspaper articles about ticks in cattle.

---

36 Sakari Pälsi, "Ukonilma," *Forssan Lehti*, July 17, 1943.

37 I., "Taisteluun kärpäsiä, itikoita ja puutiaisia vastaan," *Maatalouden Karjanhoitolehti* no. 3 (1915): 42–43; Joel Lehtonen, *Rai Jakkerintytär* (Helsinki: Otava, 1927), 192; Vappula, "Puuntäit eli puutiaiset kesäkiusana"; J. B-g., "Koiran lois-ihotaudeista," *Suomen Kennelklubin Aikakauskirja* 41, no. 3 (1936): 33–36.

38 Pirkko Kolbe, "Pieni itikka voi olla paha peto," *Helsingin Sanomat*, June 13, 1979; Klaus A. Järvinen, "Kotoiset puutiaiset voivat levittää sairautta," *Etelä-Suomen Sanomat*, July 4, 1988; "Seuraava potilas," *Länsi-Savo*, October 7, 1991.

39 Lehtonen, *Rai Jakkerintytär*, 192.

40 See, e.g., B-g., "Koiran lois-ihotaudeista"; "Koiran elämää," *Uusi Suomi*, July 2, 1965.

41 See, e.g., "Puutiaisia jäniksissä vielä myöhään syksyllä," *Metsästys ja Kalastus* 15, no. 11 (1926): 417; Vappula, "Puutiainen ja sen suhde punatautiin"; "Pääskynen lääkärissä," *Uusi Suomi*, August 23, 1939.

42 Vappula, "Puuntäit eli puutiaiset kesäkiusana".

A few texts from the early 20th century still referred to ticks as a mere bother to cattle, but by the 1920s at the latest, texts describing the relationship between cattle and ticks began to refer to the redwater fever that ticks spread to cattle. As a result, ticks were no longer understood as annoying arachnids, but as small animals that spread death among cattle herds.[43] Overall, the understanding of ticks as pesky but harmless to humans and animals declined in the second half of the 20th century, no doubt due to the fact that ticks came to be understood as animals that spread life-threatening diseases to cattle and humans.

5 Ticks "Living in Trees" Cause Diseases to Cattle

The ticks' involvement in spreading redwater fever in cattle was recognized as early as the 1890s. While the disease was already widely known in Europe and the United States in the mid-19th century, it wasn't until the 1890s that American scientists Fred L. Kilborne and Theobald Smith discovered the bacteria responsible for the disease and realized that it was transmitted to cattle through tick bites.[44] The first two decades of the 20th century saw the beginning of writing about ticks in the context of redwater fever cases, but the volume of discussion increased significantly during the 1920s and 1930s. Most of these texts described the symptoms of redwater fever and how to react when cattle get ill. These texts also mentioned the tick that spread the disease, sometimes only briefly and sometimes in more detail.

It is clear that the texts mentioning the tick as an animal that spreads redwater fever portrayed them in a negative light. In this context, it was no longer just a nuisance but an arachnid causing death and economic loss. Therefore, the central theme of ticks in the news coverage of redwater fever was how to get rid of these animals once and for all. In a few contexts, for example, it was emphasized that the tick should not be released but burned. This was justified on the grounds that the ticks would not be able to continue their life and spread redwater fever to cattle.[45]

43 See, e.g., Vappula, "Puutiainen ja sen suhde punatautiin".

44 Anderson, "The Natural History of Ticks," 205. See also Taina Syrjämaa's chapter in this volume.

45 "Toimenpiteitä punataudin ehkäisemiseksi," *Suomen Eläinvakuutusyhdistyksen Asiamieslehti* no. 2 (1930): 2–3; B-g., "Koiran lois-ihotaudeista".

However, in the case of redwater fever, it was not so much a question of destroying individual ticks but of eradicating them from entire areas. Already at the beginning of the 20th century, the tick habitat was believed to be "watery and deciduous grassland," and ticks were believed to live in alder trees, willow, and hazel bushes.[46] Therefore, texts proposing to remove all shrubs from pastures and to drain and dry wet places were published as early as the 1920s.[47] In addition to these practices, other ways people tried to get rid of ticks included spreading chalk and ants in pastures.[48] However, by far the most frequent solution identified in the public discussion was the destruction of alders and various shrubs. The idea of destroying alder trees and various shrubs to eradicate ticks was proposed in several texts until the 1950s,[49] when herbicides such as hormone-type plant killers were introduced as a way to destroy these trees and shrubs.[50] The last mention of destroying alder and willow bushes to prevent redwater fever is found in my material from 1978.[51]

Burning trees to get rid of ticks appears strange from today's perspective, as we now know that ticks do not live in trees. In a scientific context, this was noticed already in the 1950s. Biologist Christina Öhman, who studied ticks in that decade, discovered that when she collected the animals for experiments, all attempts to collect ticks from trees and bushes proved futile. They could only be found in hay.[52] However, it was not until the late 1970s that the notion of ticks falling from the branches of alder or walnut trees to their prey began to be questioned in the public discussion.[53] Even so, in 1990, for example, a news-

46 See, e.g., "Punatauti lehmistä ja sen hoito," *Hämeen Sanomat*, August 19, 1915; the Finnish etymology of the tick, "puutiainen," refers to a bug that lives in a tree. See *Suomen etymologinen sanakirja*, s.v. "Puutiainen", https://kaino.kotus.fi/ses/?p=article&etym_id=ETYM_cd5753864f0d7f141dbcdd9fb7ec3b80&word=puutiainen.

47 L.P., "Punataudin vastustaminen," *Maaseudun Tulevaisuus*, July 29, 1922.

48 Vappula, "Puuntäit eli puutiaiset kesäkiusana".

49 See, e.g., Eläinten hoitaja, "Punataudin aika lähenee!," *Pellervo* no. 19 (1926): 307; R.S., "Punataudin ehkäisytoimenpiteistä ja punatautisen lehmän hoidosta," *Karjatalous* no. 11 (1930): 388–391; "Lehmien punataudista," *Hämeen Sanomat*, July 6, 1934.

50 "Lepät pois laitumilta," *Työkansan Sanomat*, June 17, 1950.

51 Hakon Westermarck, "Punatauti uhkaa laitumilla. Vesakoiden torjuminen tärkeää," *Maaseudun Tulevaisuus*, May 23, 1978.

52 Christina Öhman, "The Geographical and Topographical Distribution of Ixodes Ricinus in Finland," *Acta Societatis pro Fauna et Flora Fennica* 76, no. 4 (1961): 4.

53 "Puutiainen levittää vaikeaa aivotulehdusta," *Ilta-Sanomat*, September 22, 1978; Riitta Kallioinen, "Pistoista oireet lievenevät," *Uusi Suomi*, July 3, 1979; "Suomen suvi hyttysiä täynnä," *Maaseudun Tulevaisuus*, July 23, 1987.

paper article on ticks was published in which ticks were said to crave "blood on the branches of deciduous trees."[54]

Overall, the importance of redwater fever as a disease of cattle began to decline in Finland, when forest grazing of livestock ceased in the 1950s. In general, the whole agricultural sector in Finland underwent a structural change; the result was that cattle farming became a business rather than a way of life for farmers. Cattle were given fenced pastures, and many farms introduced semi-detached cattle sheds, where these animals lived indoors all year round.[55] This also reduced contact with ticks and the incidence of redwater fever. For example, a 1988 newspaper article mentions ticks as a target not only for humans but also for dogs, horses, sheep, and large mammals in general.[56] The cattle were not even mentioned by name any more in relation to ticks.

6 Disease Vector to Humans

In Finland, it was slowly realised in the 1950s that ticks spread a form of brain fever to humans. The link between ticks and cases of tick-borne encephalitis (TBE) was discovered by Soviet scientists in the late 1930s.[57] In Finland, the first cases were recorded in the 1940s on the island of Kumlinge in Åland, which is why TBE was almost exclusively called "Kumlinge disease" in my data until the 1980s.[58]

Although the importance of ticks as the cause of a dangerous disease for humans entered the public discussion in the 1950s, the texts published in newspapers and magazines on the subject until the mid-1960s were almost

54 Esko Keränen, "Mökkihöperö," *Etelä-Suomen Sanomat*, August 3, 1990.

55 See, e.g., Taija Kaarlenkaski, *Kertomuksia lehmästä: tutkimus ihmisen ja kotieläimen kulttuurisen suhteen rakentumisesta* (Jyväskylä: Suomen kansantietouden tutkijain seura, 2012), 20, 207–208; Taija Kaarlenkaski and Annika Lonkila, "In Search of Invisible Cows: Collaboration, Resistance and Affection in Human-Animal Relationships on Contemporary Dairy Farms," *Ethnologia Fennica* 47, no. 2 (2020), doi:10.23991/ef.v47i2.88774.

56 Klaus A. Järvinen, "Kotoiset puutiaiset voivat levittää sairautta," *Etelä-Suomen Sanomat*, July 4, 1988.

57 Vladimir I. Zlobin, Vanda V. Pogodina and Olaf Kahl, "A Brief History of the Discovery of Tick-Borne Encephalitis Virus in the Late 1930s (Based on Reminiscences of Members of the Expeditions, Their Colleagues, and Relatives)," *Ticks and Tick-Borne Diseases* 8, no. 6 (2017), doi:10.1016/j.ttbdis.2017.05.001.

58 "Kumlingen taudin syitä etsitään," *Etelä-Suomen Sanomat*, June 11, 1956. See also Peter Wahlberg et al., "TBE in Åland Islands 1959–2005: Kumlinge Disease," *Scandinavian Journal of Infectious Diseases* 38, no. 11–12 (2006), doi:10.1080/00365540600868297. See also Suvi Rytty's chapter in this volume.

without exception either written by virologists or were short news items from research conferences. Some short and provocative texts warning of the diseases spread by ticks to humans, however, appeared already at the turn of the 1950s and 1960s. Nevertheless, on a larger scale, such texts began to appear in the 1970s and 1980s.[59]

In 1978, for example, a newspaper article was published, and it came with a warning: "Beware of ticks, or you will become their blood meal. In the worst case, you can get a severe brain infection from a tick, which takes a long time to recover from."[60] In addition, in 1980, an even more affective text appeared, telling how TBE had been found on islands other than the island of Kumlinge. An interesting detail in the text is that it emphasized that the archipelago of Hiittinen, where the disease was found, is place to many holiday resorts for people living in the Finnish capital region.[61] This was clearly intended to highlight the increased risk of exposing larger numbers of people to the virus. Also, in 1982, a text was published describing TBE as a danger to boaters and emphasizing that there is no cure for the disease.[62] In the press, concern about boaters from the capital region seemed to be clearly higher than for permanent residents of archipelago.

There was clearly an attempt to create horror scenarios about TBE at the turn of the 1970s and 1980s,[63] but the number of such texts was small, and they did not seem to create any panic, at least not in the public discussion about ticks. In general, news coverage of TBE failed, on the basis of my data, to significantly impact people's awareness and perception of the character of the disease. For example, in 1978, *Uusi Suomi* newspaper presented the question "What are ticks?" and answered with the following: "Ticks (Ixodes ricinus) live in our area, especially in alder woods. They are harmless to humans but can spread dysentery in cattle."[64]

The above quote is a good indication that not all Finns were even aware of TBE at the turn of the 1970s and 1980s, even though the disease had been known in Finland for about 25 years. In general, the public discussion suggests that TBE did not attract nearly as much public attention as it does today. The reason for people's ignorance and disinterest in the disease was very probably

---

59 See Suvi Rytty's chapter in this volume.

60 Pekka Hiekkala, "Puutiainen levittää vaikeata aivotulehdusta," *Ilta-Sanomat*, September 22, 1978.

61 "Kumlingen tauti leviää," *Uusi Suomi*, April 12, 1980.

62 Hannu Laaksonen, "Kumlingen tauti – Vaara veneilijöille," *Tekniikan Maailma* no. 7 (1982): 180.

63 See Suvi Rytty's chapter in this volume.

64 "Tiedon tikapuut," *Uusi Suomi*, November 29, 1978.

that the public discussion had created an image of the disease, especially by its name, as an exotic virus found only on the island of Kumlinge, far away in the archipelago, isolated from inland settlements.

In the public discussion, it was pointed out that TBE occurred not only in Åland but also in the Turku archipelago and on Finland's eastern border along the coast of the Gulf of Finland. Despite this, TBE was implicitly defined in the public discussion only for the island of Kumlinge and its surroundings.

Although the link between TBE and ticks in Finland in the 1950s might have been thought to have sparked fear about ticks, this did not happen until virologists discovered the link between borreliosis and ticks in the early 1980s.[65] Of course, ticks did not become frightening creatures overnight but gradually, as the media began to feed information about the link between ticks and borreliosis to the general public.

The first mention of borreliosis in my data is not until 1986, and in that text, it is mentioned as an extremely rare disease in Finland.[66] Of course, this was not the case.[67] After two years, the situation had changed and articles appeared that clarified information about borreliosis for the wider public. These new articles showed that borreliosis was associated with a central nervous system infection and that there must be hundreds of cases of borreliosis in Finland each year. Information about the wider symptoms and the number of people affected certainly contributed to people's anxiety about ticks, but I would argue that a more significant factor in people's increased fear of ticks was the implicit statement in various press articles that, in addition to the spread across Kumlinge Island, tick-borne diseases could be caught almost anywhere in Finland.[68]

Thus, suddenly, people began to understand the tick, previously known to people mainly as an irritating pest spreading the mysterious Kumlinge disease, as a frightening creature (in the public discussion) that could cause serious illness in familiar and previously safe environments. This was very quickly reflected in texts written about ticks as early as 1989–1991. Among other things, ticks began to be described as dangerous and insidious enemies and as silent

65 Burgdorfer et al., "Lyme Disease"; see also Barbour & Benach, "Discovery of the Lyme Disease Agent".

66 "Kysy lääkäriltä," *Länsi-Savo*, May 24, 1986.

67 J. Cuellar et al., "Seroprevalence of Lyme Borreliosis in Finland 50 Years Ago," *Clinical Microbiology and Infection* 26, no. 5 (2020), doi:10.1016/j.cmi.2019.10.003.

68 See, e.g., STT, "Punkista voi saada keskushermostoinfektion," *Etelä-Suomen Sanomat*, August 7, 1987; Klaus A. Järvinen, "Kotoisat puutiaiset voivat levittää sairautta," *Etelä-Suomen Sanomat*, July 4, 1988.

predators with a thirst for blood.[69] Affective material, such as images of people's bare feet in the grass, also began to be used to illustrate texts about ticks.[70]

Nevertheless, even in the 1990s, some comments still appeared in the public discussion, referring to the fact that ticks used to bite people also in the past without serious consequences.[71] One example is the conversation between a concerned mother and a doctor that I mentioned at the beginning of this chapter.[72] However, the number of these texts was much smaller than the number of articles in newspapers and magazines, which talked about ticks and the diseases they spread, creating fear and anxiety. During the 1990s, new knowledge and understanding of ticks as dangerous and terrifying creatures began to effectively take hold in Finnish culture and society. The change was so effective that few people today are actually aware that, during the previous century, the tick was mainly understood as a harmless pest.

7 Conclusion

I have shown in this chapter that the tick has not always been understood as a creature of fear, anger, and anxiety as it is today. These arachnids have, in fact, evoked the strong emotions we are familiar with today in Finland for a relatively short period of time, from the turn of the 1980s and 1990s until today. From the 1890s to the 1990s alone, there were many different ways of understanding ticks. They were used as metaphors for human activity and were widely understood as nuisances that sought to attach themselves to the skin of humans and other animals. Of course, the hatred associated with ticks emerged in public discussion as early as the early 1900s due to the redwater fever that they spread to cattle. However, the disdain that emerged in the context of the discussion on this disease did not define the whole debate on ticks. There were also many different ways of talking about ticks at the same time. Even the fact that ticks were found to spread TBE on the island of Kumlinge in the 1950s did not create fear and anxiety; however, once the connection was made between borreliosis and ticks, particularly the fact that borreliosis can be caught almost anywhere

69 Marja Haapalahti, "Vaarallinen puutiainen," *Suomen Kuvalehti* no. 30 (1989): 59–61; Keränen, "Mökkihöperö".

70 Niina Lempiäinen, "Punkin purema voi olla vaarallinen," *Länsi-Savo*, August 16, 1990.

71 See e.g. Päivi Repo, "Puutiaistaudit eivät ole lisääntyneet," *Helsingin Sanomat*, August 24, 1992; "Seuraava potilas," *Länsi-Savo*, June 12, 1995.

72 "Seuraava potilas" (1991).

in Finland, the public discussion and media formulated this knowledge to articles and news that propelled the current fear about ticks.

Thus, the tick became "a frightening enemy" in the public discussion after its danger outgrew human control. That is, the tick was not perceived as frightening until its ability to spread diseases to humans was recognized, but above all because it was discovered that ticks were not limited to a small geographical area. People realized that ticks spreading borreliosis can be found almost anywhere in Finland. This detail in particular makes the tick intimidating because we cannot simply isolate these animals to a specific area. In practice, the public discussion on ticks since the 1990s, which highlights the frightening and dangerous characteristics of these animals, is quite similar to the discussion on the danger of predator animals such as wolves, as well as their numbers and distribution. What breaks the illusion of control over our relationship with nature created by Western culture becomes frightening and monster-like. This is what ticks have become in the Finns' perception, and it is hard to say whether we will ever succeed in changing our views on these animals unless we first make a broader systematic change in our relationship with nature.

Bibliography

Newspapers and Periodicals

Apu
Eteenpäin
Etelä-Saimaa
Etelä-Suomen Sanomat
Forssan Lehti
Helsingin Sanomat
Hämeen Sanomat
Ilta-Sanomat
Kansan voima
Karjala
Karjatalous
Kotiliesi
Länsi-Savo
Maailma
Maaseudun Tulevaisuus
Maatalouden Karjanhoitolehti
Metsästys ja Kalastus
Oman kodin opas

Partio
Pellervo
Savo-Karjala
Savolainen
Sisä-Suomi
Suomen Eläinvakuutusyhdistyksen Asiamieslehti
Suomen Kennelklubin Aikakauskirja
Suomen Kuvalehti
Suomen Sosialidemokraatti
Tampereen Sanomat
Tekniikan Maailma
Turun Sanomat
Työkansan Sanomat
Uusi Aura
Uusi Suometar
Uusi Suomi
Vaasa

Literature

Aikasalo, Päivi. *Alli Wiherheimo – Uranaisen sydän*. Helsinki: Otava, 2004.

Anderson, John F. "The Natural History of Ticks." *Medical Clinics of North America* 86, no. 2 (2002): 205–218. doi:10.1016/S0025-7125(03)00083-X.

Burgdorfer, Willy, Alan G. Barbour, Stanley F. Hayes, Jorge L. Benach, Edgar Grunwaldt and Jeffrey P. Davis. "Lyme Disease – A Tick-Borne Spirochetosis?" *Science* 216, no. 4552 (1982): 1317–1319. doi:10.1126/science.7043737.

Barbour, Alan G. and Jorge L. Benach. "Discovery of the Lyme Disease Agent." *mBio* 10, no. 5 (2019). doi:10.1128/mbio.02166-19.

Cuellar, J., T. Dub, J. Sane, and J. Hytönen. "Seroprevalence of Lyme Borreliosis in Finland 50 Years Ago." *Clinical Microbiology and Infection* 26, no. 5 (2020). doi:10.1016/j.cmi.2019.10.003.

DeMello, Margo. *Animals and Society: An Introduction to Human-Animal Studies*. New York: Columbia University Press, 2012.

Gentes, Andrew A. "The Institution of Russia's Sakhalin Policy, from 1868 to 1875." *Journal of Asian History* 36, no. 1 (2002): 1–31.

Kaarlenkaski, Taija. *Kertomuksia lehmästä: Tutkimus ihmisen ja kotieläimen kulttuurisen suhteen rakentumisesta*. Jyväskylä: Suomen kansantietouden tutkijain seura, 2012.

Kaarlenkaski, Taija and Annika Lonkila. "In Search of Invisible Cows: Collaboration, Resistance and Affection in Human-Animal Relationships on Contemporary Dairy Farms." *Ethnologia Fennica* 47, no. 2 (2020): 27–53. doi:10.23991/ef.v47i2.88774.

Kivi, Aleksis. "Olviretki Schleusingenissä." In *Näytelmiä*, edited by E.A. Saarimaa. Helsinki: SKS, 1916 [1866].

Kivi, Aleksis. *Seitsemän veljestä*. Helsinki: Gummerus, 2014 [1870].

Laaksonen, Maija, Eeva Sajanti, Jani J. Sormunen, Ritva Penttinen, Jari Hänninen, Kai Ruohomäki, Ilari Sääksjärvi et al. "Crowdsourcing-Based Nationwide Tick Collection Reveals the Distribution of Ixodes Ricinus and I. Persulcatus and Associated Pathogens in Finland." *Emerging Microbes & Infections* 6, no. 1 (2017): 1–7. doi:10.1038/emi.2017.17.

Latva, Otto. *The Giant Squid in Transatlantic Culture: The Monsterization of Molluscs*. London: Routledge. doi:10.4324/9781003311775.

Lehtonen, Joel. *Rai Jakkerintytär*. Helsinki: Otava, 1927.

Lillbroända-Annala, Sanna and Oscar Winberg. "'Fästingen håller inget säkerhetsavstånd': Konkurrerande riskdiskurses om fästingar i media." *TRACE ∴ Journal for Human-Animal Studies* 9 (2023): 58–80. doi:10.23984/fjhas.121761.

Mylly, Juhani. "The Emergence of the Finnish Multi-Party System. A Comparison with Developments in Scandinavia, 1870–1920." *Scandinavian Journal of History* 5, no. 1–4 (1980): 277–293. doi:10.1080/03468758008578978.

National Library of Finland. https://digi.kansalliskirjasto.fi/stats?set_language=en. Accessed October 24, 2023.

Sajanti, Eeva, Mikko Virtanen, Otto Helve, Markku Kuusi, Outi Lyytikäinen, Jukka Hytönen and Jussi Sane. "Lyme Borreliosis in Finland, 1995–2014." *Emerging Infectious Diseases* 23, no. 8 (2017): 1282–1288. doi:10.3201/eid2308.161273.

Salmi, Hannu. *What Is Digital History?* Cambridge: Polity Press, 2021.

Wahlber, Peter, Sten-Anders Carlsson, Hans Granlund, Christian Jansson, Mogens Lindén, Clara Nyberg and Dag Nyman. "TBE in Åland Islands 1959–2005: Kumlinge Disease." *Scandinavian Journal of Infectious Diseases* 38, no. 11–12 (2006): 1057–1062. doi:10.1080/00365540600868297.

Zlobin, Vladimir I., Vanda V. Pogodina and Olaf Kahl. "A Brief History of the Discovery of Tick-Borne Encephalitis Virus in the Late 1930s (Based on Reminiscences of Members of the Expeditions, Their Colleagues, and Relatives)." *Ticks and Tick-Borne Diseases* 8, no. 6 (2017): 813–820. doi:10.1016/j.ttbdis.2017.05.001.

Öhman, Christina. "The Geographical and Topographical Distribution of Ixodes Ricinus in Finland." *Acta Societatis pro Fauna et Flora Fennica* 76, no. 4 (1961).

CHAPTER 8

Anopheline Mobilities and More-Than-Mosquito Biopolitics in Making Biotechnology

Marianne Mäkelin

1 Introduction

Mosquitoes fly. They are hard to contain. They travel across national and legislative borders. They pierce through skin, crossing boundaries between the human body and its environment, and sometimes their bites introduce unicellular *Plasmodium* parasites into human bloodstream. Both the mosquitoes and the parasites need humans in order to reproduce; malaria as a human disease emerges as a by-product of this cyclical exchange.

Because of this association with a fatal disease, the *Plasmodium*-carrying *Anopheline* mosquitoes arguably are among the most well-studied and well-known types of insects.[1] Studying mosquito biology and life-worlds has, for the past century, been driven by a desire to alter human – insect relations. Recently, scientists have been exploring ways to modify mosquito bodies in a manner that, they hope, could reduce malaria transmission. One approach is the development of "gene drives" – modifications made to the mosquito genome that increase the likelihood of the altered sequence being inherited by the next generation. By combining a gene drive with a modified genetic sequence that would render mosquitoes either infertile or resistant to the malaria parasite, wild mosquito populations could potentially be transformed in the timespan of a few generations.

In this chapter, I analyze debates on gene drive mosquitoes from the point of view of more-than-human biopolitics. The concept, introduced by Foucault, points to the ways in which knowledge of life is connected to forms of power that aim to mold life in order to protect it.[2] Biopolitics operates with knowledge practices that produce calculable populations and information about

1 Richard Jones, for example, makes such a claim in his book *Mosquito* (Chicago: The University of Chicago Press, 2012), 12.

2 Michel Foucault, *The History of Sexuality. Vol 1. An Introduction* (London: Lane, 1978), 139.

their ways of life.[3] Central to discussions on gene drives is a tension between the imperative to protect human life and concerns about the uncertain consequences of environmental interventions. This ambiguity lingers despite the growing influence of perspectives that emphasize more-than-human entanglement in public health policies, as Christos Lynteris has argued.[4] Lynteris suggests that the practical and political limitations of frameworks of multispecies intimacies become apparent in moments of epidemic crisis. He notes that these moments instead see the deployment of "a militarized apparatus" of "culling, stamping out, disinfection, disinfestation, separation, and eradication; what we may call the sovereign heart of public health in relation to animal-borne diseases."[5]

The persistence of malaria as a disease can also be framed within these kinds of crisis narratives.[6] Yet, in debates over gene drive mosquitoes, the apparatus described by Lynteris and its self-evident justification of eradication and separation seems to dissipate. I argue that the gene drive strategy reveals emergent shifts in this apparatus. These become tangible as the sacrificial rationale is extended from individual animals and local populations to more uncertain chains of influence in the more-than-human and more-than-mosquito worlds the insects become involved in. As I will show, these shifts are connected to the how the multiple mobilities of mosquitoes and their genes are articulated against visions of biotechnology and aspired control over processes of life.

The analysis is based on publicly available materials that I have collected as part of a multi-sited study on gene drive research. The gene drive strategy is discussed among the scientific community and both national-level and international organizations and regulatory bodies. The materials this chapter is based on comprise of scientific research articles produced in the projects and outside them, and policy documents produced by organizations such as United Nations Convention on Biological Diversity, WHO, the EU, the African Union and the US National Academies. I have also collected media representations, blog posts, YouTube videos, websites, conference papers, guidelines and info sheets produced in gene drive projects. In addition to two systematic reviews on the scientific literature done in 2019 and 2023, these were gathered

3 Michel Foucault, *Security, Territory, Population. Lectures at the Collège de France 1977–1978* (New York: Picador, 2007).

4 Christos Lynteris, "Introduction: Infectious Animals and Epidemic Blame," in *Framing Animals as Epidemic Villains. Histories of Non-human Disease Vectors*, ed. Chistos Lynteris (Cham: Palgrave Macmillan, 2019), 2.

5 Lynteris, "Infectious Animals," 2.

6 See Alex M. Nading, "The Lively Ethics of Global Health GMOs: The Case of the Oxitec Mosquito," *BioSocieties* 10, no. 1 (2015), doi:10.1057/biosoc.2014.16.

as corollary to the more ethnographical insights from the same study. All in all, the materials originate from 2002 to 2023.[7] The time period spans from early writings that suggest gene drive systems as an abstract possibility to the approach being developed in several multinational research consortia.

Gene drive mosquitoes are, at the time of writing, an experimental technology. Releases outside laboratory insectaries have not been carried out, and indeed the question of what the conditions for such a release would be frames much of the discussion. Correspondingly, this chapter does not aim at describing a fully-formed framework of the biopolitical implications of the gene drive mosquito. Nor do I mean to claim that genetically modified insects would be ushering in an entirely new era of understanding insects and disease. In many ways gene drive mosquitoes are a continuation of strategies and research paths that, for long, have considered diseases like malaria foremost an object of biotechnological intervention.[8] Instead, I trace points of fracture and transformation in these frameworks.

To do this, I integrate insights from science and technology studies and multispecies approaches with a framework of biopolitics to ask how are mosquito mobilities articulated and what kinds of implications they carry for disease management frameworks. Before delving into this analysis, I will first discuss juxtaposing Foucault's framework of biopolitics with theoretical approaches that give voice to the nonhuman world in the social sciences, and in the context on mosquito control and the way it historically has shaped shared human-insect worlds specifically in the next section. After this, I introduce the debates on gene drive mosquitoes as they happen in scientific articles and other materials produced by gene drive scientists, regulatory texts, and media representations. I discuss the way I examine them as a lens into making biotechnology, as parallel to the work of tinkering with mosquitoes in laboratory work. After these two sections introducing theoretical perspectives and materials and methodology of the study. After this, the empirical analysis is divided into two sections. The first one observes how mosquito mobilities are articulated and unfold amid expectations placed on biotechnology in public health. The second section discusses the mobility of the genetic matter the mosquitoes are expected to transmit. The resulting, *more-than-human* biopolitical framework consider the mosquitoes as lively agents that take part in shaping the worlds

7 270 publicly available documents written mostly in English (3 individual blog posts in French), collected between 2019–2023.

8 See Ann Kelly and Uli Beisel, "Neglected Malarias: The Frontlines and Back Alleys of Global Health," *BioSocieties* 6, no. 1 (2011), doi:10.1057/biosoc.2010.42.

they live in. Nevertheless, I argue, the biopolitics of mosquito control are also more-than-mosquito: the mosquitoes do not act alone.

2 Insect Control as Biopolitics

Thinking of insect control as biopolitics brings to light the ways in which knowing, protecting, valuing, and devaluing life takes place in multispecies networks. While Foucault's conceptualization of biopolitics was largely focussed on how and why modern states produce knowledge about *human* populations, an interplay between protecting and shaping forms of life also characterizes the myriad relationships of production, exploitation, co-habitation, avoidance, and killing between humans and other animals, including insects.

Jamie Lorimer and Clemens Driessen see more-than-human biopolitics as happening in "systematic anatomical, ethological and ecological modifications of nonhuman forms and rhythms."[9] In other words, cultivating and making use of non-human life is connected to knowledge practices that have specific ways of understanding what is at stake in the process. In disease control, as in experimental laboratory work, biopolitical governing is, according to Lorimer and Driessen,[10] concerned with securing the risky vitality of certain forms of life. This is based on notions of fixed and stable nations and bodies – organisms, populations, species, and territories that can be defended, controlled or valued.

A focus on the mosquito vector has characterized malaria research and control measures, to varying degrees, since the link between mosquitoes, the parasite, and the disease was shown.[11] Insects and arachnids of several kinds – mosquitoes as well ticks and fleas – have for long been associated with danger and death, but it was with developments in microbiology in the 19th century that nonhuman animals became categorically seen as sources of disease risk.[12] The 20th century saw the roll-out of large-scale disease control measures that were often connected to projects of colonial expansion and

9 Jamie Lorimer and Clemens Driessen, "Bovine Biopolitics and the Promise of Monsters in the Rewilding of Heck Cattle," *Geoforum* 48 (2013): 252, doi:10.1016/j.geoforum.2011.09.002.

10 Lorimer and Driessen, "Bovine Biopolitics," 253.

11 Marcos Cueto, "A Return to Magic Bullet? Malaria and Global Health in the Twenty-First Century," in *When People Come First*, ed. João Biehl and Adriana Petryna (Princeton and Oxford: Princeton University Press, 2013).

12 Lynteris, "Infectious Animals," 3.

state-building.[13] Malaria was considered an obstacle to projects of "progress" and the kind of control over natural or material worlds that it was hinged on.[14] Control measures tended to take the form of interventions on environments the mosquitoes lived in, as destruction of wetlands that provided larval breeding sites, and later, with the use of insecticides.[15] Timothy Mitchell has argued that narratives of the expansion and development of capitalism and colonial rule need to consider how their governing logics intertwine and become possible, or desired, or collapse and fail, with the mosquito, or the river along which it travels, or the *Plasmodium* parasite as it moves in and out of human bodies.[16]

Mitchell's call resonates alongside scholarship in multispecies studies that has argued for understanding social worlds as constitutively more-than-human.[17] While attention in this field has been paid to sites where life and its resilience emerge in multispecies entanglements, taking the vector of a lethal disease, the malaria mosquito, as an analytic focus points towards slightly different kinds of questions about shared lives and vulnerabilities.

3 The Gene Drive Debates

The way in which gene drive mosquitoes are worked into a potential biotechnological tool is influenced both the larger scale debates on gene drive organisms as well as the micro-politics of laboratory work.[18] These laboratory encounters also are a place where gene drive researchers enact research ethics.[19] The

13 Timothy Mitchell, *The Rule of Experts. Egypt, Techno-politics, Modernity* (Berkeley and Los Angeles: University of California Press, 2002).

14 Jones, *Mosquito*; Mitchell, *The Rule of Experts*.

15 Melissa Graboyes and Hannah Carr, "Institutional Memory, Institutional Capacity: Narratives of Failed Biomedical Encounters in East Africa," *Canadian Journal of African Studies* 3 (2016), doi:10.1080/00083968.2016.1266678.

16 Mitchell, *The Rule of Experts*, 52.

17 Thom van Dooren, Eben Kirksey and Ursula Münster, "Multispecies Studies: Cultivating Arts of Attentiveness," *Environmental Humanities* 8, no. 1 (2016), doi:10.1215/22011919-3527695; Donna Haraway, *When Species Meet* (Minneapolis and London: University of Minnesota Press, 2008).

18 Marianne Mäkelin, "Between the Lab and the Wild: Establishing the Potential of Gene Drive Mosquitoes for Malaria Control," *Science as Culture* (March 2024), doi:10.1080/09505431.2024.2325982.

19 Marianne Mäkelin, Elina Helosvuori and Mianna Meskus, "Strategic Naturalizing in the Anthropocene: Managing Cells, Bodies and Ecosystems," *The Sociological Review* (forthcoming).

focus here, however, is on the text data, which speaks of the wider context of anticipations and anxieties that are linked to the mosquitoes, or the human-mosquito-parasite nexus.[20] Scientific research and technological development in the contemporary world are phenomena that happen in many locations and modalities at once: in the laboratories where the mosquitoes are being worked with, but also in the scientific and regulatory discussions about them.[21] These are processes that are usually not separated into neatly partitioned sites but instead unfold as circulations and precarious local assemblages that inform each other.[22] Consequently, my goal is not to investigate how a specific institution or a journal discusses gene drive mosquitoes, but rather these debates are an entry point into tensions and organizing logics in the field that both emerge from and inform the research as it is being done.[23]

The gene drive debates operate with multiple and intersecting articulations of what mosquitoes, or gene drives, are and what they could (or could be made to) do. I examine these as what geographer Erik Swyngedouw has called "scalar narratives": accounts that connect entities and events across scales.[24] Narratives that link single organisms or technologies as representing wider processes have been shown to be prevalent in global health[25] as well as in biotechnology,[26] underpinning both techno-solutionist views as well as those that highlight ecological concerns. In analyzing the debates, I have paid attention to instances where mosquitoes and their attributes were connected to visions of the role of biotechnology in public health or the environmental implications of genetic modification. I have then interpreted these moments in relation to their significance to governing life in the biopolitics of public health and biotechnological interventions. The following two analytical sections discuss mosquitoes in gene drive debates from two perspectives. First,

20 Venla Oikkonen, "Affect, Technoscience and Textual Analysis: Interrogating the Affective Dynamics of the Zika Epidemic through Media Texts," *Social Studies of Science* 47, no. 5 (2017), doi:10.1177/0306312717723760.

21 G.E. Marcus, "Ethnography in/of the World System: The Emergence of Multi-Sited Ethnography," *Annual Reviews of Anthropology* 24 (1995).

22 Ayo Wahlberg, "Assemblage Ethnography: Configurations across Scales, Sites, and Practices," in *The Palgrave Handbook of the Anthropology of Technology*, ed. Maja Hojer Bruun et al. (Singapore: Palgrave Macmillan, 2022), 137, doi:10.1007/978-981-16-7084-8_6.

23 Nading, "Lively Ethics"; Oikkonen, "Technoscience and Textual Analysis".

24 Erik Swyngedouw, "Neither Global nor Local: 'Glocalization' and the Politics of Scale," in *Spaces of Globalization: Reasserting the Power of the Local*, ed. Kevin Cox (New York: Guilford Press, 1997).

25 Nading, "Lively Ethics".

26 Stefan Helmreich, *Alien Ocean. Anthropological Voyages in Microbial Seas* (Berkeley and Los Angeles: University of California Press, 2009), 172.

I examine how the mosquitoes' movements are articulated in relation to the wider processes of control and transfer of biotechnology. Second, I trace ways in which these scalar narratives about mosquitoes and the genes they carry are connected to a concern over movement and transfer over time, and a desire to guard future populations.

4 Articulations of Mosquito Mobility

Containment and geographical separation underlie the biopolitics of both biological experimentation and managing infection. Valuing forms of life and securing some forms of life from those deemed as risky is made possible by the logic of bounded bodies, populations, and environments described by Lorimer and Driessen.[27] Experiments on living organisms are possible if the experiments stay in the confined space of the laboratory. Malaria control measures such as bed nets likewise work by establishing a physical separation between human and insect bodies. What this means for an organism, an insect, that moves and is designed to move, is one of the central themes in gene drive debates. The mobilities that are ascribed to the mosquitoes however are not threatening of promising by themselves. In what follows, I examine what the mosquitoes are imagined to carry with them; how their mobile lives imply intertwined economic, ecological, and health futures.

Assessments of introducing genetically modified organisms into specific environments have been based on showing the expected effects of the introduced organism would have on the surrounding environment, conceptualized as a more or less self-contained ecosystem. Interpreted in a biopolitical framework, what is governed or being controlled are unwanted movement of these organisms. A discussion piece in Nature Biotechnology refers to the Cartagena Protocol, an international treaty signed in 1993 that governs the use and handling of what the treaty calls living modified organisms,[28] and posits that the development of gene drive mosquitoes should be taken as a moment to reconsider the terms of the agreement:

> Gene drive was not an issue that was considered when the terms of the Cartagena Protocol were first negotiated and [...] the fact that mosquitoes are a vector of human disease poses "new considerations and challenges

---

27 Lorimer and Driessen, "Bovine Biopolitics".

28 Convention on Biological Diversity, "The Cartagena Protocol on Biosafety," accessed August 30, 2023, https://bch.cbd.int/protocol.

> during the risk assessment process." Questions arise as to whether the risks of this technology should be weighed against the potential to control disease on a global scale.[29]

The framing where gene drive mosquitoes are discussed along with the need to revise the regulation of genetically modified organisms (GMOs) is often repeated in the scientific discussions. As insects, mosquitoes trouble risk management strategies that operate by producing separated bodies, populations, and milieus. In comparison with other genetically modified organisms such as plants, the way mosquitoes are associated with crossing boundaries is seen to be at the heart of this trouble. When the Cartagena protocol was devised the matter at stake was the use of genetic modification in agriculture: the GMO as a seed. The debates and controversy around agricultural GMOs concerned who was able to make legitimate claims about environmental impact, and the role of seeds as traded commodity.[30] For both those advocating for opening up the discussion about the conditions under which GMOs might be released, as in the quoted article above, as well as concerns that emphasize how releasing gene drive mosquitoes would not be compatible with the existing risk assessment frameworks, engage with an unsettling mobility the mosquitoes are seen to possess.[31]

In fact, mosquitoes do fly – but entomological research on mosquitoes that transmit malaria describes them as living rather locally situated lives. This understanding goes back to early pamphlets on mosquito eradication that describe the "war" on mosquitoes taking place in the standing puddles of water in backyards.[32] The *Anopheles* mosquitoes that have specialized to feed on humans are anthropophilic; they have evolved to feed on human blood specifically.[33] They tend to live close to human settlements, and will only fly a couple of kilometers on their own wings, and because of this, mosquito populations tend to be separated from each other. In contrast to malaria as a global

29 John M. Marshall, "The Cartagena Protocol and Genetically Modified Mosquitoes," *Nature Biotechnology* 28, no. 9 (2010), doi:10.1038/nbt0910-896.

30 Isabelle Stengers, *In Catastrophic Times. Resisting the Coming Barbarism* (Lüneburg: Open Humanities Press, 2015), 36–37.

31 See e.g. WHO, *Global Guidance Framework for the Responsible Use of the Life Sciences: Mitigating Biorisks and Governing Dual-use Research* (Geneva: World Health Organization, 2022).

32 See e.g. Ronald Ross, *Mosquito Brigades and How to Organise Them* (London & Liverpool: George Philip & Son, 1902).

33 Cédric Pennetier et al., "'Singing on the Wing' as a Mechanism for Species Recognition in the Malarial Mosquito Anopheles Gambiae," *Current Biology* 20, no. 2 (2010), doi:10.1016/j.cub.2009.11.040.

phenomenon, seen through infection rates, other accounts describe malaria as a local and rural disease.[34] In these accounts of mosquito life-worlds and their mobility, the gene drive technology is a continuation of the ongoing interplay between humans and insects.

Mosquito mobilities come to signify promise, or environmental threat, when articulated against a set of practices that would make the promise, or threat, realize.[35] The potential to "control disease on a global scale," that the quote above refers to, points at a long-standing goal defining global health, a predilection towards technological solutions that would be applicable universally, regardless of place.[36] Mosquito mobility is made to stand in for, and embody, technological transfer. As studies in science and technology studies have shown, the "global" in these imaginaries is necessarily a result of harmonization efforts, work where knowledge and practices from disparate locales is purified and converged into a coherent whole that can represent a phenomenon such as a global epidemic.[37] The global scale in this case is articulated, and gains its stakes, in relation to a global malaria, enacted as comparable sets of numbers that signify and simplify certain aspects of the human – mosquito – parasite nexus.

Another source of tension that is connected to the mosquito mobility is the transfer of scientific knowledge and technology, and the role of local and international regulatory bodies in this transfer. Even if the *Anopheles* mosquitoes themselves live stationary lives, as products of biotechnological innovation, they are transferred across borders. Narratives about the possibilities of mosquito mobility sometimes can slide between whether they are connected to agency attributed to the mosquitoes and the way in which mosquitoes can become part of scientific or trade practices and infrastructures. A UK policy report advocates for looser regulation on GMOs by referring to the way mosquitoes, too, are able to move freely:

> There is no consistent, internationally recognized, regulatory protocol or convention for the testing and release of GM insects. This appears to conflict with the fact that insects, unlike crops or other GMOs, are not contained by national boundaries. Indeed, increasing global import and export trade means that insects can readily be moved across the world.[38]

34 WHO, *World Malaria Report 2022* (Geneva: World Health Organization, 2022).

35 Carrie Friese, "Models of Cloning, Models for the Zoo: Rethinking the Sociological Significance of Cloned Animals," *BioSocieties* 4, no. 4 (2009).

36 Cueto, "A Return to Magic Bullet?".

37 Kelly and Beisel, "Neglected Malarias".

38 House of Lords, Science and Technology Select Comittee, "Genetically Modified Insects" (London: The Stationery Office Limited, 2015).

The report animates the mosquitoes with a notion of globality that here is not that of epidemic numbers but connected to an imaginary of global trade. The mosquitoes themselves are given an ambiguous role in enacting this global mobility. Mosquitoes, portrayed as uncontainable and inherently transnational, seem to stand for the biotechnology industry in the UK, which, the report argues, should similarly be free to move in and act upon the world.

The boundaries being crossed are national borders that designate the span of regulatory control. Here, however, the borders themselves seem to be an obstacle to managing disease. The biopolitics of securing borders behind existing regulation concerned itself with limiting a *risky vitality* by establishing zones of separation.[39] Instead, the UK report advances its message by connecting gene drive mosquitoes to a vitality where borders instead become a constrain to life – or a kind of liveliness that is seen as beneficial.

This imaginary of mobility, enacted by mosquitoes in concert with global trade, does often gloss over inequalities in opportunities to take part in setting health and environmental policies. Gene drive mosquitoes are developed in locations like the UK, with no local significant malaria epidemic and a robust tradition of conducting and funding medical and biotechnological research. A legacy of the colonial history of fields such as entomology and tropical medicine is the way some areas are seen as reservoirs of data that can be turned into knowledge and information elsewhere, and again transported back.[40]

Gene drive mosquito development, too, relies on knowledge of localized mosquito life-worlds.[41] The accounts of what mosquitoes could do and what kind of transfer and transformation they engender and can be made to advance also engage with international knowledge politics. A publication by the African Union connects gene drive mosquitoes to scientific progress and control over insect worlds:

> While existing interventions have significantly reduced the burden of malaria across Africa, complementary new interventions are required to […] eventually achieve malaria elimination on the continent. Africa should invest in the development and regulation of gene drive technology, whose greatest and most urgent application will be in malaria control and elimination. […] Researchers and developers should establish a network of Africa-based scientists and developers to register their

39 Lorimer and Driessen, "Bovine Biopolitics".

40 Clapperton Chakanetsa Mavhunga, "Organic Vehicles and Passengers. The Tsetse Fly as Transient Analytical Workspace," *Transfers* 6, no. 2 (2016), doi:10.3167/TRANS.2016.060206.

41 Mäkelin et al., "Strategic Naturalizing".

> studies, self-regulate, share information regarding their technology, and peer-review all ongoing developments and field testing of the technology on the continent. They should also adopt a 'co-development' approach that emphasises collaboration between the partners in the teams, from research design to the creation of standard operating procedures. [...] The Panel calls for the development of strategies that should address the challenges of the availability of African skills, the issue of regulation and ethics, education and awareness creation to prepare young people for their future role as decision-makers.[42]

For the African Union, the global and mobile mosquito is a biotechnological tool whose arrival associated with being subject to biotechnological trials that might not deploy local knowledge. To secure the right kind of proliferation of life, expertise that originates and is developed close to the sites where the technology is used is needed. The possibilities of gene drive technology, managing life are again connected to a future where scientific expertise likewise thrives. Responding to a threat of transported technology is, here, also an opportunity to develop "a network of Africa-based scientists". Unlike in the visions about the UK biotechnology sector however, the mosquito instead ties this expertise to a place. Other studies have observed how, in the global South, taking part in biotechnological advances has signified national modernization and taking part in the world economy.[43] For securing life, and securing borders, it might be desirable to conspire with the mosquitoes.

5 Mosquitoes, Genes, and the Persistence of Modified Life

Not only mosquitoes but also the genes that they carry figure in gene drive debates. A genetically modified mosquito is expected to transfer on its edited inheritance on to future mosquito generations. This meant that risk and its mitigation were also articulated against persistence over time. This section looks at the relationship between mosquitoes and genes, and the mobility and persistence that the altered genes, carried by mosquitoes, have over time. I explore how biopolitical concerns over future populations and future ecologies took shape in gene drive debates. I also observe how concerns about the

42 NEPAD, *Gene Drives for Malaria Control and Elimination in Africa* (Midrand: NEPAD, 2018), 2–3.

43 Luísa Reis-Castro and Kim Hendrickx, "Winged Promises: Exploring the Discourse on Transgenic Mosquitoes in Brazil," *Technology in Society* 35, no. 2 (2013).

mosquitoes as organisms and the cellular workings of their genes would sometimes shift scales and blend into each other.

The proposition to alter insect bodies to achieve larger-scale ecological changes is by itself not a novel one, nor is it uniquely connected to the gene drive technology.[44] An early precursor to altering the genes of insects can be traced back to the 1950s and the sterile insect technology, which refers to subjecting insects to radiation to render them sterile. The sterile insect technique has been to reduce populations, both for the purposes of agricultural pest control as well as to curb disease.[45] More recently, and concurrently with the advance of gene drive strategies, a method that uses bacteria called *Wolbachia* to disrupt mosquitoes' reproductive processes has been developed in hopes to reduce dengue fever and Zika.[46]

In comparison with these two strategies, gene drive strategies mobilize not only the mosquitoes as organisms, but the laboratory-modified genetic material. The mobility of genetic matter grounded the framing of gene drives as a novel and transformative technology in press coverage with headlines such as "The Extinction invention."[47] Different mobilities – those of mosquitoes, of biotechnology, and of genetic matter through reproduction and inheritance – could also shift into one another.

Microbiological work on mosquitoes is based on creating spaces of separation and purification between the laboratory as an experimental space and its outsides, where experiments are possible because of the containment practices that ensure that only information leaves the laboratory.[48] The *risky vitality* that made the mosquitoes a useful tool in the laboratory would, so far, be managed by enclosing it within the laboratory. What this meant in the case of gene drives, a technology that was designed to spread, was a topic that was discussed from time to time. As Lorimer and Driessen analyze, the foundational risk in the biopolitical mode of securing life from the risk of unwanted forms

44 See James E. McWilliams, *American Pests. Losing the War on Insects from Colonial Times to DDT* (New York: Columbia University Press, 2008); Marcos Cueto, *La salud internacional y la Guerra Fría Erradicación de la malaria en México, 1956–1971* (México: Universidad Nacional Autónoma de México, 2013).

45 Sandrine Dupé, "Transformer pour contrôler. Humains et moustiques à La Réunion, à l'ère de la biosécurité," *Revue d'anthropologie des connaissances* 9, no. 2 (2015), doi:10.3917/rac.027.0213.

46 Luísa Reis-Castro, "Becoming Without," *Environmental Humanities* 13, no. 2 (2021), doi:10.1215/22011919-9320178.

47 Antonio Regalado, "The Extinction Invention," *MIT Technology Review* 113, no. 3 (2016).

48 Dupé, "Transformer pour contrôler".

of life is imagined as *infection*.[49] The issues specific to working with genetically modified insects were associated with either the possibility of transfer of the genetic material outside the laboratory, or with the mobility associated with insects. A scientific article in the materials that discusses the containment of genetically modified mosquitoes also notes that managing microbial life has often served as the model case for laboratory containment measures. The article goes on to note that "[u]nlike microbes, arthropods [such as mosquitoes] crawl, fly, and hop."[50]

These might implicate the researchers themselves as a risk as their bodies move in and out of the laboratory. As most of the laboratory work on mosquitoes that involved genetically modified insects is done in places where malaria and malaria-carrying mosquitoes are not endemic – in university laboratories in Europe and the United States – the insectary spaces needed to be adjusted for *Anopheles* life. The geographical and ecological distance between where the technology was begin developed and its imagined eventual use became a form of "ecological containment."[51] Even if a solitary mosquito would find its way outside, perhaps hitching a ride in a researcher's lab coat, its altered genome would not be transferred on to other *Anopheles* outside the laboratory.

Whether the altered genes carried by mosquitoes would or would not remain in the wild after release framed many of the risks attributed to the gene drive strategy. The mosquitoes do not only crawl, fly, and hop – crucially, they also reproduce. In the light of mosquito biopolitics, this kind of mobility suggested a shift in the governing logic, from safeguarding spaces to understanding risk and security as temporal. The mobility in question is not that of the mosquitoes themselves but that of the altered genes, as they are replicated and transferred from one body to another. The staying power of the genes taps into an imaginary of genes as fundamentally informational entities that transcend the bodies they are part of.[52] A *temporal* transfer of the edited genetic sequence through mosquito generations also called into question a mode of biopolitical governing that relied on creating entities – populations and species – that can be assessed as more or less useful or harmful in some wider context.[53]

49 Lorimer and Driessen, "Bovine Biopolitics".

50 Mark Q. Benedict et al., "Recommendations for Laboratory Containment and Management of Gene Drive Systems in Arthropods," *Vector Borne and Zoonotic Diseases* 18, no. 1 (2018), doi:10.1089/vbz.2017.2121.

51 Benedict et al., "Recommendations for Laboratory Containment".

52 Lily E. Kay, *Who Wrote the Book of Life? A History of the Genetic Code* (Stanford: Stanford University Press, 2000).

53 Jamie Lorimer, *Wildlife in the Anthropocene: Conservation After Nature* (Minneapolis and London: University of Minnesota Press, 2015), 15.

Securing future populations, instead of spaces, might then mean that instead of containability and separation, a risk management strategy might concern itself with reversability. What was to be protected was temporal persistence in or displacement of mosquito generations. A prominent researcher talks of managing this as a technical question, while reflecting on what might be construed as "genetic pollution":

> If engineered genes are viewed in the popular imagination as likely to remain in the population for centuries, gene drive will be perceived as a form of lasting genetic pollution. But if it is widely understood that all engineered genes can be readily removed, [...] then even an unauthorized self-propagating gene drive is no longer dangerous contamination.[54]

Gene drives that would be able to reverse earlier technologies were indeed being developed alongside ones that aimed for mosquito control. So-called "reversal drives" were presented as an answer to concerns over unlimited persistence, or the difficulty of conducting contained field tests.

6 Conclusion

In the beginning of this chapter, I posed a question of how the framework of managing disease is shifted with the development of a biotechnological malaria control tool. Genetically modified mosquitoes represent a continuation of the approach to understanding public health, where non-human animals acting as disease vectors are targeted in health interventions and knowledge practices.[55] The mosquitoes embody a promise of a globally transferable, scalable solution that might be assessed in relation to epidemiological models that similarly see malaria as a global epidemic, abstracted from the local social and ecological worlds it emerges in.

At the same time, the novel biotechnology seems to trouble and shift the biopolitics of managing disease. The focus on bounded geographies and bodies and disposability within them does not seem to align with the gene drive strategy. These points of fracture were usually linked to a *mobility* attributed to the mosquitoes and seen as inherent to the strategy. The biopolitics of grappling

54 Kevin Esvelt, "Gene Drive Technology: The Thing to Fear is Fear Itself," accessed December 19, 2019. https://www.sculptingevolution.org/.

55 Lynteris, "Infectious Animals".

with insect-borne disease are more-than-human, and involve the mosquitoes. Yet, they are also more-than-mosquito: the mobilities attributed to the mosquitoes were plural, and arose in how the mosquitoes were valued, made usable or rendered disposable in gene drive research. The unsettling mobilities also involved how the altered genetic matter was expected or imagined to be passed on from mosquito to mosquito, and endure in wild landscapes.

At the moment of writing, the regulatory status of gene drives is still under question. It remains to be seen how the gene drive debates come to be settled, or how they continue as time goes on. Regardless of what the significance of the technology comes to be to its goal of reducing malaria, I suggest that research and debate on gene drives offer insight into how the role of biotechnology comes to be defined in an era where discussions over health, inequality and biodiversity are ever more pressing. The little mosquito draws its own line of flight across these concerns.

Bibliography

Benedict, Mark Q., Austin Burt, Margareth L. Capurro, Paul De Barro, Alfred M. Handler, Keith R. Hayes, John M. Marshall et al. "Recommendations for Laboratory Containment and Management of Gene Drive Systems in Arthropods." *Vector Borne and Zoonotic Diseases* 18, no. 1 (2018): 2–13. doi:10.1089/vbz.2017.2121.

Canguilhem, Georges. *Knowledge of Life*. New York: Fordham University Press, 2008.

Convention on Biological Diversity. "The Cartagena Protocol on Biosafety." Accessed August 30, 2023. https://bch.cbd.int/protocol.

Cueto, Marcos. "A Return to Magic Bullet? Malaria and Global Health in the Twenty-First Century." In *When People Come First: Critical Studies in Global Health*, edited by João Biehl, and Adriana Petryna, 30–53. Princeton and Oxford: Princeton University Press, 2013.

Cueto, Marcos. *La salud internacional y la Guerra Fría: erradicación de la malaria en México, 1956–1971*. México: Universidad Nacional Autónoma de México, 2013.

van Dooren, Thom, Eben Kirksey, and Ursula Münster. "Multispecies Studies: Cultivating Arts of Attentiveness." *Environmental Humanities* 8, no. 1 (2016): 1–23. doi:10.1215/22011919-3527695.

Dupé, Sandrine. "Transformer pour contrôler. Humains et moustiques à La Réunion, à l'ère de la biosécurité." *Revue d'anthropologie des connaissances* 9, no. 2 (2015): 213–236. doi:10.3917/rac.027.0213.

Esvelt, Kevin. "Gene Drive Technology: The Thing to Fear is Fear Itself." Accessed December 19, 2019. https://www.sculptingevolution.org/.

Foucault, Michel. *The History of Sexuality. Vol. 1. An Introduction.* London: Lane, 1978.

Foucault, Michel. *Security, Territory, Population. Lectures at the Collège de France 1977–1978.* New York: Picador, 2007.

Friese, Carrie. "Models of Cloning, Models for the Zoo: Rethinking the Sociological Significance of Cloned Animals." *BioSocieties* 4, no. 4 (2009): 367–390.

Graboyes, Melissa and Hannah Carr. "Institutional Memory, Institutional Capacity: Narratives of Failed Biomedical Encounters in East Africa." *Canadian Journal of African Studies* 3 (2016): 361–77. doi:10.1080/00083968.2016.1266678.

Haraway, Donna. *When Species Meet.* Minneapolis and London: University of Minnesota Press, 2008.

Helmreich, Stefan. *Alien Ocean. Anthropological Voyages in Microbial Seas.* Berkeley and Los Angeles: University of California Press, 2009.

Holt, Robert A., G. Mani Subramanian, Aaron Halpern, Granger G. Sutton, Rosane Charlab, Deborah R. Nusskern, Patrick Wincker, et al. "The Genome Sequence of the Malaria Mosquito Anopheles Gambiae." *Science* 298, no. 5591 (2002): 129–149. doi: 10.1126/science.1076181.

House of Lords, Science and Technology Select Comittee. "Genetically Modified Insects." London: The Stationery Office Limited, 2015.

Jones, Richard. *Mosquito.* Chicago: The University of Chicago Press, 2012.

Kay, Lily E. *Who Wrote the Book of Life? A History of the Genetic Code.* Stanford: Stanford University Press, 2000.

Kelly, Ann, and Uli Beisel. "Neglected Malarias: The Frontlines and Back Alleys of Global Health." *BioSocieties* 6, no. 1 (2011): 71–87. doi:10.1057/biosoc.2010.42.

Lorimer, Jamie. *Wildlife in the Anthropocene: Conservation After Nature.* Minneapolis and London: University of Minnesota Press, 2015.

Lorimer, Jamie, and Clemens Driessen. "Bovine Biopolitics and the Promise of Monsters in the Rewilding of Heck Cattle." *Geoforum* 48 (2013): 249–259. doi:10.1016/j.geoforum.2011.09.002.

Lynteris, Christos. "Introduction: Infectious Animals and Epidemic Blame." In *Framing Animals as Epidemic Villains. Histories of Non-human Disease Vectors*, edited by Chistos Lynteris, 1–26. Cham: Palgrave Macmillan, 2019.

Marcus, G.E. "Ethnography in/of the World System: The Emergence of Multi-Sited Ethnography." *Annual Reviews of Anthropology* 24 (1995): 95–117.

Marshall, John M. "The Cartagena Protocol and Genetically Modified Mosquitoes." *Nature Biotechnology* 28, no. 9 (2010): 896–897. doi:10.1038/nbt0910-896.

Mavhunga, Clapperton Chakanetsa. "Organic Vehicles and Passengers. The Tsetse Fly as Transient Analytical Workspace." *Transfers* 6, no. 2 (2016): 74–93. doi:10.3167/TRANS.2016.060206.

McWilliams, James E. *American Pests. The Losing War on Insects from Colonial Times to DDT.* New York: Columbia University Press, 2008.

Mitchell, Timothy. *The Rule of Experts. Egypt, Techno-politics, Modernity*. Berkeley and Los Angeles: University of California Press, 2002.

Mäkelin, Marianne. "Between the Lab and the Wild: Establishing the Potential of Gene Drive Mosquitoes for Malaria Control." *Science as Culture* (March 2024): 1–20. doi:10.1080/09505431.2024.2325982.

Mäkelin, Marianne, Elina Helosvuori and Mianna Meskus. "Strategic Naturalizing in the Anthropocene: Managing Cells, Bodies and Ecosystems." *The Sociological Review* (forthcoming).

Nading, Alex M. "The Lively Ethics of Global Health GMOs: The Case of the Oxitec Mosquito." *BioSocieties* 10, no. 1 (2015): 24–47. doi:10.1057/biosoc.2014.16.

Nelson, Nicole. *Model Behavior. Animal Experiments, Complexity, and the Genetics of Psychiatric Disorders*. Chicago and London: The University of Chicago Press, 2018.

New Partnership for Africa's Development (NEPAD). *Gene Drives for Malaria Control and Elimination in Africa*. Midrand: NEPAD, 2018.

Oikkonen, Venla. "Affect, Technoscience and Textual Analysis: Interrogating the Affective Dynamics of the Zika Epidemic through Media Texts." *Social Studies of Science* 47, no. 5 (2017): 681–702. doi:10.1177/0306312717723760.

Pennetier, Cédric, Ben Warren, K. Roch Dabiré, Ian J. Russell and Gabriella Gibson. "'Singing on the Wing' as a Mechanism for Species Recognition in the Malarial Mosquito Anopheles Gambiae." *Current Biology* 20, no. 2 (2010): 131–136. doi:10.1016/j.cub.2009.11.040.

Regalado, Antonio. "The Extinction Invention." *MIT Technology Review* 113, no. 3 (2016).

Reis-Castro, Luísa and Kim Hendrickx. "Winged Promises: Exploring the Discourse on Transgenic Mosquitoes in Brazil." *Technology in Society* 35, no. 2 (2013): 118–128.

Reis-Castro, Luísa. "Becoming Without." *Environmental Humanities* 13, no. 2 (2021): 323–347. doi:10.1215/22011919-9320178.

Ross, Ronald. *Mosquito Brigades and How to Organise Them*. London & Liverpool: George Philip & Son, 1902.

Smith, David L., Katherine E. Battle, Simon I. Hay, Christopher M. Barker, Thomas W. Scott, and F. Ellis McKenzie. "Ross, Macdonald, and a Theory for the Dynamics and Control of Mosquito-Transmitted Pathogens." PLoS Pathogens 8, no. 4 (2012): e1002588. doi: 10.1371/journal.ppat.1002588.

Stengers, Isabelle. *In Catastrophic Times. Resisting the Coming Barbarism*. Lüneburg: Open Humanities Press, 2015.

Swyngedouw, Erik. "Neither Global nor Local: 'Glocalization' and the Politics of Scale." In *Spaces of Globalization: Reasserting the Power of the Local*, edited by Kevin Cox, 137–166. New York: Guilford Press, 1997.

Wahlberg, Ayo. "Assemblage Ethnography: Configurations across Scales, Sites, and Practices." In *The Palgrave Handbook of the Anthropology of Technology*, edited by Maja Hojer Bruun, Ayo Wahlberg, Rachel Douglas-Jones, Cathrine Hasse, Klaus

Hoeyer, Dorthe Brogård Kristensen and Brit Ross Winthereik, 125–144. Singapore: Palgrave Macmillan, 2022. doi:10.1007/978-981-16-7084-8_6.

World Health Organization (WHO). *World Malaria Report 2022*. Geneva: World Health Organization, 2022.

World Health Organization (WHO). *Global Guidance Framework for the Responsible Use of the Life Sciences: Mitigating Biorisks and Governing Dual-use Research*. Geneva: World Health Organization, 2022.

PART 4

Bodies at Risk

CHAPTER 9

Clandestine Agents in Meadows: Ticks, Cattle and Redwater Fever in Finland, 1860s–1930s

Taina Syrjämaa

1 Introduction

A notable outbreak of Redwater fever was registered in central Finland in the summer of 1869. According to a newspaper published by the financial section of the Senate of the Grand Duchy, not only had cattle fallen ill in a number of villages in Kuopio province, but the disease had also killed well-kept cattle.[1] Consequently, discussion of the disease, as well as its spread and the potential ways to combat it, was inserted into the agenda of a provincial agricultural meeting and the governor gave orders to the provincial veterinarian to examine the situation.[2] From the perspective of posterity, it is uncontroversial that the sheep tick (*Ixodes ricinus*) triggers Redwater fever by transmitting Babesia protozoans, which destroy red blood cells in the body of a bovine.[3] At the time, however, the potential causes of the disease were unknown and was a major cause of perplexity among the peasantry, as well as estate owners, the authorities and scientists. Many potential causes were considered and a plethora of potential remedies were administered, but at no point were ticks suspected of transmitting the disease.

In this chapter I examine the relations between ticks, bovines and humans through the prism of Redwater fever. I explore an early stage of tick – human relations when the ticks gradually became seen a dangerous threat, not yet to human health, but to animal health and consequently human wealth. This process, which I explore from the 1860s to the 1930s, is not solely a question of

1 "Kuopiosta," *Suomalainen Wirallinen Lehti*, July 29, 1869.

2 Johtokunta, "Kuopion läänin maanwiljelysseuran kokous ja näyttelö w. 1869," *Tapio*, August 21, 1869; "Mainittavia tapauksia kuwernööreiltä tulleiden ilmoitusten mukaan," *Suomalainen Wirallinen Lehti*, September 9, 1869.

3 On the current definition of bovine Rewater fever, see Finnish Food Authority, "Punatauti," accessed November 29, 2023, https://www.ruokavirasto.fi/elaimet/elainten-terveys-ja-elaintaudit/elaintaudit/naudat/punatauti/; World Organization for Animal Health, "Bovine Babesiosis," Accessed November 29, 2023, https://www.woah.org/fileadmin/Home/eng/Animal_Health_in_the_World/docs/pdf/Disease_cards/BOVINE_BABESIOSIS.pdf.

human perception and understanding vis-à-vis what ticks were supposed to be like. Instead, it is closely connected to practices and daily actions in a shared and lived multispecies environment.

I focus on Redwater fever in order to grasp the interactions of ticks, cattle and humans, while also some other actors or suspected actors, such as red-starts and alders. Tick-borne cattle disease, which received its name from the reddish discoloring of urine that is a symptom of the illness, was an annual and widespread threat in Finland. It could kill the sole cow of a poor peasant family as well as a great number of pedigree cattle belonging to the affluent owner of a manor house. The spread of the disease in the Finnish context coincided with the gradual growth of animal husbandry, especially of dairy farming.[4]Redwater fever was a known disease in Europe from the mid-nineteenth century and it had a wide geographical distribution from France to Romania and from Southern Europe to the Nordic countries. Similar tick-borne diseases were also a menace in other continents.[5] Especially notorious – in terms of its economic consequences – was the so-called Texas fever as it took hold in the context of the massively industrialized North American meat industry, which required transportation of millions of bovine over long distances.[6] The era was characterized by scientific breakthroughs in discovering micro-organisms.[7] Research on the causes of bovine haemoglobinuria, for example, proved to be an important step in the novel field of veterinary epidemiology.[8] Numerous scientists in the United States and Europe were employed in this field and they advanced diverse results and hypothesis. The existence of Babesia protozoans and the role of ticks as vectors was gradually established by scientists in the 1880s and 1890s.[9] This, however, did not signify an immediate change in everyday practices and conceptions, as we will see in this chapter.

---

4 On Finnish agriculture and the intensification of animal husbandry, see, for example, Teppo Vihola, *Leipäviljasta lypsykarjaan. Maatalouden tuotantosuunnan muutos Suomessa 1870-luvulta ensimmäisen maailmansodan vuosiin* (Helsinki: SHS, 1991).

5 Michael Köhler and Werner Köhler, "Zentralblatt für Bakteriologie – 100 years ago. Victor Babeş and enzootic haemoglobinuria of cattle," *International Journal of Medical Microbiology* 293, no. 4 (2003): 233.

6 Joshua Specht, *The Red Meat Republic: A Hoof-to-Table History of How Beef Changed America* (Princeton University Press, 2019), for example, 166; Cecil Kirk Hutson, "Texas Fever in Kansas, 1866–1930," *Agricultural History* 68, no. 1 (1994); Norman F. Cheville, *Pioneer Science and the Great Plagues: How Microbes, War, and Public Health Shaped Animal Health* (West Lafayette, IN: Purdue University Press, 2021), 47, 80–81.

7 Bruno Latour, *The Pasteurization of France* (Cambridge, MA: Harvard University Press, 1988).

8 Cheville, *Pioneer Science*, 47.

9 Köhler and Köhler, "Zentralblatt für Bakteriologie".

The essential theoretical premise of this research has been an overarching comprehension of the inevitability of continuous human – non-human interaction and consequent relativity of all agency, whether human or more-than-human.[10] Small ticks and even smaller Babesia have had a major influence on cattle and humans and have directly and indirectly impacted a number of other species that shared the same environment. They have forcefully shaped, for example, human – cattle practices. It is also important to bear in mind that this occurred over the course of many decades when their agency was unknown to humans.[11] This research is inspired by the Latourian actor-network theory, which highlights more-than-human agency and the interdependency of all actors in a network.[12] However, I emphasize that it is only an inspirational starting point. I will not carry out a Latourian analysis of how a network is constituted. Indeed, I find Latour's equation of animate and inanimate actors to be very problematic. Here, my emphasis is on animate actors. Methodologically, this study represents a contextualizing and qualitative form of historical research, with an emphasis on the history of the everyday life of multispecies. I refer to current bioscientific research in regard to the physiology of ticks.

In this study I mainly examine three types of historical source: newspapers, other print media and oral history sources. The outbreaks of Redwater fever were newsworthy and the rapidly expanding press sector in Finland hurried to inform the public of the emergence of the disease. Newspapers also reprinted reports produced by agricultural associations and other organizations, as well as the communiques and recommendations of the authorities. Furthermore, they enhanced public discussion on animal husbandry and cattle diseases by publishing the contributions of those who wanted to share their views and expertise, whether this was professional or based on practical experience. Other print media examined in this chapter consists of the annual reviews of the National Board of Health, as well as booklets published by veterinarians to

10 Donna Haraway, *When Species Meet* (Minneapolis: University of Minnesota Press, 2008); Susan Nance, "Introduction," in *The Historical Animal*, ed. Susan Nance (Syracuse, NY: Syracuse University Press, 2015); Vinciane Despret, "From Secret Agents to Interagency," *History and Theory* 52 (2013); Chris Pearson, "Dogs, History, and Agency," *History and Theory* 52, no. 4 (2013); Tuomas Räsänen and Taina Syrjämaa, eds., *Shared Lives of Humans and Animals. Animal Agency in the Global North* (London: Routledge, 2017).

11 On multispecies practices, see, Nora Schuurman and Taina Syrjämaa, "Shared Spaces, Practices and Mobilities: Pet–Human Life in Modern Finnish Homes," *Home Cultures* 18, no. 2 (2021), doi:10.1080/17406315.2021.1963611.

12 Bruno Latour, *Reassembling the Social: An Introduction to Actor-Network-Theory* (Oxford: Oxford University Press, 2005).

advise farmers and popular science publications designed to present fauna to the general public.

Oral history collections also form an important source as they contain descriptions of unwritten daily practices and beliefs. The Finnish Literature Society Archive hosts a remarkable collection in which material on folk healing includes numerous references to cattle, grazing, herding and explicitly to Redwater fever. The oral history material used in this research was collected between the 1880s and the 1950s. It reveals quite a diversity in the conception of nature and non-human animals as well as practices to treat and mitigate cases of Redwater fever.

In this chapter, I first trace the presence of ticks – unknown agents – and their interactions with humans and other animals. Written and oral descriptions of outbreaks of Redwater fever indicate and reveal practices and spaces – to use Donna Haraway's concept of contact zones – of multispecies co-living. They also show human attempts to defend the cattle by methods that range from modern medicine and technologies to traditional folk healing and magic. I then explore the search for the cause of the disease in the context of daily life and subsequently among scientists, which eventually initiated the process of uncovering the agency of ticks. This chapter shows the complexity and hybridity of the long road in human understanding of ticks, whereby they were initially viewed as a mere nuisance to humans and their livestock to the more recent comprehension of the arachnids as a menace to a modernizing society.

2 Impact of Tick Bites

Ticks have probably lived for centuries in the territory that today is Finland. However, their exact distribution was not mapped before the 1950s.[13] Sheep ticks can attach to almost any warm-blooded vertebrate and usually attach to hosts of different species in their three different life stages. In each stage they need blood. Ticks quest for potential hosts among plants. In the second and third stages nymphs and adults ascend higher in thick undergrowth where they react to odor, exhaled carbon dioxide and movement. Adult ticks prefer larger hosts and when on them they are able to crawl around looking for a

13 Niko Kulha et al., "Does Environmental Adaptation or Dispersal History Explain the Geographical Distribution of Ixodes Ricinus and Ixodes Persulcatus Ticks in Finland?," *Ecology and Evolution* 12, no. 12 (2022), doi:10.1002/ece3.9538.

thinner spot of skin on which to attach.[14] With the increase of animal husbandry during the latter part of the nineteenth century, a growing number of cattle – as well as their human herders – walked in Finnish meadows, an environment shared with ticks, from spring to autumn thereby multiplying their mutual encounters.

Animal husbandry at this time was intensifying internationally and was being transformed into a large-scale business. In Finland, animal production was very limited in comparison to many other countries, but it was seen as a promising way to break out of the prevailing poverty. It was thought that animal husbandry offered a more stable livelihood than traditional agriculture, which was heavily reliant on grain production and was notoriously risky as ground frost often destroyed crops and led to famine. In fact, when the abovementioned outbreak of Redwater fever occurred in Kuopio province in the summer of 1869, Finland was just coming out of a disastrous famine that had killed approximately eight per cent of the human population over the course of three years.[15] Yet, animal husbandry alone was not expected to offer nourishment to the population. It was also hoped that Finland would be able to join the growing international animal business markets and that it would export animal-based products.[16] Indeed, dairy products did become an important sector of the Finnish economy.[17]

Until the 1870s, Finnish animal husbandry was, however, still in its infancy. Finland was pre-eminently a country of smallholders, who concentrated on growing of grain for home consumption and whose sparse resources limited the opportunity to invest in the modernization of agriculture. It was quite common, for example, that the scarce farm animals were not properly fed. It was not customary in Finland to cultivate foraging. Animals were given leaves and branches that had been gathered in woods and as soon as the weather

14 Jani Sormunen, *Questing Ticks, Hidden Causes: Tracking Changes in Ixodes Ricinus Populations and Associated Pathogens in Southwestern Finland* (Turku: University of Turku, 2018), 7–9.

15 Antti Häkkinen and Henrik Forsberg, "Finland's Famine Years of the 1860s: A Nineteenth-Century Perspective," in *Famines in European Economic History: The Last Great European Famines Reconsidered*, ed. Declan Curran et al. (London: Routledge, 2015), 99, 108–109.

16 Taina Syrjämaa, "Eläimet, eläinperäiset tuotteet ja edistysusko," *Tunteva tuote. Kuinka eläimistä tuli osa teollista tuotantoa?*, ed. Taija Kaarlenkaski and Otto Latva (Tampere: Vastapaino, 2022); Taina Syrjämaa, "Multispecies Mobilities and Human Belief in Progress," in *Animal Industries. Nordic Perspectives on the Exploitation of Animals since 1860*, ed. Taina Syrjämaa et al. (Berlin: De Gruyter 2024).

17 Matti Peltonen, "Uudet kaupallistumisen muodot," in *Suomen maatalouden historia II*, ed. Matti Peltonen (Helsinki: SKS, 2004).

FIGURE 10.1 A smoky fire expelling flying insects when cows are milked in the evening, Northern Savonia, 1917. Unknown photographer.
SOURCE: PRESS PHOTO ARCHIVE JOKA/FINNISH HERITAGE AGENCY. HTTPS://FINNA.FI/RECORD/MUSEOVIRASTO.71620E24-D405-4582-A9F9-D1E9DD427A7B%20, CC BY 4.0

permitted it in the spring, the pinched cattle were taken out to eat whatever they could find by themselves in meadows and woodland.[18] Ticks were waiting for them in these places and could also relocate to new locations should the cattle move.

18 Ann-Catrin Östman, "Mekanisoinnin ensimmäinen aalto," in *Suomen maatalouden historia II*, ed. Matti Peltonen (Helsinki: SKS, 2004); Teppo Vihola, "Pärjääkö pienviljelys?," in *Suomen maatalouden historia II*, ed. Matti Peltonen (Helsinki: SKS, 2004), 164.

Human and non-human animals usually react to skin injuries and irritations, but the saliva of ticks suppresses such reactions. As tick bites do not cause immediate pain or itchiness, the arachnids can continue to feed for days without being interrupted by their hosts.[19] In this respect, many other small creatures in meadows, such as horseflies and other midges could be perceived as worse pests.[20] Horseflies were especially persistent in their manner of assault and their bites hurt immediately. Indeed, it was customary to light a smoky fire in the evening when cows were milked, which was meant to keep flying insects away from both cows and their milkers.[21]

Tick bites can also become inflamed over time and when large numbers became attached to the muzzles, udders and genitals of grazing cattle, they became an irritant. Definitely, humans at this time also experienced unpleasantness from tick bites. Yet, not much attention was paid to ticks as they were not considered to be particularly harmful. However, ticks that carried the invisible Babesia protozoans caused for much more severe illness for cattle than mere local inflammations.

Bovines who fell ill because of ticks and Babesia protozoans became tired and they lost their appetite. Walking to the meadow in the morning could become difficult and they fell behind other members of the herd. At this point, a sharp-eyed herder could suspect that a cow was suffering from Redwater fever and let it rest. After the first symptoms became noticeable, the health of a diseased cow could rapidly decline over the course of a few days: urine – and milk if any was still being produced – would turn red and the animal would suffer from fever, diarrhea and later constipation. Descriptions of the main external symptoms are highly consistent whether described by persons with practical experience or by those with veterinary education throughout the time period under investigation in this study.[22]

---

19 Ladislav Šimo et al., "The Essential Role of Tick Salivary Glands and Saliva in Tick Feeding and Pathogen Transmission," *Frontiers in Cellular and Infection Microbiology* 7 (2017): article 281, doi:10.3389/fcimb.2017.00281.

20 "Asikkalasta," *Suomalainen Wirallinen Lehti*, July 6, 1872.

21 Antti Vallius, *Kaunis Suomi. Maaseutumaisemakuvaston historiaa 1800-luvulta EU-Suomeen* (Helsinki: Tammi, 2015), 144–154. See also, for example, Porajärvi. Helmi Helminen 1496. 1943 – Ivan Hermonen, aged 75. FLSA (Finnish Literature Society Archive), 76. Folk religion II, cattle raising 11, cattle diseases, bovine diseases (FR, CA, CD).

22 J.A. Wegelius (Häm), "Helppo keino punatautia ja werenkusemista wastaan elukoissa," *Suomalainen Wirallinen Lehti*, August 16, 1867; "Karjarutosta," *Suomalainen Wirallinen Lehti*, May 20, 1873; O.B., "Raawaskarjan punataudista," *Tampereen Sanomat*, February, 7, 1884; Oskar von Hellens, *Verenheitosta (punataudista) raavaseläimissä* (Helsinki: Lääkintöhallitus, 1906), 7–8; Kaarlo A. Zinck, *Nautakarjan tarttuva punatauti. Toimenpiteet sen levenemisen ehkäisemiseen* (Kuopio: Itä-Suomen karjanjalostusyhdistys, 1929), 6–7.

It was difficult for humans, including veterinarians, to evaluate whether a bovine would recover or whether the illness would be fatal. The interests of cattle owners could lead to a desperate attempt to slaughter an animal in order to also avoid losing the meat and hide of the livestock.[23] Farmers were officially obligated to inform the authorities when any epidemic was suspected and had to bury the corpses of diseased animals. Failure to comply with these regulations could result in a fine.[24] Redwater fever was not among the most catastrophic diseases that affected livestock, such as cattle plague or rabies.[25] Nonetheless, Redwater fever had a major negative impact at a regional level, not to mention on individual farms and families, as it could break out every summer. It is also important to note that it was quite difficult for farmers to be sure of any diagnosis and to exclude other diseases. Moreover, it was not clear how contagious Redwater fever actually was as long as the mechanism of the infection was not known. However, in general, it was noted that bovines did not directly infect each other.

The distribution of ticks and Babesia seems to have expanded during the last decades of the nineteenth century. Whilst there were annual differences in the intensity of Redwater fever outbreaks, there were numerous localities that regularly suffered from the disease. Heinävesi, in the Finnish lake district, for example, became notorious for its outbreaks. By the mid-1880s the disease had already been prevalent in the region for approximately thirty years.[26] Every now and then new outbreaks were reported in villages where the disease had previously not been experienced, thereby providing evidence of the gradual spread of Redwater fever.[27] In the early twentieth century, cases of the disease were reported almost each year in all Finland's provinces. The highest concentration of Redwater fever continued to be in Kuopio province and in other

23 "Karjan-rutto," *Karjalatar*, July 27, 1877; "Lyhykäisiä kirjoituksia maataloudessa, toimittanut Uudenmaan ja Hämeenläänien maanwiljelysseura," *Hämäläinen*, June 11, 1881.

24 Keisarillisen Majesteetin Armollinen Julistus, koskewa mitä tarttuvain eläintautien estämiseksi ja häätämiseksi pitää waariin otettaman. Annettu Helsingissä, 6 p:nä kesäkuuta 1864. *Suomen Suuriruhtinanmaan Asetus-Kokous* (Helsinki: Keisarillisen Senaatin kirjapaino, No. 18, 1864), 1–7. Redwater fever is not mentioned but the statute has a generic comment that includes also other than named diseases.

25 On cattle diseases, see, Anneli Mäkelä-Alitalo, "Karjataudit ja eläinlääkintä," in *Suomen maatalouden historia I*, ed. Viljo Rasila, Eino Jutikkala and Anneli Mäkelä-Alitalo (Helsinki: SKS, 2003).

26 "Punatauti," *Savonlinna*, June 17, 1886.

27 "Punatauti raivoaa Oriwedellä," *Hämeen Sanomat*, July 13, 1895.

FIGURE 10.2 Grazing cattle on a lake shore in Viitasaari, central Finland, 1900–1915
SOURCE: UNKNOWN PHOTOGRAPHER, THE MUSEUM OF CENTRAL FINLAND, HTTPS://FINNA.FI/RECORD/KSM.158386785528900, CC BY-ND 4.0

areas in central Finland and Karelia with hundreds of cases. Only a few cases were recorded in the most northerly province.[28]

This is logical as more favorable conditions for ticks in Finland were to be found near the water systems in southern and central Finland. The climate in the north was less propitious for ticks. Yet, the greater prevalence of the disease in central Finland than in coastal areas of western Finland, for example, may be partly explained by differences in agricultural practices.

The need for a veterinarian was often cited during outbreaks, but in most cases no-one was available. The number of veterinarians in Finland was extremely small: in 1882 there were only thirteen in the entire country and there were still only thirty-five in 1892.[29] Another obstacle faced by farmers was that the administrative procedure to commission a veterinarian was slow.

---

28 See, for example, *Bidrag till Finlands officiella statistik XI. Medicinalverket, Medicinalstyrelsens berättelse för år 1900* (Helsingfors, 1901), 146. [1901: 154; 1905: 270; 1908: 458; 1909: 458; 1910: 482].

29 Katri Helminen, *Eläinlääkärinä kolmella vuosisadalla. Suomen eläinlääkäriliitto 1892–2017* (Helsinki: Fennovet, 2017), 7, 84.

Only after a regional governor provided authorization could a provincial veterinarian set off on what was likely a cumbersome journey to the disease hot spot.[30] Thus, many bovine patients either succumbed to Redwater fever or had recovered by the time a veterinarian finally arrived.

The treatment offered by veterinarians could not save all of the bovine patients that had been bitten by ticks. Veterinarians could assist in handling symptoms and in sustaining some bodily functions, but they did not have a patent solution regarding how to save diseased animals. It is therefore hardly surprising that peasants were often disappointed in the treatment offered by veterinarians and their medicaments. Furthermore, some highlighted significant social differences, including linguistic problems: Finnish-speaking peasants could not understand the instructions given to them by Swedish-speaking veterinarians.[31]

Various remedies were used to try and cure diseased animals in the 1870s, many of which must have been quite an ordeal for the patient. Here diverse human conceptions of how to react to illnesses – and to challenges posed by nature in general – collided, but also co-existed: from the scientific approach of veterinarians to the pragmatic experiments carried out by cattle owners and even to traditional magical practices.[32] In subsequent decades many medicinal formulas were utilized, often consisting of vinegar, linseed oil, certain kinds of lichen and a watery sort of gruel. Sometimes plants and herbs, such as marsh tea and juniper berries, were used. Some formulas also contained red ochre, spirits, various acids and salts. These remedies constituted a balance between what was considered as being potentially useful and what was at hand.[33] The resulting mixture was forced down the throat of a sick cow, who hopefully swallowed it rather than it getting caught in the trachea.

---

30 "Kovin hankalaa," *Uusi Suometar*, August 9, 1889; "Punatauti lehmissä," *Keski-Suomi*, July 7, 1896. This much-criticized system prevailed until 1937. Helminen, *Eläinlääkärinä kolmella vuosisadalla*, 54.

31 "Mikkelin läänin maanwiljelyskokous Heinäwedellä 2 ja 3 päiwä Syyskuuta 1886," *Wiipurin Sanomat*, September 25, 1886; A.L., "Hätähuuto Heinävedeltä!," *Uusi Suometar*, July 11, 1888.

32 Cf. co-existence of human medicine and folk healing: Ulla Piela, "'Konsti elää kauwwan.' Parantaminen Suomessa varhaismodernilta ajalta nykypäivään," in *Kiistellyt tiet terveyteen. Parantamisen monimuotoisuus globaalihistoriassa*, ed. Markku Hokkanen and Kalle Kananoja (Helsinki: SKS, 2017).

33 J.A. Wegelius (Häm), "Helppo keino punatautia ja werenkusemista wastaan elukoissa," *Suomalainen Wirallinen Lehti*, August 16, 1867; "Asikkalasta," *Suomalainen Wirallinen Lehti*, July 7, 1872; "Kuopiosta 20 p. heinäk.," *Suomalainen Wirallinen Lehti*, July 26, 1873. See, for example, J.W. Kotikoski, Korpilahti (1908) and J.E. Tuomala, Porvoo parish (1890) FLSA, FH, RF (Folk healing, Redwater fever in animals).

Newspapers rarely hinted that magical techniques were employed and when they did so the tone was instructive. Such remedies were presented as old-fashioned superstitions.[34] Yet, the brief mention of such treatments reveals the existence of locally-known and respected individuals who were invited to treat diseased animals by means of folk healing. What was deemed to be an obsolete set of beliefs and practices in newspaper articles appear in a totally different light when one examines oral history sources. These sources reveal a lively variety of folk practices, rituals and spells. An oft-repeated cure was to force a diseased cow to swallow a living frog, frog spawn, a dor beetle or a tarred roach.[35] Other complicated and detailed rituals were also performed. In 1889, for example, a man in his sixties in south-eastern Finland related that a nog, made of rowan, should be driven into the ground at the spot where a cow is seen to urinate for the first time after she has fallen ill.[36] This must refer to red urine as an evident symptom of disease. This interviewee, as well as many others, describes various magic rituals and gives the impression that such practices were common and that they themselves had first-hand experience of using or witnessing their use. However, people did not only try to heal animals, but there were also attempts to ward off sickness altogether. This required a more precise diagnosis in terms of suspected culprits.

3 In Search of Culprits

In September 1890 – twenty-one years after the agricultural meeting in Kuopio province mentioned at the beginning of this chapter – the same issues continued to puzzle the attendees at a similar gathering: what caused Redwater fever and how could it be stopped? Participants at the meeting informed others of their personal experiences. For example, Adolf von Wright, a state dairyman, explained how he had previously suffered losses due to Redwater fever every summer, but when he kept his cattle inside a corral on an open upland area none of his herd fell ill.[37] Von Wright's method pointed to the most widely accepted explanation for the cause of Redwater fever: damp meadows.

---

34 "Taikausko ja taikatemput," *Savo* June 27, 1883; "Punatauti," *Aamulehti*, July 8, 1890.

35 J.E. Tuomala, Porvoo parish (1890); The farmer Teuteri Laitinen, aged 63, Rauhajärvi 1932–1933. FLSA, FR, CA, CD.

36 Kymi. Vihtori, Alava. IV. A. 39. 1889 – Otto Matinpoika, aged 59, native of Sysmä, FLSA, FH, RWF.

37 "Kuopion läänin Maanwiljelyskokous Suonenjoella," *Savo*, September 9, 1890.

Over the decades, countless references were made to wet pastures by cattle owners, veterinarians and by peasant interviewees. Such meadows were described as consisting of thick bushes and fast-growing trees, such as alders, willows and aspens.[38] The theories regarding the dangers of meadows coincide perfectly with the favorite habitat of ticks. Hard ticks, including the sheep tick, require a relatively high degree of humidity and suffer from desiccation if there is not enough water vapor in the air. They do not prosper in strong sunlight and in windy locales. Furthermore, ticks need moss or dead leaves on the ground to develop when they are not feeding and over the course of winter. They also need grass or other suitable plants to climb when questing potential hosts.[39]

While contemporaries agreed on the dangerous nature of moist meadows, it was rarely possible to avoid them. Cattle owners rarely possessed better quality pastures. Drainage was another oft-suggested solution. This was a technique also applied to low-lying fields that were prone to high levels of frost. Such initiatives, however, required notable labor resources. It continued to be unavoidable for many farmers to take their cattle to the same meadows in the spring despite the known risk of Redwater fever.[40]

Ticks were not suspected of being carriers of Redwater fever for a long time. Other species that lived in the same environment bore the brunt of blame for outbreaks of the epidemic. It was recognized early on that the outbreaks of the disease occurred between spring and autumn and the first cases could be expected soon after the cattle had started to graze in meadows.[41] Indeed, contemporary biologists have ascertained that sheep ticks require a temperature of +7c to be active.[42]

In the nineteenth century, blame was understandably cast upon animals, plants and conditions that were connected with meadows and grazing. On occasions it was suspected that hungry bovines had eaten unsuitable plants and weeds that thrived in humid areas, or that they had drunk unsanitary water.[43] Miasmas were also cited following the widely believed idea that

38 "Punatauti," *Keski-Suomi*, July 23, 1891; "Maanwiljelys- ja talousosasto," *Keski-Suomi*, July 6, 1901.

39 Sormunen, *Questing Ticks*, 8–10.

40 See, for example, "Punatauti karjassa wedessä kulkemisen seurauksena," *Uusi Suometar*, March 1, 1890.

41 O.B., "Raawaskarjan punataudista," *Tampereen Sanomat*, February 7, 1884.

42 Sormunen, *Questing Ticks*, 7.

43 For example, "Lyhykäisiä kirjoituksia maataloudessa, toimittanut Uudenmaan ja Hämeenläänien maanwiljelysseura," *Hämäläinen*, June 11, 1881; "Muutama sana pernataudista," *Hämäläinen*, August 6, 1881.

gases emanating from putrid soil were harmful.[44] Cowherds who worked in the meadows on a daily basis had their own suspects. Redstarts, for example, were mentioned in numerous oral history recollections. It was believed that if a redstart flew beneath the belly of a cow, the latter would be contaminated. Redstarts were consequently killed and their nests were destroyed.[45]

The reason why these birds were blamed seems to depend on a logic of equivalence in folk healing.[46] As Redwater fever turned urine and milk red, anything red was seen to be connected with the disease; either as a cause or as potentially being capable of preventing contamination. Hence, cloths stained with women's menstrual blood were also utilized. The longevity and dynamicity of such beliefs can be seen in the curious hybrid practice of giving red lemonade, instead of red ochre, to cows.[47] While many traditional practices continued to exist – for example, making a cow swallow a living frog was still attested in the 1930s – it can be noted that these traditions gradually became less widely employed. Younger interviewees of ethnographical surveys seem to refer to magical practices as rarities or as traditions that had already ended and of which they only had indirect knowledge. A woman born in the late 1890s and interviewed in 1936, for example, explained that bovines who had died of Redwater fever in the center of Heinävesi had their heads taken to a nearby sacred spring. Apparently, she had not witnessed it herself, but as a proof she described how a pole thrusted in the spring clattered into the skulls of the dead cattle.[48] Such a story connects with age-old beliefs in sacred places in nature that had their roots in pre-Christian culture. This story hints that such a practice may have existed in the late nineteenth century in Finland but had become obsolete in the early twentieth century.

4 The Slowly Exposed Agency of Ticks

In the late nineteenth century scientific theories relating to the cause of Redwater fever and Texas fever were thoroughly transformed. This was an era that witnessed remarkable changes in how the natural environment was perceived.

---

44 O.B., "Raawaskarjan punataudista," *Tampereen Sanomat*, February 7, 1884; Peter Thorsheim, *Inventing Pollution: Coal, Smoke, and Culture in Britain Since 1800* (Athens, OH: Ohio University Press, 2017).

45 See, for example, Sortavala, Otsoinen. Matti Moilanen 3187. 1937. Jaakko Patja, aged 74, FLSA, FR, CA, CD.

46 Cf. humans suffering from jaundice ate yellowhammers to recover. Piella, "'Konsti elää kauwwan'," 120.

47 Koivisto, Lauri Laiho 1935, FLSA, FH, RWF.

48 Heinävesi, L. Karhu 128. 1936. – Ida Koponen, aged 39, FLSA, FR, CA, CD.

Louis Pasteur's discovery of a new actor – the microbe – is the most notable example. It was soon turned into a heroic narrative of a supposedly lonely scientific genius, which was not the case, as Bruno Latour has shown. Nonetheless, the revelation of the existence of microbes not only dramatically transformed human conceptions of more-than-human networks, but also offered new means to treat diseases.[49] The quest for the cause of Redwater fever formed part of the booming field of research into micro-organisms. This was not a straightforward process. It included several changing hypotheses, for example, as well as practical experiments in laboratories and in pastures in different continents. It also involved academic publications and a good amount of scientific rivalry in the late 1880s and the 1890s. American researchers, of whom Theobald Smith is the most widely known, tried to uncover the cause of Texas fever, which was fatal to northern cattle herds grazing in pastures traversed by southern cattle. Concurrently, the Romanian researcher Victor Babeş sought out the cause of endemic Redwater fever that menaced the lowlands near the Danube. His efforts led to the discovery of the microscopic parasites that caused haemoglobinuria in cattle. These parasites came to be named after Babeş, although his initial identification of them as bacteria in 1888 proved to be an error and they soon became classified as protozoans.[50]

The need for scientific knowledge in Finland was increasingly emphasized. In September 1890, for example, the participants of the Kuopio provincial agricultural meeting proved to have their fingers on the pulse as they decided to ask the Senate to hasten Finnish research into Redwater fever.[51] During the summers of 1893 and 1894 two young researchers – Oscar von Hellens (1867–1948) and Ali Krogius (1864–1939) – studied the disease and focused on the Kuopio province and Heinävesi. Von Hellens also visited Parainen on the southwestern coast and Impilahti and Salmi in Karelia.[52] At the time, von Hellens was a recent graduate in veterinarian science and was employed at the National Board of Health, where he was able to continue his studies in medicine, while Krogius had already completed his doctorate in medicine and surgery.[53] Their

49 Latour, *The Pasteurization of France*.

50 Köhler and Köhler, "Zentralblatt für Bakteriologie"; Mircea-Ioan Popa, "Where Pathology, Microbiology and Virology Converge: Professor Victor Babeş," *Romanian Archives of Microbiology and Immunology* 80, no. 2 (2021).

51 "Kuopion läänin Maanwiljelyskokous Suonenjoella," *Savo*, September 9, 1890.

52 F., "Blodstallningssjukdomen," *Nya Pressen*, July 1, 1896.

53 Ilkka Alitalo, "Hellens, Oskar von," Kansallisbiografia online publication, *Studia Biographica* 4 (Helsinki: SKS, 1997), accessed March 20, 2023, http://urn.fi/urn:nbn:fi:sks-kbg-006842; Theodor Mikael Scheinin, "Ali Krogius," Kansallisbiografia online publication. *Studia Biographica* 4 (Helsinki: SKS, 1997), http://urn.fi/urn:nbn:fi:sks-kbg-006168.

observations reinforced recent results of international researchers, who had demonstrated that small parasites were able to destroy the red blood cells of diseased bovines.[54]

Another crucial question concerned how the parasites were transmitted to bovines. In America, some stockmen and veterinarians had noted as early as the 1860s that Texas fever was only prevalent in areas inhabited by ticks, but for a long time these observations were disregarded.[55] However, various scientific experiments began to support the connection between ticks, parasites and the disease. This claim was supported by the studies of Theobald Smith and F.L. Kilborne from the 1880s to the 1890s,[56] and promulgated, for example, in a talk at a biological society meeting in Washington D.C. in February 1890.[57] Yet, it took years before a scientific consensus was reached vis-à-vis the role of ticks as vectors. In this regard, Robert Koch's experiments in East Africa in 1897 proved to be essential.[58]

When the Finnish National Board of Health published a booklet by von Hellens on Redwater fever in 1901, the readers were offered a detailed description of the parasites and their actions. It was demonstrated how they penetrated red blood cells and destroyed them, thereby causing haemoglobin to dissociate and exit from the body in urine. Even the appearance of the microscopic parasites was described: they were said to resemble roundish cysts or to be the shape of a pear. Thus, readers were able to have a visual image of the microscopic creatures that they could not see. With some caution, von Hellens informed his readers that ticks probably infect bovines. It was well known that cows often had a great number of ticks that sucked their blood. Now scientific experiments confirmed that the bodies of ticks contained similar parasites as in the blood of diseased cows.[59] Thus, the agency of ticks was gradually revealed – in principle – as not only was new knowledge disseminated in booklets published by the National Board of Health, but also newspapers and journals circulated abbreviated versions of this information. Yet, in many instances the menace of ticks went undetected and they were able to continue their lives undisturbed for decades.

54 F., "Blodstallningssjukdomen," *Nya Pressen*, July 1, 1896.

55 Cheville, *Pioneer Science*, 80–81.

56 Hutson, "Texas Fever in Kansas," 93.

57 Cooper Curtice, "The Biology of the Cattle Tick," *The Journal of Comparative Medicine and Veterinary Archives* 12, no. 7 (1891): 317, https://www.ncbi.nlm.nih.gov/pmc/articles/PMC9261720/pdf/jcmvetarch132938-0001.pdf.

58 Köhler and Köhler, "Zentralblatt für Bakteriologie," 237.

59 Oskar von Hellens, *Verenheitosta*, 5–7; O. von Hellens, "Verenheitto eli punatauti raawaseläimissä," *Mikkelin Sanomat*, August 10, 1901.

For example, when the agronomist and teacher of agriculture Evert von Konow summarized the content of von Hellens' booklet in *Pellervo*, the journal of a society promoting cooperative action, he listed the two established explanations for the cause of the disease: namely, grass in wet meadows and puddles of water in swamps and bogs. Only after these factors did he briefly mention ticks. It seems that he was not totally convinced of von Hellens' views regarding ticks or the requisite medicines needed to treat Redwater fever. Von Hellens recommended modern quinine, but von Konow preferred old medicinal solutions. He justified his views by recourse to practical experience: he had taken care of a herd of 180 bovines for seven years in a disease-stricken region.[60] Thus, he placed more value on his own experience than in von Hellens' scientific experiments and preferred long-time suspects, that is, damp meadows and unfit water, to ticks.

There is no reason to belittle the practical experiences and observations of contemporaries. They may often have led to conclusions that were later reversed with the help of new research instruments and methods. Yet, the validity of some older beliefs are still being confirmed. For example, the perception that young animals suffer less severely from Redwater fever and were subsequently more resistant than those adult individuals who arrived from disease-free regions was often repeated in contemporary texts and is nowadays part of the confirmed characteristics of the disease.[61]

If a teacher of agriculture was slow to accept the role of ticks as crucial vectors, then maybe it should not be surprising that smallholders did not readily accept the idea. Among dozens of references to Redwater fever in the oral history collection of the Finnish Literature Society, only one points to ticks. In 1910, a collector of folk culture was told in Korpilahti, in central Finland, that Redwater fever appears only where there are ticks.[62] In the following decades, more information regarding ticks as vectors was directed to the population in numerous newspaper articles and booklets,[63] but with a limited degree of success. Still in 1929, decades after scientific consensus on the topic had been established, Kaarlo A. Zinck, a veterinarian of the breeding association of Eastern Finland, complained that many cattle owners believed that

60 E. v. K-w. [Evert von Konow], "Punatauti nautakarjassa," *Pellervo* 2, no. 6 (1901): 179–183.

61 Finnish Food Authority, "Punatauti."

62 Korpilahti, J.W. Kotikoski 755. 1910. FLSA, FH, RWF.

63 See, for example, Ingvald Lieberkind, *Vaarallisia vieraita. Eläimet tartunnanlevittäjinä ja taudinaiheuttajina*, trans. Heikki Väänänen (Helsinki: Otava, 1929), 151–153; J.E. Aro, *Kuvauksia eläinten elämästä* (Helsinki: Valistus, 1922), 37–39.

Redwater fever was caused by poisonous plants, putrid drinking water or poison used to kill foxes.[64]

Alder trees serve to show the complexity and hybridity of the search for culprits. In assessing their role, quite different theories merged. In Finnish folk culture, alders had traditionally been connected with Redwater fever because of their reddish color. Professionals who did not share a belief in folk healing also suspected alders, but for different reasons: in the watery places in which Redwater fever appeared, there were often plenty of alders. It is not always clear whether it was the alder itself that was considered to be dangerous, or whether the tree merely signaled a hazardous environment. Whatever the case, there was long-lasting suspicion and a dislike of alders. When scientists and veterinarians began to trumpet the key part played by ticks in Redwater fever infections, a new reason to abhor alders appeared as it was believed that the ticks were especially fond of living in alders.[65] Nowadays, scientists point to tall grass and bushes as the environments in which ticks quest a host, not trees. However, the Finnish word for ticks used to be *puuntäi*, which literally means "a louse from a tree." Furthermore, there was yet another reason to dislike alders. They were accused of destroying meadows as they are particularly fast-growing and produce large quantities of suckers. They are also capable of surviving quite extreme cutting, which made it laborious to maintain open meadows. Model state pastures were used as encouraging examples of how to dispose of alders.[66]

Zinck crystallized the conceptions of veterinarians and the authorities when he demanded that cattle owners should pay more attention to ticks. He maintained that it was the fault of farmers if their cows fell sick. He instructed them to drain meadows and to "destroy willows and alders," thereby making the terrain uninhabitable for ticks. When the ticks were eradicated, he argued that Redwater fever would also disappear. Zinck also referred to the modernization of animal husbandry practices as the safest solution: to avoid meadows altogether and to feed cows in cowsheds or to keep them tethered. In the meantime, when both ticks and Redwater fever still appeared, Zinck recommended the careful removal of all ticks from cows and that that the former

---

64 Zinck, *Nautakarjan tarttuva punatauti*, 2–5.

65 "Punatautiin sairastuu maassamme vuosittain," *Uusi Suomi*, August 21, 1935; "Eläintauteja on tänä kesänä ollut Satakunnassa vähän," *Satakunnan Kansa*, August 20, 1939; Ällä Pee, "Kellotkin," *Maaseudun Tulevaisuus*, August 26, 1923; E.K. Keränen, "Punatautiin on kuluvana kesänä kuollut erittäin paljon nautaeläimiä," *Suomen Sosiaalidemokraatti*, July 26, 1935. See also the chapter by Otto Latva in this book.

66 "Laitumen raivauksesta," *Liitto*, April 30, 1927; F.J.T., "Laitumen raivauksesta," *Loviisan Sanomat*, July 5, 1929; "Valtion näytelaiduntilat," *Savo*, November 23, 1930.

should then be incinerated.[67] In this authoritative view, ticks were the downright enemies of cows and humans and his exhortation was to eliminate them and their supposedly favorite habitat – alders.

5 Conclusion

An examination of Redwater fever in Finland permits us to discover the presence of ticks and their agency in the past at a time when humans were not very interested in them. Before the role of ticks as vectors of Redwater fever became known, they were not considered perilous and explicit references to them were rare. However, their presence and their actions are disclosed by outbreaks of Redwater fever, which, in turn, have been documented. Despite human authors not placing blame on ticks, these descriptions open a view to multispecies co-living at the time when ticks were clandestine agents in meadows.

In the late nineteenth- and early twentieth century, ticks had an excellent opportunity to latch on to large mammals as the number of cattle increased and they tended to be fed in meadows, which were often wet. Countless things happened because of tick bites. If ticks became infected by Babesia parasites they could subsequently kill cows if they became attached to them. Their influence on humans was also notable, although less direct. No doubt humans could also suffer from troublesome inflammations, but Redwater fever was a bovine disease. Nonetheless, ticks also affected human lives in many ways as the loss of cattle could be disastrous for a farmer. Ticks made humans act. This could take the form of casting spells in the cowshed, treating cattle with different mixtures, draining meadows, reporting on outbreaks of the disease to governors, writing newspaper articles and arranging meetings, as well as undertaking scientific research in laboratories and in the field.

In the history of the interaction between ticks, cattle and humans, there are no straightforward, simple paths. Even though the mechanism of Redwater fever infection was discovered by scientists in the 1880s and 1890s, thereby revealing the agency of ticks and Babesia protozoans, other assumptions and practices continued to exist for decades. From the point of view of everyday history, uncovering the agency of ticks was a slow process with intriguing hybrids of beliefs and conceptions that derived from divergent ways of perceiving the world. Giving red lemonade to a cow bitten by a tick resonated with age-old traditions in a modernizing world.

---

67 Zinck, *Nautakarjan tarttuva punatauti*, 7, 11.

Bibliography

Archival Sources

Finnish Literature Society Archive (FLSA)

Newspapers and Periodicals

Aamulehti
Hämeen Sanomat
Hämäläinen
Karjalatar
Keski-Suomi
Liitto
Loviisan Sanomat
Maaseudun Tulevaisuus
Mikkelin Sanomat
Nya Pressen
Pellervo
Satakunnan Kansa
Savo
Savonlinna
Suomalainen Wirallinen Lehti
Suomen Sosialidemokraatti
Tampereen Sanomat
Tapio
Uusi Suometar
Uusi Suomi
Wiipurin Sanomat

Literature

Alitalo, Ilkka. "Hellens, Oskar von." Kansallisbiografia online publication. *Studia Biographica* 4. Helsinki: SKS, 1997. Accessed November 29, 2023. http://urn.fi/urn:nbn:fi:sks-kbg-006842.

Aro, J.E. *Kuvauksia eläinten elämästä.* Helsinki: Valistus, 1922.

Bidrag till Finlands officiella statistik XI. Medicinalverket, Medicinalstyrelsens berättelse för år 1900. Helsingfors, 1901 [and the subsequent yearbooks 1901, 1905, 1908, 1909, 1910].

Cheville, Norman F. *Pioneer Science and the Great Plagues: How Microbes, War, and Public Health Shaped Animal Health.* West Lafayette, IN: Purdue University Press, 2021.

Curtice, Cooper. "The Biology of the Cattle Tick." *The Journal of Comparative Medicine and Veterinary Archives* 12, no. 7 (1891): 313–319. https://www.ncbi.nlm.nih.gov/pmc/articles/PMC9261720/pdf/jcmvetarch132938-0001.pdf.

Despret, Vinciane. "From Secret Agents to Interagency." *History and Theory* 52 (2013): 29–44.

Finnish Food Authority. "Punatauti." Accessed November 29, 2023. https://www.ruokavirasto.fi/elaimet/elainten-terveys-ja-elaintaudit/elaintaudit/naudat/punatauti/.

Häkkinen, Antti and Henrik Forsberg, "Finland's Famine Years of the 1860s: A Nineteenth-Century Perspective." In *Famines in European Economic History: The Last Great European Famines Reconsidered,* edited by Declan Curran, Lubomyr Luciuk and Andrew G. Newby, 99–123. London: Routledge, 2015.

Haraway, Donna. *When Species Meet.* Minneapolis: University of Minnesota Press, 2008.

Hellens, Oskar von. *Verenheitosta (punataudista) raavaseläimissä.* Helsinki: Lääkintöhallitus, 1906.

Helminen, Katri. *Eläinlääkärinä kolmella vuosisadalla. Suomen eläinlääkäriliitto 1892–2017*. Helsinki: Fennovet, 2017.

Hutson, Cecil Kirk. "Texas Fever in Kansas, 1866–1930." *Agricultural History* 68, no. 1 (1994): 74–104.

Kulha, Niko, Kalle Ruokolainen, Eero J. Vesterinen, Maija Lamppu, Tero Klemola and Jani J. Sormunen. "Does Environmental Adaptation or Dispersal History Explain the Geographical Distribution of Ixodes Ricinus and Ixodes Persulcatus Ticks in Finland?" *Ecology and Evolution* 12, no. 12 (2022). doi:10.1002/ece3.9538.

Keisarillisen Majesteetin Armollinen Julistus, koskewa mitä tarttuvain eläintautien estämiseksi ja häätämiseksi pitää waariin otettaman. Annettu Helsingissä, 6 p:nä kesäkuuta 1864. *Suomen Suuriruhtinanmaan Asetus-Kokous*. Helsinki: Keisarillisen Senaatin kirjapaino, No. 18, 1864.

Köhler, Michael and Werner Köhler. "Zentralblatt für Bakteriologie – 100 years ago. Victor Babeş and enzootic haemoglobinuria of cattle." *International Journal of Medical Microbiology* 293, no. 4 (2003): 233–239.

Latour, Bruno. *The Pasteurization of France*. Cambridge, MA: Harvard University Press, 1988.

Latour, Bruno. *Reassembling the Social: An Introduction to Actor-Network-Theory.* Oxford: Oxford University Press, 2005.

Lieberkind, Ingvald. *Vaarallisia vieraita. Eläimet tartunnanlevittäjinä ja taudinaiheuttajina*, translated by Heikki Väänänen. Helsinki: Otava, 1929.

Mäkelä-Alitalo, Anneli. "Karjataudit ja eläinlääkintä." In *Suomen maatalouden historia I*, edited by Viljo Rasila, Eino Jutikkala and Anneli Mäkelä-Alitalo, 579–597. Helsinki: SKS, 2003.

Nance, Susan. "Introduction." In *The Historical Animal*, edited by Susan Nance, 1–16. Syracuse, NY: Syracuse University Press, 2015.

Östman, Ann-Catrin. "Mekanisoinnin ensimmäinen aalto." In *Suomen maatalouden historia II*, edited by Matti Peltonen, 38–43. Helsinki: SKS, 2004.

Pearson, Chris. "Dogs, History, and Agency." *History and Theory* 52, no. 4 (2013): 128–145.

Peltonen, Matti. "Uudet kaupallistumisen muodot." In *Suomen maatalouden historia II*, edited by Matti Peltonen, 92–107. Helsinki: SKS, 2004.

Piela, Ulla. "'Konsti elää kauwwan.' Parantaminen Suomessa varhaismodernilta ajalta nykypäivään." In *Kiistellyt tiet terveyteen. Parantamisen monimuotoisuus globaalihistoriassa*, edited by Markku Hokkanen and Kalle Kananoja, 83–129. Helsinki: SKS, 2017.

Popa, Mircea-Ioan. "Where Pathology, Microbiology and Virology Converge: Professor Victor Babeş." *Romanian Archives of Microbiology and Immunology* 80, no. 2 (2021): 179–188.

Räsänen, Tuomas and Taina Syrjämaa, eds. *Shared Lives of Humans and Animals. Animal Agency in the Global North.* London: Routledge, 2017.

Scheinin, Theodor Mikael. "Ali Krogius," Kansallisbiografia online publication. *Studia Biographica* 4. Helsinki: SKS, 1997. Accessed November 29, 2023. http://urn.fi/urn:nbn:fi:sks-kbg-006168.

Schuurman, Nora and Taina Syrjämaa. "Shared Spaces, Practices and Mobilities: Pet – Human Life in Modern Finnish Homes." *Home Cultures* 18, no. 2 (2021): 173–194. doi:10.1080/17406315.2021.1963611.

Šimo, Ladislav, Maria Kazimirova, Jennifer Richardson and Sarah I. Bonnet. "The Essential Role of Tick Salivary Glands and Saliva in Tick Feeding and Pathogen Transmission." *Frontiers in Cellular and Infection Microbiology* 7 (2017): article 281. doi:10.3389/fcimb.2017.00281.

Sormunen, Jani. *Questing Ticks, Hidden Causes: Tracking Changes in Ixodes Ricinus Populations and Associated Pathogens in Southwestern Finland.* Turku: University of Turku, 2018.

Specht, Joshua. *The Red Meat Republic: A Hoof-to-Table History of How Beef Changed America.* Princeton University Press, 2019.

Syrjämaa, Taina. "Eläimet, eläinperäiset tuotteet ja edistysusko." *Tunteva tuote. Kuinka eläimistä tuli osa teollista tuotantoa?*, edited by Taija Kaarlenkaski and Otto Latva, 27–47. Tampere: Vastapaino, 2022.

Syrjämaa, Taina. "Multispecies Mobilities and Human Belief in Progress." *Animal Industries. Nordic Perspectives on the Exploitation of Animals since 1860*, edited by Taina Syrjämaa, Marja Jalava, Taija Kaarlenkaski, Otto Latva, Eeva Nikkilä and Tuomas Räsänen, 25–43. Berlin: De Gruyter 2024.

Thorsheim, Peter. *Inventing Pollution: Coal, Smoke, and Culture in Britain Since 1800.* Athens, OH: Ohio University Press, 2017.

Vallius, Antti. *Kaunis Suomi. Maaseutumaisemakuvaston historiaa 1800-luvulta EU-Suomeen.* Helsinki: Tammi, 2015.

Vihola, Teppo. *Leipäviljasta lypsykarjaan. Maatalouden tuotantosuunnan muutos Suomessa 1870-luvulta ensimmäisen maailmansodan vuosiin.* Helsinki: SHS, 1991.

Vihola, Teppo. "Pärjääkö pienviljelys?" In *Suomen maatalouden historia II*, edited by Matti Peltonen, 135–216. Helsinki: SKS, 2004.

World Organization for Animal Health. "Bovine Babesiosis." Accessed November 29, 2023. https://www.woah.org/fileadmin/Home/eng/Animal_Health_in_the_World/docs/pdf/Disease_cards/BOVINE_BABESIOSIS.pdf.

Zinck, Kaarlo A. *Nautakarjan tarttuva punatauti. Toimenpiteet sen levenemisen ehkäisemiseen*. Kuopio: Itä-Suomen karjanjalostusyhdistys, 1929.

CHAPTER 10

Social Construction of Tick-Borne Diseases from the 1950s to the Twenty-First Century: a View from the History of Medicine

Suvi Rytty

1 Introduction

This chapter examines changes in the public perceptions of tick-borne diseases as a health risk to humans since the 1950s. The topic is approached from the perspective of the history of medicine and health in the Finnish context. As stated in the introduction to this book, the most important tick-borne diseases in Finland are borreliosis (Lyme disease), a bacterial infection affecting the skin, nervous system, joints, and heart, and tick-borne encephalitis (TBE), a viral infectious disease that attacks the central nervous system and can even lead to death.[1] These are emerging infectious diseases that are increasing in number and geographic range. This is mostly because climate change and global warming have allowed ticks to spread further north.[2]

Both TBE and borreliosis are caused by pre-existing pathogens that were previously unknown or very little was known about them. As the previous chapter shows, the role of ticks as vectors of animal diseases had already been established at the end of the nineteenth century,[3] but it was not until the 1950s that Finns came to understand that ticks were also to blame for the mysterious brain inflammation afflicting humans that is known today as TBE.[4] Tick-borne diseases have been seen as an example of new risks generated by technologies and human activities, which Anthony Giddens and Ulrich Beck have argued threaten modern societies.[5] However, the health risk they pose has not always

1 Jukka Hytönen et al., "Puutiainen ja sen levittämät taudit," *Duodecim Terveyskirjasto,* June 9, 2021, https://www.terveyskirjasto.fi/kpp00002.

2 I. Rochlin and A. Toledo, "Emerging Tick-Borne Pathogens of Public Health Importance: A Mini-Review," *Journal of Medical Microbiology* 69, no. 6 (2020), doi:10.1099/jmm.0.001206.

3 See Taina Syrjämaa's chapter in this volume.

4 "Kumlingesjukan är fortfarande ett mysterium," *Hufvudstadsbladet,* December 6, 1950; "Punkit levittävät yhä enemmän sairauksia", *Iltalehti,* July 8, 2019.

5 P. Peretti-Watel, J. Ward, R. Lutaud and V. Seror, "Lyme Disease: Insight from Social Sciences," *Médecine et Maladies Infectieuses* 49 (2019), doi:10.1016/j.medmal.2018.12.005.

been perceived as high as the public currently sees it.[6] Moreover, the public and scientific experts do not necessarily share the same perception of the given risks.[7] For example, according to the Finnish Institute for Health and Welfare, the actual risk of contracting a tick-borne disease in Finland today is lower than the general public fears.[8]

The obvious explanation for the increased fear of tick-borne diseases among the Finns is the proliferation of both ticks and tick-borne diseases.[9] However, this may not be the whole picture, for although diseases are basically biological conditions, our understanding of them is also influenced by social and cultural meanings attached to them.[10] According to Charles E. Rosenberg, disease does not exist as a social phenomenon in some ways until we agree that it does. This process involves perceiving, naming, and responding to it.[11] The mode of historical analysis that views non-biological factors, such as beliefs, economic relationships or societal institutions, influencing our understanding of particular illnesses has been referred to as the social construction of disease.[12]

The purpose of this chapter is to examine, how these kinds of non-biological factors have contributed to making tick-borne diseases appear more frightening from the 1950s up to the early 2000s. The main shaper of public opinion has always been the media and, in earlier decades, newspapers in particular. This chapter analyses the image of tick-borne diseases that has been conveyed to the public by the Finnish newspaper press since the 1950s, and the social factors, ranging from scientific knowledge and individual experience to health policy priorities, which have influenced public risk perceptions and subsequent actions concerning tick-borne diseases.

---

6 "Virusten leviämistä tutkitaan Tvärminnessä: Ahvenanmaan puutiainen levittää Kumlingentautia," *Helsingin Sanomat*, August 25, 1976.

7 Peretti-Watel et al., "Lyme Disease," 135.

8 Jussi Sane, "Riski saada punkilta tauti on pieni, muttei merkityksetön," *THL Blog*, May 17, 2017, accessed September 21, 2023, https://blogi.thl.fi/riski-saada-punkilta-tauti-on-pieni-muttei-merkitykseton/.

9 See for example Otto Latva's chapter in this volume.

10 Sigrun Olafsdottir, "Social Construction and Health," in *Medical Sociology on the Move: New Directions in Theory*, ed. William C. Cockerham (Dordrecht: Springer, 2013), 43.

11 Charles E. Rosenberg, "Disease in History: Frames and Framers," *The Milbank Quarterly* 67, Supp. 1 (1989): 1–2.

12 Robert Aronowitz, "Framing Disease: An Underappreciated Mechanism for the Social Patterning of Health," *Social Science & Medicine* 67, no. 1 (2008): 2. In later socio-historical scholarship "social construction" has often been replaced by the less programmatically charged metaphor "framing." This is due to the unwanted connotations sometimes associated with constructionist arguments, such as dated cultural relativism and reflexive opposition to biomedicine. See Rosenberg, "Disease in History," 3.

The main source material consists of newspapers digitized by the National Library of Finland. This newspaper collection contains an almost uninterrupted series of newspapers published in Finland from the late 1700s.[13] The electronic archive of *Helsingin Sanomat*, Finland's largest subscription newspaper in terms of circulation, has been utilized for those issues that are missing from the National Library collection (1980–2016).[14] Relevant source material has been located by using search words related to ticks and tick-borne diseases. The research period begins in 1950, when the first news about ticks being able to spread a human disease was published, and extends to the early 2000s, by which time the public debate on tick-borne diseases had begun to take its current form.

2 From a Mystery to a Disease

In 1950 a mysterious illness made the headlines in Finland after it affected eleven people on the small island of Kumlinge in the Åland archipelago. This previously unknown disease, in which those afflicted suffered from a variety of symptoms, including fever, headache, nausea, and a stiff neck, seemed to have appeared out of nowhere.[15] In reality, the first two cases had already been detected in 1942, and by 1950 thirty-nine Åland islanders had contracted the disease.[16] As the cause of the disease was not known, it was named after the place it was first detected; hence Kumlinge disease.[17] Due to the increase in cases in 1950, the disease was initially feared to be something of an epidemic – especially as it caused symptoms similar to polio. The regional doctor in Kumlinge speculated in a newspaper interview in 1953 that the disease could be some kind of new variant or subtype of poliomyelitis,[18] which caused widespread epidemics in western countries, including Finland, at this time.[19]

13 The National Library of Finland, "Search from digital materials."

14 *Helsingin Sanomat*, "Arkisto"; *Helsingin Sanomat*, "HS Aikakone."

15 "Kumlingesjukan är fortfarande ett mysterium"; "Epidemisk magsjuka i Kumlinge," *Hufvudstadsbladet*, December 31, 1950.

16 "Lärt och populärt," *Hufvudstadsbladet*, February 12, 1956; Nils Oker-Blom, "Kumlinge Disease: A Meningo-Encephalitis Occurring in the Aaland Islands," *Annales Medicinae Experimentalis et Biologiae Fenniae* 34 (1956): 309–310.

17 "Lärt och populärt"; Oker-Blom, "Kumlinge Disease," 309–310.

18 "Kumlingesjukan är fortfarande ett mysterium"; "Epidemisk magsjuka i Kumlinge"; "Kumlinge-sairaus on lapsihalvauksen tapainen sairaus," *Helsingin Sanomat*, September 10, 1953; "Kumlingesjuka – polioart?," *Hufvudstadsbladet*, September 10, 1953; "Kumlingesjukan avart av barnförlamning," *Ny Tid*, September 10, 1953.

19 Jacob Heller, *The Vaccine Narrative* (Nashville: Vanderbilt University Press, 2008), 3–5.

The true nature of Kumlinge disease was discovered by the virologist Nils Oker-Blom (1919–1995) and his research team from the Department of Virology at the University of Helsinki. Because no children were affected, and the incidence of paralysis was very low, Oker-Blom suspected the disease to be viral encephalitis (inflammation of the brain) spread by ticks,[20] known today as tick-borne encephalitis (TBE). TBE was first recognized and medically described in Austria in 1931 as "meningitis serosa epidemica" of unknown origin.[21] From the early 1930s, an acute central nervous system disease with a high death rate was also recorded in the Far East of the Soviet Union. The first scientific expedition team was sent to the Far-Eastern taiga in 1937 and it elucidated the etiology of this new disease, discovered a previously unknown virus, and determined the vector as being the tick. The second expedition team in 1938 developed an inactivated vaccine against the disease. The Soviet expedition team called the disease "spring epidemic encephalitis" or "tick-borne encephalitis".[22]

Investigating a previously unknown virus was not easy. The work of the Soviet expedition teams was undertaken under basic conditions in the remote taiga, and several team members were infected either through tick bites or while working with the virus in the laboratory. Some even died.[23] The slow accumulation of scientific knowledge is also reflected in the fact that the classification of the virus and its subtypes has varied over time, as have the names of both the disease and the virus. The knowledge gleaned in the 1950s about the disease, which formed the basis of Oker-Blom's work, differs from that of the present day.[24]

---

20 Markus Brummer-Korvenkontio, *Virusten ja prionien luonnonhistoriaa: Myyräkuumeesta SARS:iin, Ebolasta AIDS:iin ja arboviruksista lintuinfluenssaan* (Helsinki: Helsinki University Press, 2007), 238.

21 Christian Kunz, "TBE vaccination and the Austrian experience," *Vaccine* 21, no. 1 (2003): 50–51, doi:10.1016/S0264-410X(02)00813-7.

22 Vladimir I. Zlobin, Vanda V. Pogodina and Olaf Kahl, "A Brief History of the Discovery of Tick-Borne Encephalitis Virus in the Late 1930s (Based on Reminiscences of Members of the Expeditions, Their Colleagues, and Relatives)," *Ticks and Tick-Borne Diseases* 8, no. 6 (2017): 813–817, doi:10.1016/j.ttbdis.2017.05.001.

23 Zlobin, Pogodina and Kahl, "A Brief History," 813–817.

24 Today it is known that tick-borne encephalitis virus (TBEV) is a member of the virus genus Flavivirus, which includes approximately eighty species. Most of these viruses are primarily transmitted by infected arthropods (mosquitos or ticks), and therefore classified as arboviruses. In 1962, Clarke proposed the division of TBEV into two subtypes: Russian spring-summer encephalitis (RSSE) and Central European encephalitis (CE or CEE). In 1999, TBEV was segregated into three subtypes according to their primary geographical distribution: European, Siberian, and Far-Eastern (former RSSE), although in recent years additional subtypes have been proposed. Due to geographical, pathogenetic, and environmental peculiarities, Louping ill virus (LIV) is not classified as a subtype, but a related tick-borne flavivirus, although the European subtype of TBEV is closer to LIV than to other

For example, due to the pioneering work of the Soviet researchers, the disease and the virus were initially known as "Russian spring-summer encephalitis" (RSSE).[25] In 1943 it was demonstrated that RSSE had a close antigenic relationship with Louping ill (LI), which is a tick-borne viral disease affecting mainly sheep that had been described in Scotland since the eighteenth century.[26] The first European tick-borne encephalitis virus was isolated in Czechoslovakia in 1948 and in 1950 this Central European encephalitis (CE/CEE) was shown to be closely related to RSSE. In the 1950s RSSE, LI and CEE were understood to form a "Russian spring-summer complex" or "Russian spring-summer encephalitis – Louping ill group".[27]

In the 1950s, Finland was still recovering from the Second World War and a lack of funds made it difficult at first to undertake virological research. Only with the help of grants from the Sigrid Juselius Foundation and the Samfundet Folkhälsan was it possible to initiate the investigation of Kumlinge disease in the summer of 1954.[28] Several medical-biological expeditions were made to Kumlinge Island to collect both blood samples and ticks. The first research results, based among other things on antibody tests, were published in 1956. According to this data, Kumlinge disease seemed to belong to the Russian spring-summer encephalitis – Louping ill group. Further confirmation came from the fact that encephalitides belonging to this group had been detected in the neighboring countries Soviet Union and Sweden.[29] Thus, a fundamentally

---

TBEV subtypes in terms of genetic distance. See Andrei A. Deviatkin et al., "TBEV Subtyping in Terms of Genetic Distance," *Viruses* 12, no. 11 (2020), doi:10.3390/v12111240; S.I. Belikov et al., "The Relationship Between the Structure of the Tick-Borne Encephalitis Virus Strains and Their Pathogenic Properties," *PLoS ONE* 9, no. 4 (2014): 1–4, https://doi.org/10.1371/journal.pone.0094946.

25 Paul Heyman et al., "A Clear and Present Danger: Tick-Borne Diseases in Europe," *Expert Review of Anti-Infective Therapy* 8, no. 1 (2010), doi: 10.1586/eri.09.118.

26 William L. Pond, Russ B. Sudie and Joel Warren, "The Russian Spring-Summer Encephalitis and Louping Ill Group of Viruses: Relationship of European and Asiatic Strains of Russian Spring-Summer Encephalitis Viruses and Louping Ill Virus," *The Journal of Infectious Diseases*, vol. 93, no. 3 (1953); C.L. Jeffries et al., "Louping Ill Virus: An Endemic Tick-Borne Disease of Great Britain," *Journal of General Virology* 95, no.5 (2014), doi:10.1099/vir.0.062356-0.

27 Oker-Blom, "Kumlinge Disease," 317; H. Williams and H. Thorburn, "The Serological Response of Sheep to Infection with Louping-Ill Virus," *The Journal of Hygiene* 59, no. 4 (1961); Heyman et al., "A Clear and Present Danger," 33–50.

28 "Lärt och populärt".

29 Oker-Blom, "Kumlinge Disease," 313–317; "Lääkäri saapuu lasten luo saaristoon 'sokerilaivalla'," *Suomen Sosiaalidemokraatti*, July 22, 1956; Hex, "Ihmisiä työnsä ääressä," *Suomen Sosiaalidemokraatti*, December 16, 1956; Pond et al., "The Russian Spring-Summer Encephalitis," 294–300.

new concept of disease had been introduced in Finland – although the exact nature of the disease remained obscure as no virus had been isolated.[30] Thousands of ticks had been collected in Åland, but the problem was that most of them were still virus-free at that time. This was a pity, in Oker-Blom's view, because "the villainous role of the tick was considered proven."[31]

Once Kumlinge disease was identified, it was discovered that it probably also occurred elsewhere, such as in the Turku archipelago nearer to the mainland. This gave the impression that the disease was spreading eastwards.[32] However, the reports about encephalitis cases from hospitals and municipal doctors showed that tick-borne encephalitis might also be common in the Lappeenranta area in Eastern Finland near the border with the Soviet Union. This suggested that TBE might be spreading across the border from the USSR.[33] As long as there was no isolated virus, there was no certainty as to whether the tick-borne encephalitis in Finland was of the eastern ("Russian") or western ("Swedish") type. The distinction was crucial in that the eastern form was known to be much more severe and deadly, with estimated mortality rates of up to 10–30%, whereas the estimated mortality rate of the western form was only 1–2%.[34]

Throughout the 1950s, Kumlinge disease was primarily a concern for health authorities and virologists. It was initially unknown how and with what speed the disease spread, as well as how widespread it had already become. Hence, the severity it posed as a public health threat remained unknown. The press, however, seemed to be less interested in Kumlinge disease itself than in its scientific research. The discovery that the TBE-virus belonged to the group of arthropod-borne viruses (arboviruses) responsible for many dangerous global diseases gave research into Kumlinge disease international significance. For a young nation whose independence had been at stake in the recent war against the Soviet Union, international attention was much sought after and important.[35] In 1958 Oker-Blom received a large grant from the U.S.-based Rockefeller

30 Oker-Blom, "Kumlinge Disease," 317.

31 "Fästing ger sjukdom," *Nya Pressen*, June 25, 1958.

32 Hex, "Ihmisiä työnsä ääressä".

33 "Fästing ger sjukdom"; "Aivotulehdustutkimusta punkkien avulla," *Länsi-Savo*, August 16, 1958; "Punkkiretkikunnat etsivät aivokuumeen aiheuttajaa," *Ilta-Sanomat*, May 30, 1959.

34 "Forskningsunderstöd till Nils Oker-Blom," *Hufvudstadsbladet*, October 31, 1954; "Lärt och populärt"; "Punkkiretkikunnat".

35 Hex,"Ihmisiä työnsä ääressä"; "13 mmk till N. Oker-Bloms virusforskning," *Hufvudstadsbladet*, February 8, 1958; "Punkkien levittämää aivokuumetta tutkitaan," *Helsingin Sanomat*, July 6, 1958.

Foundation to study the viral inflammations of the brain in Finland, which the press interpreted as a sign of international recognition for Finnish virological research.[36]

With the help of this grant, Oker-Blom's team made a scientific expedition to the Lappeenranta area in Eastern Finland in July 1958 in order to investigate the geographical distribution of TBE. Finnish newspapers at the time presented the expedition as a sign that the dangerous Russian spring-summer encephalitis was spreading across the border from USSR into Finland.[37] Finns had historically become accustomed to the fact that serious epidemics, such as smallpox, usually spread to Finland from Russia.[38] Nevertheless, the public was reassured that the further west one went, the milder RSSE seemed to become, that is, the disease was less dangerous in Finland than in the Soviet Union.[39] This observation was proven correct when Oker-Blom's research group finally managed to isolate a tick-borne encephalitis virus (Kumlinge A52) from *Ixodes ricinus* tick in Åland in 1959. In the early 1960s, D.H. Clarke demonstrated that it represented a strain of Central European encephalitis (CE/CEE) and not of RSSE – both of which she classified as the subtypes of the tick-borne encephalitis virus (TBEV).[40]

3 A Multispecies Disease

Initially the knowledge that Kumlinge disease was spread by ticks and that the virus primarily circulated in the animal kingdom, instead of transmitting from one person to another, was presented as being reassuring when discussed in the public domain. It meant that the virus spread slowly and the disease

36 "13 miljoonan apuraha virustutkimuksiin Rockefelleriltä," *Ilta-Sanomat*, February 7, 1958; "Rockefeller-säätiö avustaa Suomen virustutkimusta," *Helsingin Sanomat*, February 8, 1958. See also Downs, "The Rockefeller Foundation Virus Program," *Annu Rev Med* (1982): 7, doi:10.1146/annurev.me.33.020182.000245.

37 "Aivotulehdustutkimusta punkkien avulla"; "Punkkiretkikunnat"; "Venäläinen kevätkesä leviämässä Eurooppaan," *Uusi Suomi*, September 18, 1958.

38 Uuno Winter, "Isosta rokosta ja rokotuksesta," *Terveydenhoitolehti* 24, no. 2 (1912): 19; Ensio Alho, "Onko rokotuspakko poistettava?," *Terveydenhoitolehti* 41, no. 9 (1929): 120–121.

39 "Fästing ger sjukdom"; "Punkkien levittämää aivokuumetta tutkitaan"; "Venäläinen kevätkesä leviämässä Eurooppaan".

40 Delphine H. Clarke, "Further Studies on Antigenic Relationships Among the Viruses of the Group B Tick-borne Complex," *Bulletin of the World Health Organization* 31, no. 1 (1964); Richard Moreland Taylor (ed.), *Catalogue of Arthropod-Borne Viruses of the World: A Collection of Data on Registered Arthropod-Borne Animal Viruses* (U.S. Government Printing Office, 1967), 825–828; Brummer-Korvenkontio, *Virusten ja prionien luonnonhistoriaa*, 238.

occurred only sporadically in areas where infected ticks were abundant. By the mid-1960s, the relatively remote and sparsely-populated areas of Kumlinge and the rest of the Åland Islands, as well as the Turku archipelago and the eastern border region had been identified as the main areas of the outbreak.[41] Hence, most of the Finns seemed to be safe from the disease. Nevertheless, in a newspaper article in 1958 the public was for the first time directly warned about ticks that were able to spread certain kind of encephalitis, and especially city dwellers on vacation in Åland were recommended to avoid thickets and alder forests, where ticks were believed to thrive at the time.[42]

In the 1950s and 1960s, Kumlinge disease was constantly compared to the severe form of tick-borne encephalitis then occurring in the Soviet Union, which made Finnish tick-borne encephalitis appear very mild. The estimated mortality rate was "only" 1–2%, which was about the same as for polio.[43] On the one hand, this reflects an era when generally hazardous communicable diseases, such as polio, tuberculosis and measles, still formed part of everyday life, although their mortality rates had begun to decline in the 1950s with the expanding national vaccination program.[44] On the other hand, evaluating the risk of Kumlinge disease as minor also reflects the health policy of the time, which assessed the risk at the level of the population as a whole. The population losses after the Second World War had given rise to a debate on demographic policy in Finland, in which a large, healthy and productive population was seen as important for the existence of the nation. The severity or danger of diseases was weighed against their demographic significance, that is, the mortality they caused in the productive population.[45]

---

41 "Lärt och populärt"; "Punkkiretkikunnat etsivät aivokuumeen aiheuttajaa"; Juhani Lyy, "Kumlingen kumma tauti," *Suomen Kuvalehti*, December 5, 1964.

42 "Fästing ger sjukdom".

43 "Punkkiretkikunnat etsivät aivokuumeen aiheuttajaa".

44 Helene Laurent, *Asiantuntijuus, väestöpolitiikka, sota. Lastenneuvoloiden kehittyminen osaksi kunnallista perusterveydenhuoltoa 1904–1955* (Helsinki: Unigrafia, 2017), 270–284. In Finland, vaccination against tuberculosis began in 1941, but even in the 1950s tuberculosis was a common disease among all age groups. However, the number of deaths from the disease fell rapidly, thanks in part to new anti-tuberculosis drugs. Polio vaccination was introduced in Finland in 1957. Initially, many parents were reluctant to use the vaccine, as a manufacturing error in the United States in 1955 had resulted in 260 children being infected with polio. A measles vaccination was not introduced in Finland until 1975. Laurent, *Asiantuntijuus, väestöpolitiikka, sota*, 270–280; The Finnish Institute for Health and Welfare, "Milloin eri rokotukset ovat alkaneet Suomessa?," accessed November 7, 2023, https://thl.fi/fi/aiheet/infektiotaudit-ja-rokotukset/tietoa-rokotuksista/kansallinen-rokotusohjelma/milloin-eri-rokotukset-ovat-alkaneet-suomessa-.

45 Minna Harjula, *Terveyden jäljillä: Suomalainen terveyspolitiikka 1900-luvulla* (Tampere: Tampere University Press, 2007), 55–65.

Moreover, the news coverage of Kumlinge disease at the time was largely based on the opinions of scientific experts. They thought about a risk in terms of probability and severity, rather than in terms of the suffering the disease might cause to an individual.[46] Oker-Blom's team had discerned that even in the risk areas only a small proportion of ticks carried the virus, so the chance of infection was small for someone who had been bitten. The majority of people also had a very mild form of the disease. Indeed, in the 1960s the most severe form of TBE was thought to affect only about 20 people a year.[47] In addition, some researchers were reluctant "to make a big noise" about Kumlinge disease in order to prevent unnecessary panic.[48] As one researcher noted as late as 1976:

> Finnish Kumlinge disease is not very dangerous. Usually, it can be treated within 2–3 weeks in hospital and it leaves no lasting effects. [...] Only one in 200 ticks spreads the virus. Around one in four people who are bitten will become ill, the rest will simply get a lifelong vaccination against the disease.[49]

Nevertheless, the very fact that TBE was a zoonosis also raised new questions and concerns. Farmed animals were known to carry the virus even though they often did not have a clinically proven disease. There was a possibility that the virus could be excreted in milk and transmitted to humans, as had happened in Central Europe via unpasteurized goat's milk. In Finland, concern was directed toward dairy cows, for cattle farming still played an important role in the 1950s in the country's economic structure.[50] As the previous chapter in this volume demonstrates, cows were traditionally grazed in woodland pastures, where they could become riddled with a lot of ticks.[51] At the time, however, alimentary TBE infections in Finland could not be proven.[52]

46 Peretti-Watel et al., "Lyme Disease," 135.

47 "Punkkiretkikunnat"; "Sokeritaudista hypotermiaan," *Helsingin Sanomat*, September 28, 1959; "Möss föda för farliga fästingar," *Hufvudstadsbladet*, August 13, 1961; Lyy, "Kumlingen kumma tauti".

48 "Tutkijat eri mieltä vaarasta," *Uusi Suomi*, April 12, 1980.

49 "Virusten leviämistä tutkitaan Tvärminnessä".

50 Arvo M. Soininen, "Suomen maataloushistorian päälinjat," *Agricultural and Food Science* 38, no. 2 (1966): 103, doi:10.23986/afsci.71655.

51 Erkki Räikkönen, "Ei punkeista tainnut ennen tauteja tulla," *Helsingin Sanomat*, July 5, 2007.

52 "Fästing ger sjukdom"; "Lärt och populärt".

From the 1960s, the public debate expanded from Kumlinge disease to other arthropod-borne diseases. Attention was drawn to how arthropods, such as mosquitoes and ticks, spread dangerous diseases not only in tropical regions, but also in more northerly climes. In Finland, *Uukuniemi* virus had been found in ticks, and mosquitoes had been discovered to spread *Inkoo* virus.[53] Ticks do not travel very long distances on their own, but one of the questions discussed internationally was whether dangerous viral diseases could be carried from one area to another by ticks arriving with migratory birds. In Finland, the number of ticks arriving with migratory birds was estimated to be around 2,300,000 per year. However, on the basis of a random sample, it was concluded that the birds left their African ticks in Central Europe, and brought mainly European castor-bean ticks (*Ixodes ricinus*) to Finland. Nevertheless, it was feared that migratory birds arriving from the area of the Soviet Union, would bring with them taiga ticks (*Ixodes persulcatus*), which spread the dangerous Russian Spring Summer Encephalitis.[54]

From the 1960s ticks and the microbes they carried started to fuse in the public discussion making the former appear more dangerous than previously thought.[55] For example, *Suomen Kuvalehti* magazine reported that one tick carrying the encephalitis virus had a viral concentration high enough to kill 100,000 mice.[56] This made a tick sound like a deadly weapon. Sometimes ticks were even described as "microbial reservoirs" or "syringes loaded with pathogens."[57] In the public debate of the 1990s, a tick bite alone became deemed as being dangerous and was reflected in headlines, such as "a tick bite can be

53 Pekka Nuorteva, "Vaaralliset hyönteiset," *Kansan Uutiset*, November 1, 1966; "Punkeista voi saada aivokuumeenkin," *Helsingin Sanomat*, June 5, 1968; Pekka Saikku, "Hyttyset ja punkit virustautien levittäjinä," *Uusi Suomi*, July 14, 1968. Inkoo virus is one of the world's northernmost and most common arboviruses. It can cause flu-like illness, nausea, vomiting, confusion, stiff neck, headaches, drowsiness and convulsions, especially in children. Most infections are asymptomatic. Maija Rummukainen, "Hyttysten levittämät taudit lisääntyvät," *Lääkärilehti* 75, no. 24–33 (2020), https://www.laakarilehti.fi/tieteessa/katsausartikkeli/hyttysten-levittamat-taudit-lisaantyvat-3289/. In general, Uukuniemi virus has not been considered to be of public health significance. Gustavo Palacios et al., "Characterization of the Uukuniemi virus group (Phlebovirus: Bunyaviridae): evidence for seven distinct species," *Journal of Virology* 87, no. 6 (2013), doi:10.1128/JVI.02719-12.

54 Pekka Nuorteva, "Muuttolinnut tautien levittäjinä," *Uuden Suomen Viikkolehti*, June 7, 1964; "Punkeista voi saada aivokuumeenkin"; Saikku, "Hyttyset ja punkit".

55 "Punkki vaarallisten tautien levittäjänä," *Helsingin Sanomat*, July 9, 1960; Urpo Iivarinen, "Tunnettuja suomalaisia tiedemiehiä XXIX," *Maaseudun Tulevaisuus*, December 10, 1964.

56 Lyy, "Kumlingen kumma tauti".

57 Nuorteva, "Muuttolinnut tautien levittäjinä"; Tom Ståhlberg, Anssi Junnila ja Erkki Leppäkoski, *Apua! Punkki ja sen levittämät taudit* (Turku: Arkipelagia-seura, 1994), 17.

dangerous" or "a tick can bite you sick."[58] Interestingly, the Finnish Literary Society's survey on ticks (2019) also contains a few responses that mention a poisonous tick bite.[59]

4 Lyme Borreliosis Enters the Scene

In 1964, a Finnish professor of dermatology and venereology mentioned in a newspaper interview how a tick bite could result in a skin infection. This was a reddening ring starting around the tick bite and expanding month by month as the center healed at the same time.[60] The medical name for the rash was *erythema chronicum migrans*, and until the 1980s it was sometimes mentioned in newspaper articles as a disease that could follow a tick bite in addition to tick-borne encephalitis.[61] It was only in 1986 that a doctor in a medical column of the *Länsi-Savo* newspaper linked this ring-shaped skin infection with meningitis, nerve and joint pain and myocarditis and disclosed to the public that it was called Lyme disease. He thought the ring-shaped rash was a relatively common phenomenon in Finland, but Lyme disease itself was rare, with only a few cases having been diagnosed.[62]

However, recent research (2020) based on a subset of historical serum samples collected by the Finnish Institute for Health and Welfare between 1968–1972 has shown that the proportion of antibody-positive people for Lyme borreliosis was considerably higher in Finland in the late 1960s and early 1970s than in 2011. Borreliosis has therefore been a common infection in Finland for at least half a century.[63] The problem was that symptoms of borreliosis were not recognized until U.S. researchers diagnosed it as a separate condition for the first time in the mid-1970s in Lyme, Connecticut. At first, it was mistaken for juvenile rheumatoid arthritis. The identification of the pathogen as

58 "Puutiaisen purema aiheuttaa pysyviä oireita yli 200 suomalaiselle," *Helsingin Sanomat* April 14, 1990; Niina Lempiäinen, "Punkin purema voi olla vaarallinen," *Länsi-Savo*, August 16, 1990; Kati Marjakangas, "Jos metsään haluat mennä nyt…," *Apu*, June 16, 1995.

59 Finnish Literature Society, Questionnaire 2019, "Punkit tulevat," SKS 015, 018, 092b.

60 Iivarinen, "Tunnettuja suomalaisia tiedemiehiä XXIX".

61 Nuorteva, "Vaaralliset hyönteiset"; "Pakeneminen ei auta, vain karkote pelottaa hyttysiä," *Helsingin Sanomat*, July 5, 1974; Pirkko Kolbe, "Pieni itikka voi olla paha peto," *Helsingin Sanomat*, June 13, 1979.

62 "Kysy lääkäriltä," *Länsi-Savo*, May 24, 1986.

63 J. Cuellar et al., "Seroprevalence of Lyme Borreliosis in Finland 50 Years Ago," *Clin Microbiol Infect.* 26, no. 5 (2020), doi:10.1016/j.cmi.2019.10.003.

a spirochete (*Borrelia burgdorferi*) was made by Willy Burgdorfer and his team of researchers in 1981.[64]

As a matter of fact, the cutaneous manifestation of Lyme disease, erythema chronicum migrans (ECM) was already described by the Swedish dermatologist Arvid Afzelius (1857–1923) in 1910, and the Austrian dermato-venereologist Benjamin Lipschütz (1878–1931) in 1913. By 1955, clinical and epidemiological evidence showed that ECM was caused by a penicillin-susceptible bacterial agent transmitted by the *Ixodes ricinus* tick. But as ECM had earlier been studied by European dermatologists and "Lyme arthritis" later by U.S. rheumatologists, it was only after prospective studies that the U.S. investigators acknowledged that ECM and the subsequent rheumatological and neurological abnormalities were all related to *Borrelia burgdorferi* infection and were all manifestations of the same disease.[65]

In the late 1980s and early 1990s, it became clear that there were many more cases of borreliosis in Finland than had initially been thought. By the late 1990s the estimated number of cases had already risen to 2000–3000 per year.[66] Ticks carrying the causative pathogens of borreliosis started to be found almost everywhere in Finland except Lapland,[67] including the Helsinki metropolitan area, where about a fifth of Finns live.[68] In 1999, borreliosis was even referred to in the media as a new Finnish common disease (*kansantauti*).[69] Scientific experts tried to reassure the public that the proliferation of cases had more to do with improved diagnostics and increased public awareness than with the increase of the disease itself. Thus, there should be no cause for alarm

64 Robert A. Aronowitz, "Lyme Disease: The Social Construction of a New Disease and Its Social Consequences," *The Milbank Quarterly* 69, no. 1 (1991): 83–92.

65 Aronowitz, "Lyme Disease," 83–92; Franc Strle, Gerold Stanek and Klemen Strle, "Lyme Borreliosis: The European Perspective," in *Lyme Disease: An Evidence-Based Approach*, 2nd ed., ed. John J. Halperin (Wallingford & Boston, MA: CABI, 2018), 105–107.

66 Klaus A. Järvinen, "Kotoiset puutiaiset voivat levittää sairautta," *Etelä-Suomen Sanomat*, July 4, 1988; "Punkki aiheuttaa vuosittain tuhansia borrelioositartuntoja," *Etelä-Suomen Sanomat*, May 9, 1996.

67 Tiina Ylitalo, "Outojen vaivojen takana voikin olla punkinpurema," *Etelä-Suomen Sanomat*, March 23, 1997; Hannele Tulonen, "Punkkipuhelin valistaa, jos puutiainen puraisee pyhinä," *Helsingin Sanomat*, June 19, 1998; Mervi Turunen, "Punkit taas liikkeellä," *Helsingin Sanomat*, July 7, 1999.

68 Katri Himma, "Punkkia kärsästä kiinni," *Suomen Kuvalehti*, July 25, 1997; Juha Junttila and Miikka Peltomaa, "Punkki on kavala salamatkustaja," *Helsingin Sanomat*, July 13, 1998.

69 "TV1," *Helsingin Sanomat*, August 10, 1999.

or panic.[70] However, these attempts were not successful in preventing the tick scare of the 1990s.[71]

5 Symptoms Remain the Same – the Disease Becomes More Frightening

As tick-borne diseases are basically biological conditions, the symptoms and danger of the diseases have remained unchanged from the 1950s to the early 2000s. What has changed, however, is the perception of how serious and frightening the diseases are and how they are described at different times. For instance, in the 1950s Kumlinge disease was not considered dangerous because very few people contracted the severe form of the disease or died from it.[72] However, at the beginning of the 2000s, scientific experts considered tick-borne encephalitis to be more dangerous than borreliosis because TBE could cause paralysis and leave permanent after-effects. Moreover, about 1% of those infected died from the disease.[73] This partly reflected a refinement of the clinical picture of TBE, but it was also a result of a new focus on the harm that the disease caused to the individual instead of society as a whole. It reflects the change in health policy thinking that took place in the 1970s, when views that emphasized the interests of the nation started to decrease, while individualistic perspectives emphasizing the citizen became more common.[74]

The change in health policy thinking was also reflected in the news coverage of tick-borne diseases, which from the 1970s began to present TBE from an increasingly individual perspective that drew attention to the painful and devastating experiences of patients.[75] A growing trend in the 1990s involved

70 Lempiäinen, "Punkin purema"; Päivi Repo, "Puutiaistaudit eivät ole lisääntyneet," *Helsingin Sanomat*, August 24, 1992; Anna Paljakka, "Todellinen punkki kesä," *Helsingin Sanomat*, July 10, 1994; Pyry Lapintie, "Borrelioosirokotteen tehosta tietoja loppuvuonna," *Helsingin Sanomat*, November 17, 1995.

71 Juhani Partanen, "Seuraava potilas," *Länsi-Savo*, July 27, 1992; Riitta-Eliisa Laine, "Punkin pelko on uusi ilmiö," *Helsingin Sanomat*, July 26, 1994; Repo, "Puutiaistaudit eivät ole lisääntyneet".

72 "Punkkiretkikunnat".

73 Maria Annala, "Kostea sää voi aiheuttaa pahan punkkikesän," *Helsingin Sanomat*, June 23, 2004; Pyry Lapintie, "Ennusteet lupaavat jälleen ennätyksellistä punkkikesää," *Helsingin Sanomat*, June 24, 2005.

74 Minna Harjula, *Hoitoonpääsyn hierarkiat: Terveyskansalaisuus ja terveyspalvelut Suomessa 1900-luvulla* (Tampere: Tampere University Press, 2015), 245.

75 Maija-Liisa Peltonen, "Punkista voi saada aivotulehduksen," *Helsingin Sanomat*, August 1, 1979.

patients telling horror stories about the suffering they had endured from borreliosis. For example, a man with an exceptionally bad case of Lyme disease reported that he had been almost immobilized at the height of his illness. The disease destroyed retinal cells in his eyes and left him with permanent deterioration of hearing and a much poorer sense of balance.[76] Reports of personal tragedies increased the public's fear of tick-borne diseases, because the impact of a given piece of information on risk perception depends on how it is presented. For instance, striking personal stories have a greater impact on an audience than the more "objective" pieces of information disseminated by health authorities.[77]

Aronowitz, in his study on Lyme borreliosis, has drawn attention to how newspaper accounts of Lyme disease are typically very straightforward, thereby leaving the risk of serious symptoms unqualified. Instead of describing severe symptoms as occurring rarely or in a minority of patients, the media usually gives extensive coverage to more worrisome features of the disease, for a new and serious disease is a better selling point.[78] Accordingly, the Finnish newspaper *Helsingin Sanomat* presented a rather gloomy picture of tick-borne diseases in 2007:

> Two tick-borne diseases, Lyme disease and tick-borne encephalitis, can cause meningitis, heart problems, paralysis, eye inflammation, arthritis, permanent tremors, memory loss, loss of balance, profound fatigue, and concentration problems even years after infection. Tick-borne encephalitis can lead to death. There is no cure when bitten by a tick carrying tick-borne encephalitis. The vaccination must be given before the tick bites. There is no vaccine for Lyme disease, only antibiotics taken afterwards are effective.[79]

Personal stories drew the readers' attention to the fact that even if the risk of being infected by a tick bite was relatively small in Finland, the consequences for the infected individual could be tragic. This was highlighted in the 1990s by Ilkka Vartiovaara (1946–2010), a doctor who presumably contracted Lyme borreliosis from a tick bite during a business trip to Canada in 1987. As borreliosis was not yet known in Finland at the time, his treatment was delayed.

76 "Puutiaisen purema"; Katri Hannele Broman, "Kroonisesti sairaalla heikko sosiaaliturva," *Helsingin Sanomat*, July 6, 2008.

77 Peretti-Watel et al., "Lyme Disease," 135–136.

78 Aronowitz, "Lyme Disease," 97.

79 Mari Manninen, "Varoitus punkkiratsiasta," *Helsingin Sanomat*, 22 April, 2007.

Consequently, the pain caused by the disease became chronic and left him unable to work. Vartiovaara used his own suffering as an example of how important it was to identify borreliosis in a timely manner when it was still easily treatable with antibiotics. He suggested that there may have been tens of thousands of people with Lyme disease in Finland who felt they were ill, but who were deprived of medical treatment because their ailments and disease could not be defined.[80]

Vartiovaara's experience drew attention to medical uncertainty, which was able to create a sense of health risk or crisis even if borreliosis from a medical point of view appeared to be a mild and non-fatal disease.[81] The public debate in the 1990s demonstrated how medical knowledge about borreliosis had increased so rapidly that not all doctors had kept up and recognized the disease. In particular, it was difficult to identify the symptoms of advanced Lyme disease, which were similar to many other diseases, such as rheumatism, multiple sclerosis, neuropathic pain, facial nerve strokes and cardiac arrhythmias. Therefore, borreliosis was called "the great imitator." Even antibody tests did not necessarily confirm the presence of borreliosis in the body.[82]

6 Russian Roulette

Although it was generally known that only some ticks carried pathogens, the problem was that one could not observe this on the surface of the tick. Contracting the disease was like playing Russian roulette, which meant that every tick attached to the skin was a potential threat.[83] Tick-borne diseases present an example of an invisible risk, which can only be understood with scientific tools, such as serological tests. The invisible nature of these risks contributes to the volatility and proliferation of risk perceptions.[84] In Finland this

80 Kipsimies (pseudonym), "Tasa-arvoa kuuroutumiseen!," *Helsingin Sanomat*, July 17, 1990; "Lymen taudin vankina," *Suomen Kuvalehti*, May 19, 1995; "Miksi juuri minä?" *Apu*, February 21, 1997; Ylitalo, "Outojen vaivojen takana"; "Ilkka Vartiovaara," *Helsingin Sanomat*, April 10, 2010.

81 Peretti-Watel et al., "Lyme Disease," 134–135.

82 Järvinen, "Kotoiset puutiaiset"; Lempiäinen, "Punkin purema"; Seija Lamberg, "Punkkien hyökkäys," *Suomen Kuvalehti*, May 19, 1995; Kirsi Vuorinen, "Borrelioosista selkeät ohjeet," *Helsingin Sanomat*, August 8, 1995; Ylitalo, "Outojen vaivojen takana"; Himma, "Punkkia kärsästä kiinni".

83 Jorma Rotko, "Punkkikauhua," *Helsingin Sanomat*, August 23, 1987; Paljakka, "Todellinen punkki kesä"; Himma, "Punkkia kärsästä kiinni".

84 Peretti-Watel et al., "Lyme Disease" (2019): 134.

uncertainty gave rise to the phenomenon of going for a course of antibiotics after a tick bite "just in case".[85]

The search for a sense of security from the antibiotic treatments taken to be on the safe side was not strange in the sense that newspapers had for a long time repeated the simplistic narrative that early-stage Lyme disease was "very easily" and "completely" cured with a couple of weeks of antibiotics. Even in the late stages, a slightly longer course of intravenous antibiotics was presented as a sure cure. In some early-stage news reports, an "antibiotic shot" was recommended even in doubtful cases,[86] but as this was not recommended from a medical point of view, the public was later advised that a mere tick bite without any symptoms suggestive of Lyme disease was not a reason to prescribe antibiotics.[87]

In the early 2000s, tick-borne encephalitis re-entered the headlines because it seemed to be spreading rapidly to new areas, such as the Helsinki archipelago. Moreover, cases caused by the more dangerous Siberian subtype of the TBE virus had been detected in the Kokkola region on the west coast.[88] In 2005, the number of serious cases of TBE was estimated to have tripled in a decade and the number of Lyme disease infections had also doubled. This time the increase was estimated to be real and not just due to improved diagnostics and increased testing.[89] Renewed concern about tick-borne encephalitis was reflected in increased interest in TBE vaccinations.

An Austrian TBE vaccine had been used in Finland since 1979. As TBE was only sporadic in certain high-risk areas, the vaccine was not intended to provide herd protection, but was a "personalised vaccine", that people living or

85 Rotko, "Punkkikauhua"; Paljakka, "Todellinen punkki kesä"; Laine, "Punkin pelko"; Juhani Partanen, "Seuraava potilas," *Länsi-Savo*, June 12, 1995; Päivi Repo, "Borrelioosiin sairastuu vuodessa parisen tuhatta," *Helsingin Sanomat*, August 9, 1995; Himma, "Punkkia kärsästä kiinni".

86 Järvinen, "Kotoiset puutiaiset"; "Puutiaisen purema"; Laine, "Punkin pelko"; Lamberg, "Punkkien hyökkäys"; "Kymmeniä borrelioositapauksia vuosittain," *Pargas Kungörelser*, August 17, 1995; "Punkki rakastaa kosteaa eikä pelkää viluakaan," *Etelä-Suomen Sanomat*, July 23, 1996; Annala, "Kostea sää".

87 Partanen, "Seuraava potilas" (1995); Repo, "Borrelioosiin sairastuu"; Himma, "Punkkia kärsästä kiinni".

88 Päivi Repo, "Punkkien levittämä aivotulehdus leviää nopeasti Suomessa," *Helsingin Sanomat*, June 20, 2001; Minna Pölkki, "Kumlingen tautia havaittiin Kokkolassa," *Helsingin Sanomat*, June 18, 2003; Tapani Tikkakoski et al., "Kuolemaan johtanut puutiaisaivokuume," *Lääketieteellinen Aikakauskirja Duodecim* 127, no. 10 (2011).

89 Päivi Repo, "Ahvenanmaalaisille suunnitellaan rokotusta puutiaisaivokuumetta vastaan," *Helsingin Sanomat*, January 12, 2004; Marjut Lindberg, "Borrelioosin voi saada citypunkistakin," *Helsingin Sanomat*, May 3, 2004; Lapintie, "Ennusteet lupaavat".

vacationing in high-risk areas could purchase to protect themselves.[90] It was described in newspapers as a "cheap life insurance" despite the high price.[91] As the highest incidence of tick-borne encephalitis in Finland, relative to the population, was found in Åland, the local residents began to be offered free TBE vaccinations as part of the national vaccination program from 2006.[92] The same year, a private medical center, Saaristolääkärit, based in the Turku archipelago, launched an ongoing "tick bus" operation bringing TBE vaccinations close to people, in town squares and in front of shopping malls. In the first year, around 8,500 people purchased the vaccine.[93]

In the spring of 2007, the media reported that vaccines against tick-borne encephalitis were running out in Finland due to high demand. During that year, a total of about 15,000 people had received the TBE vaccine in both Helsinki and the Turku region. The problem was that people who did not need the vaccine had also taken it.[94] The phenomenon sparked a lively public debate. Health professionals thought that the fear of tick-borne encephalitis was disproportionate and that people's protection against ticks had already gone too far. An editorial in *Helsingin Sanomat* even referred to "tick hysteria".[95]

Obviously the TBE vaccine was seen as a risk-reducing practice or product, intended to provide safety, reassurance, fear reduction, and control over

---

90 As a personalized vaccine, TBE vaccine is comparable to the vaccines against Lyme borreliosis. See Robert A. Aronowitz, "The Rise and Fall of the Lyme Disease Vaccines: A Cautionary Tale for Risk Interventions in American Medicine and Public Health," *The Milbank Quarterly* 90, no. 2 (2012): 251, 271.

91 Riitta Kallioinen, "Pistoista oireet lievenevät," *Etelä-Suomen Sanomat*, July 3, 1979; "Tutkijat eri mieltä vaarasta"; Laaksonen, "Kumlingen tauti – vaara veneilijälle"; "Punkista voi saada vaarallisen taudin," *Maaseudun Tulevaisuus*, July 1, 1993; Lapintie, "Ennusteet lupaavat"; Pyry Lapintie, "Punkkibussin rokotuksiin jonotettiin Turun torilla," *Helsingin Sanomat*, May 31, 2006.

92 Repo, "Ahvenanmaalaisille suunnitellaan rokotusta"; Lapintie, "Ennusteet lupaavat"; Pyry Lapintie, "Punkit kiusaavat huhtikuusta lokakuuhun," *Helsingin Sanomat*, May 31, 2006; Peter Wahlberg et al., "TBE på Åland 1959–2005 – Kumlingesjukan," *Finska Läkaresällskapets Handlingar* 166, no. 1 (2006): 61.

93 Nils-Eric Backman, "Puutiaisaivokuumeen uskotaan leviävän," *Helsingin Sanomat*, April 19, 2006; "Punkkibussi starttaa ensi perjantaina," *Pargas Kungörelser*, April 20, 2006; Lapintie, "Punkkibussin rokotuksiin jonotettiin".

94 Timo Siukonen, "Punkkirokotteet loppumassa Suomestakin," *Helsingin Sanomat*, May 5, 2007; "Nyt kannatta syynätä ihoa tarkkaan," *Etelä-Suomen Sanomat*, May 21, 2007; "Punkkihysteria levisi Suomeen," *Helsingin Sanomat*, June 30, 2007; Salla Tuomola, "Kansanterveyslaitos paheksuu rokotebussien markkinointia," *Turun Sanomat*, November 13, 2007.

95 Timo Rautava, "Punkki iskee iholle," *Helsingin Sanomat*, June 4, 2007; "Punkkihysteria levisi Suomeen"; Päivi Repo, "Vain kuusi sai alkuvuonna puutiaisaivokuumeen," Helsingin Sanomat, July 29, 2007.

uncertainty.[96] For vaccine manufacturers and distributors, this was a selling point, while at the individual level, a vaccination brought relief for an individual who felt liberated from the responsibility of having to constantly monitor his/her body, environment and behavior for ticks.[97] However, according to a later survey by the Finnish Institute of Health and Welfare (THL), it was unclear to the public what the TBE vaccine was effective against. Many assumed that it also provided protection against Lyme borreliosis or prevented ticks from attaching themselves to human skin. The public also estimated the risk of tick-borne encephalitis to be up to ten times higher than it actually was. According to a THL official, public perceptions may have been confused by the fact that the TBE vaccine was referred to in the media and popular language simply as the "tick vaccine."[98]

According to Jacob Heller, Western culture has long viewed vaccines as simple and versatile tools for contending with all kinds of as-yet-unforeseen public health problems. He argues that this kind of strong cultural acceptance of vaccines is maintained by so-called vaccine narratives, that is, success stories about selfless physician-researchers inventing cheap, safe, and effective vaccines that rescue us from dreaded infectious diseases. These narratives elide and overwhelm contradictions and simplify our understanding of vaccines so that we can more easily make sense of the whole.[99]

The Finnish TBE vaccine narrative started in the 1950s, when researchers reassured the public that if Kumlinge disease proved to be highly contagious, the population could be vaccinated against it. The fact that a vaccine suitable for protecting the whole population was not yet available in Finland in the 1950s or 1960s was not seen as a problem, as it was considered perfectly feasible to develop a vaccine if necessary.[100] In the mid-1990s, the vaccine narrative was extended to borreliosis. News coverage reported optimistically that although there was not yet a vaccine against Lyme disease, several vaccines were being developed in both the USA and Europe,[101] and one was even tested

96 Aronowitz, "Lyme Disease Vaccines," 270.

97 Lapintie, "Punkkibussin rokotuksiin jonotettiin"; "Punkkihysteria levisi Suomeen".

98 Henna Mäkelä, "Puutiaisaivotulehdus ja borrelioosi, kumpi olikaan kumpi?" *THL Blog*, March 16, 2021, accessed November 14, 2023, https://blogi.thl.fi/puutiaisaivotulehdus-ja-borrelioosi-kumpi-olikaan-kumpi/.

99 Heller, *The Vaccine Narrative*, 1–2, 5, 8.

100 Hex, "Ihmisiä työnsä ääressä"; "Fästing ger sjukdom"; "Möss föda för farliga fästingar"; "Punkkiretkikunnat"; Lyy, "Kumlingen kumma tauti".

101 "Punkinpuremaan kannattaa suhtautua vakavasti," *Etelä-Suomen Sanomat*, July 21, 1994; Lamberg, "Punkkien hyökkäys"; Lapintie, "Borrelioosirokotteen tehosta tietoja loppuvuonna".

on humans in the Åland Islands in 1996.[102] Ultimately, no Lyme disease vaccine was launched on the market. This was mainly down to economic reasons.[103] Nonetheless, news coverage of it might well have obscured the public's understanding of what the vaccine on the market was for.

7 Conclusion

The Finnish social debate on tick-borne diseases highlights several factors that have influenced the risk perceptions of both experts and the public. Scientific knowledge of tick-borne diseases has been of great importance. It eased experts' concerns about the public health threat posed by the new and mysterious Kumlinge disease in the 1950s. On the other hand, increased awareness of tick-borne encephalitis as zoonosis raised new concerns among researchers about other arboviruses that could be spread by ticks and mosquitoes in northern regions. The public has perhaps been most influenced by studies on the increase in the number of infected ticks and their spread to new areas, and by the knowledge of the increase in tick-borne diseases, which in the 1990s and early 2000s led to an outright tick scare.

Changes in disease spectrum, as well as in the health policy emphases and perceptions of health and illness have also played a role. Health policy in the 1950s and 1960s focused on the lethality of the disease, in relation to which Kumlinge disease was not seen as being very dangerous. In terms of its disease pattern, it was "no more severe" than polio, the other common disease of the era. However, as part of the Western trend towards individualization, the pain and damage that the disease caused to the individual began to be emphasized in notions of health and illness from the 1970s. With the disappearance of dangerous infectious diseases, the suffering caused by illness was no longer seen as a natural part of life, which made the severe forms of Lyme disease and TBE seem frightening.

The level of risk perceived by the public has also been influenced by the way in which tick-borne diseases have been reported in the press. The emphasis on the worrisome features of the tick-borne diseases and the personal accounts of

---

102 Merja Räihä, "Ahvenanmaalla testataan rokotetta borrelioosiin," *Helsingin Sanomat*, July 18, 1995; "Ahvenanmaan lääkärit borrealia [sic.] -rokotteen koekaniineiksi," *Etelä-Suomen Sanomat*, December 5, 1995; "Borrelioosia torjutaan rokotteilla Ahvenanmaalla," *Helsingin Sanomat*, April 3, 1996.

103 In the United States the vaccines against borreliosis faced market challenges because Lyme disease was geographically limited, treatable with antibiotics, non-deadly, and preventable by other means. There was also a fear of harmful side effects and lawsuits. Aronowitz, "Lyme Disease Vaccines," 250–251, 267–270.

people with severe forms of the disease has probably contributed to the fear of tick-borne diseases. They have drawn the audience's attention to the fact that even if the risk of being infected by a tick bite might be relatively small, the consequences for the infected individual could be tragic. Experts and the press have sought to allay public concern by highlighting how tick-borne encephalitis can be prevented by vaccination and Lyme disease can be treated with antibiotics. These simplified narratives provide fear reduction and control of uncertainty, but can sometimes lead to exaggerations, such as the 2007 tick vaccine enthusiasm.

Bibliography

Archival Material

The Finnish Literature Society (SKS). Questionnaire 2019. "Punkit tulevat." SKS, Helsinki.

Helsingin Sanomat. "Arkisto." Accessed November 13, 2023. https://www.hs.fi/arkisto/.

Helsingin Sanomat. "HS Aikakone." Accessed November 2, 2023. https://www.hs.fi/aikakone/.

The National Library of Finland. "Search from digital materials." Accessed April 5, 2023. https://digi.kansalliskirjasto.fi/etusivu?set_language=en.

Newspapers and Periodicals

Apu
Etelä-Suomen Sanomat
Helsingin Sanomat
Hufvudstadsbladet
Iltalehti
Ilta-Sanomat
Kansan Uutiset
Länsi-Savo
Maaseudun Tulevaisuus
Ny Tid
Nya Pressen
Pargas Kungörelser
Suomen Kuvalehti
Suomen Sosialidemokraatti
Turun Sanomat
Uuden Suomen Kuvalehti
Uusi Suomi

Literature

Alho, Ensio. "Onko rokotuspakko poistettava?" *Terveydenhoitolehti* 41, no. 9 (1929): 119–121.

Aronowitz, Robert A. "Lyme Disease: The Social Construction of a New Disease and Its Social Consequences." *The Milbank Quarterly* 69, no. 1 (1991): 79–112.

Aronowitz, Robert A. "The Rise and Fall of the Lyme Disease Vaccines: A Cautionary Tale for Risk Interventions in American Medicine and Public Health." *The Milbank Quarterly* 90, no. 2 (2012): 250–277.

Aronowitz, Robert. "Framing Disease: An Underappreciated Mechanism for the Social Patterning of Health." *Social Science & Medicine* 67, no. 1 (2008): 1–9.

Belikov, S.I., I.G. Kondratov, U.V. Potapova and G.N. Leonova. "The Relationship between the Structure of the Tick-Borne Encephalitis Virus Strains and Their Pathogenic Properties." *PLoS ONE* 9, no. 4 (2014): 1–16. https://doi.org/10.1371/journal.pone.0094946.

Brummer-Korvenkontio, Markus. *Virusten ja prionien luonnonhistoriaa: Myyräkuumeesta SARS:iin, Ebolasta AIDS:iin ja arboviruksista lintuinfluenssaan*. Helsinki: Helsinki University Press, 2007.

Clarke, Delphine H. "Further Studies on Antigenic Relationships Among the Viruses of the Group B Tick-borne Complex," *Bulletin of the World Health Organization* 31, no. 1 (1964): 45–56.

Cuellar, J., T. Dub, J. Sane and J. Hytönen. "Seroprevalence of Lyme borreliosis in Finland 50 years ago." *Clin Microbiol Infect.* 26, no. 5 (2020): 632–636. doi:10.1016/j.cmi.2019.10.003.

Deviatkin, Andrei A., Galina G. Karganova, Yulia A. Vakulenko and Alexander N. Lukashev. "TBEV Subtyping in Terms of Genetic Distance." *Viruses* 12, no. 11 (2020): 1240. doi:10.3390/v12111240.

Downs, W.G. "The Rockefeller Foundation Virus Program: 1951–1971 with Update to 1981." *Annu Rev Med* (1982): 1–29. doi:10.1146/annurev.me.33.020182.000245.

The Finnish Institute for Health and Welfare. "Milloin eri rokotukset ovat alkaneet Suomessa?" Accessed November 7, 2023. https://thl.fi/fi/web/infektiotaudit-ja-rokotukset/tietoa-rokotuksista/kansallinen-rokotusohjelma/milloin-eri-rokotukset-ovat-alkaneet-suomessa-.

The Finnish Institute for Health and Welfare. "Puutiaisaivotulehdus." Accessed November 21, 2023. https://thl.fi/fi/web/infektiotaudit-ja-rokotukset/taudit-ja-torjunta/taudit-ja-taudinaiheuttajat-a-o/puutiaisaivotulehdus.

Harjula, Minna. *Hoitoonpääsyn hierarkiat: Terveyskansalaisuus ja terveyspalvelut Suomessa 1900-luvulla*. Tampere: Tampere University Press, 2015.

Harjula, Minna. *Terveyden jäljillä: Suomalainen terveyspolitiikka 1900-luvulla*. Tampere: Tampere University Press, 2007.

Heller, Jacob. *The Vaccine Narrative*. Nashville: Vanderbilt University Press, 2008.

Heyman, Paul, Christel Cochez, Agnetha Hofhuis, Joke van der Giessen, Hein Sprong, Sarah Rebecca Porter, Bertrand Losson et al. "A Clear and Present Danger: Tick-Borne Diseases in Europe." *Expert Review of Anti-Infective Therapy* 8, no. 1 (2010): 33–50. doi: 10.1586/eri.09.118.

Hytönen, Jukka, Jarmo Oksi, Jani Sormunen and Olli Vapalahti. "Puutiainen ja sen levittämät taudit." *Duodecim Terveyskirjasto.* June 9, 2021. Accessed on November 21, 2023. https://www.terveyskirjasto.fi/kpp00002.

Jeffries C.L., K.L. Mansfield, L.P. Phipps, P.R. Wakeley, R. Mearns, A. Schock, S. Bell et al. "Louping Ill Virus: An Endemic Tick-Borne Disease of Great Britain." *Journal of General Virology* 95, no.5 (2014): 1005–1014. doi:10.1099/vir.0.062356-0.

Kunz, Christian. "TBE vaccination and the Austrian experience." *Vaccine* 21, no. 1 (2003): 50–55. doi:10.1016/S0264-410X(02)00813-7.

Laurent, Helene. *Asiantuntijuus, väestöpolitiikka, sota: Lastenneuvoloiden kehittyminen osaksi kunnallista perusterveydenhuoltoa 1904–1955.* Helsinki: Unigrafia, 2017.

Mäkelä, Henna. "Puutiaisaivotulehdus ja borrelioosi, kumpi olikaan kumpi?" *THL Blog.* March 16, 2021. Accessed November 14, 2023. https://blogi.thl.fi/puutiaisaivotulehdus-ja-borrelioosi-kumpi-olikaan-kumpi/.

Oker-Blom, Nils. "Kumlinge Disease: A Meningo-Encephalitis Occurring in the Aaland Islands." *Annales Medicinae Experimentalis et Biologiae Fenniae* 34 (1956): 309–318.

Olafsdottir, Sigrun. "Social Construction and Health." In *Medical Sociology on the Move: New Directions in Theory*, edited by William C. Cockerham, 41–59. Dordrecht: Springer, 2013.

Palacios, Gustavo, N. Savji, A. Travassos da Rosa, H. Guzman, X. Yu, A. Desai, G.E. Rosen et al. "Characterization of the Uukuniemi Virus Group (Phlebovirus: Bunyaviridae): Evidence for Seven Distinct Species." *Journal of Virology* 87, no. 6 (2013): 3187–3195. doi:10.1128/JVI.02719-12.

Peretti-Watel, P., J. Ward, R. Lutaud and V. Seror. "Lyme Disease: Insight from Social Sciences." *Médecine et Maladies Infectieuses* 49 (2019): 133–139. doi:10.1016/j.medmal.2018.12.005.

Pond, William L., Russ B. Sudie and Joel Warren. "The Russian Spring-Summer Encephalitis and Louping Ill Group of Viruses: Relationship of European and Asiatic Strains of Russian Spring-Summer Encephalitis Viruses and Louping Ill Virus." *The Journal of Infectious Diseases* 93, no. 3 (1953): 294–300.

Rochlin, I. and A. Toledo. "Emerging Tick-Borne Pathogens of Public Health Importance: A Mini-Review." *Journal of Medical Microbiology* 69, no. 6 (2020): 781–791. doi:10.1099/jmm.0.001206.

Rosenberg, Charles E. "Disease in History: Frames and Framers." *The Milbank Quarterly* 67, Supp. 1 (1989): 1–15.

Rummukainen, Maija. "Hyttysten levittämät taudit lisääntyvät". *Lääkärilehti* 75, no. 24–33 (2020): 1516–1521. https://www.laakarilehti.fi/tieteessa/katsausartikkeli/hyttysten-levittamat-taudit-lisaantyvat-3289/

Sane, Jussi. "Riski saada punkilta tauti on pieni, muttei merkityksetön." *THL Blog*, May 17, 2017. Accessed September 21, 2023. https://blogi.thl.fi/riski-saada-punkilta-tauti-on-pieni-muttei-merkitykseton/.

Soininen, Arvo M. "Suomen maataloushistorian päälinjat." *Agricultural and Food Science* 38, no. 2 (1966): 101–108. doi:10.23986/afsci.71655.

Strle, Franc, Gerold Stanek and Klemen Strle. "Lyme Borreliosis: The European Perspective." In *Lyme Disease: An Evidence-Based Approach*, 2nd ed, edited by John J. Halperin, 105–123. Wallingford & Boston, MA: CABI, 2018.

Ståhlberg, Tom, Anssi Junnila and Erkki Leppäkoski. *Apua! Punkki ja sen levittämät taudit*. Turku: Arkipelagia-seura, 1994.

Taylor, Richard Moreland (ed.). *Catalogue of Arthropod-Borne Viruses of the World: A Collection of Data on Registered Arthropod-Borne Animal Viruses*. U.S. Government Printing Office, 1967.

Tikkakoski, Tapani, Tadeusz Kaminski, Sinikka Ingo, Seppo Tuisku, Hannu Tuominen, Anu Jääskeläinen and Olli Vapalahti. "Kuolemaan johtanut puutiaisaivokuume." *Lääketieteellinen Aikakauskirja Duodecim* 127, no. 10 (2011): 1041–1045.

Wahlberg, Peter, Sten-Anders Carlsson, Hans Granlund, Christian Jansson, Mogens Lindén, Clara Nyberg and Dag Nyman, "TBE på Åland 1959–2005 – Kumlingesjukan." *Finska Läkaresällskapets Handlingar* 166, no. 1 (2006): 61–68.

Williams, H. and H. Thorburn. "The Serological Response of Sheep to Infection with Louping-Ill Virus." *The Journal of Hygiene* 59, no. 4 (1961): 437–47.

Winter, Uuno. "Isosta rokosta ja rokotuksesta." *Terveydenhoitolehti* 24, no. 2 (1912): 17–19.

Zlobin, Vladimir I., Vanda V. Pogodina and Olaf Kahl. "A Brief History of the Discovery of Tick-Borne Encephalitis Virus in the Late 1930s (Based on Reminiscences of Members of the Expeditions, Their Colleagues, and Relatives)." *Ticks and Tick-Borne Diseases* 8, no. 6 (2017): 813–820. doi:10.1016/j.ttbdis.2017.05.001.

CHAPTER 11

Tick Smart: Practices and Materializations in Human – Tick Entanglements

Sanna Lillbroända-Annala

1 Introduction

> Since ticks have started to spread serious diseases, the attitude toward them has, of course, changed. From being harmless (in my childhood in the 70s), they have become a terrifying and menacing insect species, comparable with malaria mosquitoes or venomous spiders.[1]

This chapter sheds light on the health-related concerns surrounding ticks by studying how people have experienced the increase in ticks, the diseases they spread and the fear of ticks in their own lives in Finland. It discusses the emotionally charged, contradictory, and multifaceted relationships between ticks and humans[2] – as described in the quote above – from the perspective of everyday practices and materiality. When tick season begins in early spring, the number of tick-related newspaper articles is multiplied, along with posts on social media, where ticks are discussed as troublemakers interrupting and affecting peoples' lives in many ways. Our relationship with ticks is described as problematic and dysfunctional, impacting, e.g., peoples' outdoor activities and relationships to nature.[3] Ticks are considered disgusting[4] and are portrayed in peoples' perceptions as dangerous and threatening to human and non-human well-being.

---

1 SLS questionnaire answer number 2342.

2 See e.g. Viktor Zöldi et al., "Knowledge, Attitudes, and Practices Regarding Ticks and Tick-Borne Diseases," *Ticks and Tick-Borne Diseases* 8, no. 6 (2017), doi:10.1016/j.ttbdis.2017.07.004.

3 See Sanna Lillbroända-Annala, "When Nature Becomes a Risk: Solastalgia and Entanglement in Human-Tick Relations," *Cultural Analysis* 21, no. 2 (2023), 76, https://www.ocf.berkeley.edu/~culturalanalysis/volume21_2/vol21_2_Nature.html.

4 Sanna Lillbroända-Annala, "Fy, en fästing! Äckel i kroppsliga och affektiva möten mellan fästing, sällskapsdjur och människa," *Budkavlen*, 101 (2022), https://journal.fi/budkavlen/article/view/115761.

The health-related concerns concerning ticks are commonly acknowledged. Ticks can be a serious health risk for many animals, including humans, transmitting diseases such as Lyme borreliosis and tick-borne encephalitis (TBE).[5] Both are considered public health concerns in Finland, and if not treated properly, both diseases can cause serious, even life-threatening, conditions.

With the growing awareness that ticks can possess a health risk, the approach to ticks has gradually changed. Tick encounters and imagined encounters with ticks have brought about protective and preventive measures, *tick-related practices and materializations*, evoked by strong emotions and causing people to act defensively to avoid tick bites and tick-borne diseases. These proactive and protective measures include, for example, tick-preventing clothing, regularly performed tick checks, tick removals with certain tools, and using tick repellents on companion animals. In encounters with ticks, practices and materializations become control mechanisms for avoiding diseases and managing health risks.

The practices are linked to the tick season[6], which in Finland lasts from spring until autumn, and to tick-dense areas of the country. The practices are also anchored in a range of products intended to keep ticks at a distance and to remove them. These products are sold in grocery stores, pet stores, and pharmacies. The product range includes many products intended for companion animals, and by protecting them from ticks, we also indirectly protect ourselves.

Tick-related practices and materializations are legitimized through repetitions, experiences, and best practices emerging from expected and real encounters with ticks. By practices and materializations, I refer to what they *do* and how they are *done* in different spatial and temporal contexts. Danish ethnologists Tine Damsholt and Dorthe Gert Simonsen do not see materiality as something passive, but as something relational, processual, and performative. Materiality is ascribed agency and the opportunity to influence the world in different ways.[7]

5 See Introduction and chapters by Suvi Rytty and Taina Syrjämaa in this volume.

6 Ruut Uusitalo et al. "Predicting Habitat Suitability for Ixodes ricinus and Ixodes persulcatus Ticks in Finland." *Parasites & Vectors* 15, no. 310 (2022), doi:10.1186/s13071-022-05410-8.

7 Tine Damsholt and Dorthe Gert Simonsen, "Materialiseringer, processer, relationer og performativitet," in *Materialiseringer: Nye perspektiver på materialitet og kulturanalyse*, ed. T. Damsholt, D.G. Simonsen and C. Mordhorst (Århus: Aarhus Universitetsforlag, 2009).

Therefore, the focus is on describing how risk is experienced and practiced by analyzing how practices and materializations correlate with the awareness of ticks and tick-borne diseases. How do things and everyday practices engage in human – tick and human – thing entanglements, which could also be referred to as multispecies and multi-item entanglements.[8] Together, these entanglements refer to situations where different actors, when encountering each other, influence one another so that something or some things change.[9]

This chapter contributes to multispecies studies, human – animal studies, and cultural animal studies[10] by contextualizing and problematizing how human relationships with other species are intertwined with material-driven practices.[11] Within these research fields, questions about how we create practices in encounters with animals – often also from the viewpoints of tensions and conflicting emotions arising from these relationships – are scrutinized. Postcolonial and intersectional influences, as well as anthropocentric perspectives, are also discussed in interspecies relations.[12] Studies within all these research fields highlight human and non-human entanglements and encounters.

This chapter takes its starting point from an understanding that humans and other animals are in constant interaction[13] and that humans' relationships

8 Thom van Dooren, Eben Kirksey and Ursula Münster, "Multispecies Studies: Cultivating Arts of Attentiveness," *Environmental Humanities* 8, no. 1 (2016), doi:10.1215/22011919-3527695.

9 Tim Ingold, "Binding Against Boundaries: Entanglements of Life in an Open World," *Environment and Planning A* 40, no. 8 (2008); Damsholt and Simonsen, "Materialiseringer, processer, relationer".

10 See, e.g., Stefan Helmreich and Eben Kirksey, "The Emergence of Multispecies Ethnography," *Cultural Anthropology* 25, no. 4 (2010); Laura Ogden, Billy Hall and Kamiko Tanita, "Animals, Plants, People and Things: A Review of Multispecies Ethnography," *Environment and Society: Advances in Research* 4 (2013); Donna Haraway, *When Species Meet* (Minneapolis and London: University of Minnesota Press, 2008); van Dooren, Kirksey and Münster, "Multispecies Studies".

11 See, e.g., Tuomas Räsänen and Nora Schuurman (eds.), *Kanssakulkijat: Monilajisten kohtaamisten jäljillä* (Helsinki: Suomalaisen Kirjallisuuden Seura, 2020); Simon Ekström and Lars Kaijser, *Djur: Berörande möten och kulturella smärtpunkter* (Göteborg: Makadam, 2018); Garry Marvin and Susan McHugh (eds.), *Routledge Handbook of Human-Animal Studies* (Abingdon: Routledge, 2014).

12 Ekström and Kaijser, *Djur*; Sonja Hagelstam and Sanna Lillbroända-Annala, "Människan och andra djur," *Budkavlen* 99 (2020); Räsänen & Schurman, *Kanssakulkijat.*

13 Haraway, *When Species Meet*; Tim Ingold, "Anthropology Beyond Humanity," *Suomen Antropologi: Journal of the Finnish Anthropological Society* 38, no. 3 (2013).

with other animals are shaped by versatility and complexity.[14] This chapter can therefore be seen as a contribution to the growing research field of cultural animal studies, which analyzes human relationships with other animals and species, as well as their relationship with us. Cultural animal studies bring to the fore questions about how animals are classified and valued, i.e., how we relate to animals, how we talk about animals, and what practices arise in encounters with animals. The studies also discuss tensions and conflicting emotions that arise in relationship with animals. Cultural animal studies problematize issues of proximity and distance and how boundaries between humans and animals arise, how they are maintained, and how they are sometimes crossed.[15]

2 Research Material and Theoretical Approaches

The research material used in this chapter consists of answers to the questionnaires "Tick" (Fästingen 2019) and "The ticks are coming!" (Punkit tulevat! 2019).[16] These questionnaires were sent out and made available to answer online by the Swedish Literature Society (SLS) and the Finnish Literature Society (SKS) during the summer of 2019. The questionnaires were prepared by our research team in collaboration with the archives. The SLS questionnaire was answered by a total of 42 people and the SKS questionnaire by a total of 103 people. Answers were sent from all parts of Finland, except from the most northern parts of the country, where ticks are not to be found – yet.

In the questionnaires, information about how ticks have been treated in the past and how they are treated today was collected alongside information about the respondents' experiences with ticks and the names they use for ticks. Questions about the relationship between ticks and climate change, i.e., how changes in the climate and environment may impact ticks, were also asked, as well as whether the respondents have seen ticks on other animals and how these have been removed. We were also keen to know more about tick-related diseases and how respondents feel about vaccines. The questionnaires also point out an interest in knowing how ticks are portrayed in media of various

14 Nickie Charles and Charlotte Aull Davies, "My Family and Other Animals: Pets As Kin," in *Human and Other Animals*, ed. B. Carter and Nickie Charles (London: Palgrave, 2011).

15 Ekström and Kaijser, *Djur*; Hagelstam and Lillbroända-Annala, "Människan och andra djur"; Räsänen and Schurman, *Kanssakulkijat*.

16 The questionnaire answers used have been anonymized but can be read with the license provided by the archives. The anonymization has been done in accordance with the National Board on Research Integrity, https://tenk.fi/en/research-misconduct/responsibleconduct-research-rcr.

kinds and how the relationship between ticks, humans, and other animals can be predicted in the future.

The questionnaires contain a lot of memories, experiences, and stories about ticks as matters of health concerns. The Finnish-speaking questionnaire sent out by the SKS contains health-related testimonies to a very large extent compared with the Swedish-speaking questionnaire sent out by the SLS. When using ethnographic material such as questionnaires, it is important to remember that, as a material, they are the result of the questions asked. Looking at the questions now, a few years later, even if our aim was to collect diverse material about tick-related memories, experiences, and attitudes, some of the questions seem to be focusing on the problems and challenges ticks might bring forth. Therefore, the questionnaires should be considered in a critical light where the shortcomings as well as the advantages of a rich, qualitative material are acknowledged.[17]

The health concerns linked to ticks are born from the feeling of insecurity and fear of diseases. The risk of being physically exposed to ticks and the threat of being infected with a tick-borne disease act as strong motivators for managing risks in different ways. Ticks are also embedded in a risk discourse based on scientific research[18] and personal experiences, a common theme debated in Finnish public discourse.[19]

Ticks are considered a risk to well-being for humans and non-humans, as they challenge our integrity. Therefore, ticks are intertwined in emotions, perception of risk, and the materialization of risk. Early on, British anthropologist Mary Douglas drew our attention to the fact that peoples' experiences of risk and danger are relative and contextual.[20] Douglas emphasizes that risks

---

17 See e.g. Charlotte Hagström and Lena Marander-Eklund (eds.), *Frågelistan som källa och metod* (Lund: Studentlitteratur, 2005).

18 Institutet för hälsa och välfärd TH, "Infektionssjukdomar och vaccinationer," accessed April 6, 2021, https://thl.fi/sv/web/infektionssjukdomar-och-vaccinationer/sjukdomar-och-bekampning/sjukdomar-och-sjukdomsalstrare-a-o/fastingburen-hjarninflammation; Huldén, Lena: "Uusien vektorivälitteisten tautien mahdollinen saapuminen Suomeen ilmastonmuutoksen ja ihmisten liikkuvuuden kylkiäisinä," accessed September 15, 2023, https://julkaisut.valtioneuvosto.fi/handle/10024/163158. See also the chapter by Suvi Rytty in this volume.

19 Sanna Lillbroända-Annala & Oscar Winberg, "Fästingen håller inget säkerhetsavstånd." Konkurrerande riskdiskurser om fästingar i media, *TRACE ∴ Journal for Human-Animal Studies* 9 (2023). In our article we focused on different constructions of risk discourses and the relationship between scientific, media and public risk discourses around ticks. See also the chapter by Otto Latva in this volume.

20 Mary Douglas, *Purity and Danger: An Analysis of Concepts of Pollution and Taboo*, (London: Routledge and Kegan Paul, 1976 [1966]); Mary Douglas, *Risk Accessibility*

include a socio-cultural context that influences how individuals view risk: which risks we fear the most, which risks we are prepared to take, and which risks we ignore.

Cultural anthropologist Birgitta Hellmark Lindgren further claims, in accordance with sociologist Niklas Luhmann, that we should distinguish between risk and danger. Risks can, to some extent, be anticipated and avoided. We may be held responsible for exposing ourselves to risks or not doing so. On the other hand, dangers of various kinds cannot always be foreseen or avoided. Thoughts about responsibility are central to how we experience and handle issues related to risks and dangers.[21]

Health risks are often calculable based on an epidemiological understanding of risk.[22] The diseases ticks can transmit can be problematic for the health and well-being of humans and other animals. In contrast, what is identified as risk in society by both the general public and experts is not only a question of identifying risk in relation to pure facts; it is also influenced by socio-cultural processes, which have been explored in several studies where the interest is not focused on actual risks but rather on how definitions and valuations of risks relate to broader socio-cultural and political contexts.[23] In addition to the fact that ticks in a real sense pose a health risk to both humans and pets, the socio-cultural context in this chapter focuses on the questionnaire answers in which risk is experienced and practiced.

Materializations and practices for health-related concerns can be analyzed as *human – tick entanglements*[24] and as *human – thing entanglements.*[25]

According to the Social Sciences (London: Routledge and Kegan Paul, 1986); Mary Douglas, *Risk and Blame. Essays in Cultural Theory* (London: Routledge, 1992).

21 Birgitta Hellmark Lindgren, *Pregnoscape: Den gravida kroppen som arena för motstridiga perspektiv på risk, kön och medicinsk teknik* (Uppsala: Department of Cultural Anthropology and Ethnology Institutionen för kulturantropologi och etnologi, 2006), 28; Niklas Luhmann, *Risk: A Sociological Theory* (Berlin: Walter de Gruyter, 1993).

22 Hellmark Lindgren, *Pregnoscape*, 27, according to Deborah Lupton, "Introduction: Risk and Sociocultural Theory," in *Risk and Sociocultural Theory: New Directions and Perspectives*, ed. Deborah Lupton (Cambridge: Cambridge University Press, 1999), 1.

23 Hellmark Lindgren, *Pregnoscape*, 27; Douglas, *Risk Accessibility*; Ulrich Beck, *Risk Society: Towards a New Modernity* (London: Sage, 1992); Anthony Giddens, *Modernitet och självidentitet: Självet och samhället i den senmoderna epoken* (Göteborg: Daidalos, 1999); Luhmann, *Risk*, and Frank Furedi, *Culture of Fear. Risk-taking and the Morality of Low Expectation* (United Kingdom: Continuum, 2002 [1997]).

24 See e.g., Helmreich and Kirksey, "Multispecies Ethnography"; Ogden, Hall and Tanita, "Animals, Plants, People"; Haraway, *When Species Meet*; van Dooren, Kirksey and Münster, "Multispecies Studies".

25 Ian Hodder, *Entangled: An Archaeology of the Relationships Between Humans and Things* (Malden, MA: Wiley-Blackwell, 2012).

Humans, non-humans, and things are relationally produced, where humans, ticks, and materials form entanglements through adopted and implemented practices. Humans and ticks get entrapped in their relations with things. Archeologist Ian Hodder distinguishes two forms of dependence. The first and more general focus on dependence recognizes enabling, where the human use of things allows humans to be, live, socialize, think, etc. Hodder uses the term "dependence" in the sense of "reliance on." Dependency also involves some form of constraint, as dependences might limit some of our abilities. Therefore, dependence and dependency create a dialectical struggle within entanglement. On one hand, humans and non-humans depend on or rely on things to achieve and enable something. On the other hand, dependency and codependency occur when humans and things cannot manage without each other, and in doing so, they constrain and limit what each can do.[26]

The idea of humans and things as inseparable has been strengthened by posthumanism and the "ontological turn" across sciences and humanities, radically reimagining human – object relationships.[27] These new perspectives urge us to think of objects and people as ontologically inseparable, entangled

26 Ian Hodder, "The Entanglements of Humans and Things: A Long-Term View," *New Literary History* 45, no. 1 (2014): 20.

27 See for e.g., Karen Barad, Posthumanist Performativity: Towards an Understanding of How Matter Comes to Matter. *Signs* 28(3) (2003):801– 831; Karen Barad, *Meeting the Universe Halfway: Quantum Physics and the Entanglement of Matter and Meaning* (Durham, NC: Duke University Press, 2007); Jane Bennett, *Vibrant Matter: A Political Ecology of Things* (Durham, NC: Duke University Press, 2010); Bill Brown, "Thing Theory," In *Things*, ed. Bill Brown, 1–16 (Chicago: University of Chicago Press, 2004); Levi Bryant, Nick Smicek and Graham Harman (eds.), *The Speculative Turn: Continental Materialism and Realism* (Melbourne: RE Press, 2010); Diana Coole and Samantha Frost (editors), *New Materialisms: Ontology, Agency, and Politics* (Durham, NC: Duke University Press, 2010); M. DeLanda, *A New Philosophy of Society: Assemblage Theory and Social Complexity* (London and New York: Continuum, 2006); Gilles Deleuze and Félix Guattari, "Rhizome: Introduction," In *A Thousand Plateaus: Capitalism and Schizophrenia*, 3–28 (London and New York: Continuum, 2007 [1980]); Alfred Gell, "The Technology of Enchantment and the Enchantment of Technology," in *Anthropology, Art, and Aesthetics*, ed. J. Coote and A. Shelton, 40– 67 (Oxford: Clarendon Press, 1992); Alfred Gell, *Art and Agency: An Anthropological Theory* (Oxford: Clarendon Press, 1998); Tim Ingold, *The Perception of the Environment: Essays on Livelihood, Dwelling, and Skill* (New York and London: Routledge, 2000); Tim Ingold, *Making: Anthropology, Archaeology, Art, and Architecture* (Routledge, London, 2013); Bruno Latour, *We Have Never Been Modern* (Cambridge: Harvard University Press, 1993); Bruno Latour, *Pandora's Hope: Essays on The Reality of Science Studies* (Cambridge: Harvard University Press, 1999); Bruno Latour, *Reassembling the Social: An Introduction to Actor-Network Theory* (Oxford: Oxford University Press, 2005); Frank Trentmann, "Materiality in the Future of History: Things, Practices, and Politics," *Journal of British Studies* 48 (2009).

within mutually constitutive, yet ever-shifting relationships.[28] According to anthropologist Daniel Miller, just as we create objects, objects create us through a circular process he calls *objectification.*[29]

The human – tick and human – thing entanglements, as well as the perceptions of ticks as risk, are affected by emotions. British-Australian scholar Sara Ahmed writes that emotions are created in contact with someone or something, and are influenced by our cultural beliefs about this someone or that something.[30] Thus, emotions are produced not only in the interaction between people but also in interspecies encounters; and they are rooted in different environments, situations, and matters.[31] Emotions are a driving force in tick encounters as feelings of discomfort and fear make us do things like protect ourselves with tick-proof clothing, perform regular tick checks of our bodies, and keep ticks away from our companion animals and ourselves by using tick repellents of various kinds. These ongoing practices can also become a liability and bring forth feelings of frustration as the tick season continues.

Regarding emotions, an understanding of what emotions *do* rather than what they *are* is essential. This understanding is also important and useful in tick – human relations. Ahmed believes emotions are more than psychological states and should be understood as performative social and cultural practices. She argues that emotions can be both moving and sticky, where the subject is formed in encounters with others and where the emotions create surfaces and boundaries between subject and object. The mobility of emotions can have different effects: for some, an emotion can be fleeting and temporary; for others, it can mean a fixation[32] – it sticks firmly, as in many cases of ticks.

3 Ticks: a Growing Health Concern

As other articles about ticks in this volume confirm, ticks and tick-borne diseases have increased during the last decades. How are people experiencing this

---

28 Ruth M. van Dyke (ed.), *Practicing Materiality* (Tucson: The University of Arizona Press, 2015).

29 Daniel Miller, *Material Culture and Mass Consumption* (Oxford: Blackwell, 1987), 19–33; Daniel Miller, "Materiality: An Introduction," in *Materiality*, ed. Daniel Miller (Durham, NC: Duke University Press, 2005).

30 Sara Ahmed, *The Cultural Politics of Emotion* (Edinburgh: Edinburgh University Press, 2004), 79.

31 Ole Riis and Linda Woodhead, *A Sociology of Religious Emotion* (Oxford: Oxford University Press, 2010), 7.

32 Ahmed, *The Cultural Politics of Emotion*, 7–13.

and how do they respond to these changed circumstances in their everyday lives? How are they comparing childhood experiences of ticks with their experiences of today?

Many memories connected with ticks are strongly influenced by a "time before and after ticks," referring to a change in which the awareness of ticks, the number of tick encounters, and the knowledge of tick-transmitted diseases have increased. This marks a transition from something that was before to something that is now and lies ahead of us. This shift can be seen in the growth of tick-related practices and in the more serious attitude toward tick-borne diseases.

> Many people I know are afraid of these little devils. A man I know had noticed something black on his shoulder in the sauna and tried to wipe it away. It was soon discovered to be a tick. The man immediately went to the nearest hospital, where it was removed. Afterwards, he went to the workplace doctor to take a blood test in fear of Lyme disease. Perhaps this was an overreaction, as it is easy to remove the tick yourself, and only a fraction [of ticks] cause diseases. The man had heard horror stories about ticks from relatives and was frightened.[33]

Ticks become active and look for blood meals when the outdoor temperature stays above +5 degrees. Thus, in many parts of the country where winters are becoming less cold due to climate change, the living conditions for ticks and their host species have improved. Ticks thrive wherever they have access to blood meals. Therefore, their living environment includes nature across a wide area. However, grass and bushes, above all, are usually highlighted as the tick's habitat, as ticks are considered to be "lurking in the grass" and waiting for a passing animal. The coastal areas and the Finnish archipelago are tick-dense, but nowadays, ticks can be found in most parts of the country except for the northern parts of Lapland. When it is hot or too cold, ticks retreat into the soil to wait for more suitable conditions.[34]

The fear and real-life concern for tick encounters, and thereby transmitted diseases, is recurring in many questionnaire answers. People engage in various but seemingly similar practices in dealing with ticks in their different tick encounters. Ticks induce action: actions to remove ticks and to proactively

---

33 SKS questionnaire answer number 028.

34 Jukka Hytönen et al., "Puutiaisten elämä: Kuka pelkää punkkia?," Terveyskirjasto (2021), accessed September 15, 2023, https://www.terveyskirjasto.fi/kpp00003/puutiaisten-elama.

fight the diseases they may transmit. Vaccination against tick-borne encephalitis is one quite often used precaution, especially in tick-dense areas of Finland.

> In the last ten years or so, the number of ticks in our summer place on the island has been constantly increasing. First, we had to start checking them morning and evening, and about five years ago, it seemed wise to get vaccinated. Last year, I started checking for ticks on my skin throughout the day; there are so many of them. In the family, the attitude has remained the same, i.e., we think that Lyme disease does not occur when ticks are removed daily, and the vaccination series most likely protects against brain fever. That's enough to be careful. However, my mother had Lyme disease about 10–15 years ago. The cottage neighbors have started to keep the lawns short, and that's something I'm considering doing myself.[35]

With time, ticks have become more common compared with the past and with the increase in the number of ticks; also, different practices have become more regular and ambitious. One of the tick-preventing practices many people refer to is becoming aware of ticks outdoors, ticks as part of nature.[36] One practice to handle the risks of being outdoors is to keep lawns short and to make surroundings unavailable to ticks, as in the quotes mentioned above and below, in order to prevent ticks from coming too close to us.

> In the late 1980s and 1990s, it [the cabin] became a place we visited more and more often. The forest groves had to give way, and the paths were trimmed regularly. The mosquitoes were expelled by removing undergrowth and lower branches of trees. Construction timber was also taken. Grass and groves no longer provide significant protection, but still, ticks occasionally appear.[37]

In many questionnaire answers, the idea of nature as a place where caution is also required comes from lessons from childhood. These childhood memories and experiences originate from the 1950s until the 1990s. In childhood memories, ticks are often described as ordinary. Ticks live among us in various surroundings, and for many, they were everyday companions in their

35 SKS questionnaire answer number 082.
36 Lillbroända-Annala, "Solastalgia and Entanglement".
37 SKS questionnaire answer number 070.

childhood. The carefree yet controlled co-existence with ticks is present in many memories.

> We were three siblings all summer long at a summer cottage in Sammatti (nowadays part of Lohja) in the late 1950s. My memory of ticks was that you got them on your skin, especially from alders near the shore. If the tick got attached, you just covered it with butter and waited until it cut loose. Most of the time, the butter melted in the summer heat before the tick gave up. Apparently, however, the butter had such an effect that it was somehow easier to remove the tick from the skin by pulling it with your fingers. As far as I know, no one has had any consequences from ticks. They belonged to summer![38]

In many memories, children were supposed to avoid tick-infested natural surroundings, especially alders.[39] They were considered tick magnets as it was in the alders and in their immediate vicinity that ticks were believed to reside. By avoiding alders, one could avoid running into ticks. This protective practice of keeping children away from alders was one way of dealing with the fear of ticks in nature. It was later proven wrong, as the increasing knowledge of where ticks live shifted the tick-infested area around the alders to grass and damp lands. Ticks can even reside in our backyards, very close to where humans live.

> In my childhood, I wasn't warned about the dangers of ticks. I was just told to watch out for alders because ticks jump from them. In particular, gray alders were considered to be trees where ticks live. Now that is no longer true, either, and ticks have gone from being annoying to being really dangerous. That is why the checks continue. I have no dog, but the grandchildren are always checked for ticks. They are also aware of this risk themselves; they move around covered up, and they do not walk barefoot. It is a little annoying that they cannot fully enjoy the summer. Fortunately, ticks do not threaten when you are swimming![40]

4 From Ordinary to Extraordinary Practices

With the awareness of the health risk, peoples' attitudes toward ticks has shifted. From being perceived as ordinary and relatively harmless in childhood

38 SKS questionnaire answer number 025.
39 See also Taina Syrjämaa, Tuomas Räsänen and Heta Lähdesmäki in this volume.
40 SKS questionnaire answer number 78b.

and during previous decades, ticks are now considered a risk to the well-being of humans and other animals. This change from ordinary to extraordinary has evoked practices of protection and prevention. It has also generated adaptation, where the human – thing entanglement during the tick season brings about certain practices for clothing and covering the body to protect it from the ticks.

> Ticks are nowadays very common, and there are plenty of them at our summer place in Ruokolahti.... When I go out in nature, I protect myself with long sleeves as well as with tight socks that I pull over my trousers. And I wear rubber boots. Despite these precautions, these small animals sometimes find their way to the skin.[41]

The practice of checking and constantly monitoring ourselves and our four-legged friends is visible in the questionnaires. Stuff related to detecting and removing ticks is strongly related to bodily practices during tick season. These practices are not only induced by medical and biological facts – the diseases ticks can transmit – but are very much culturally constructed.[42] We are socialized into a tick culture, which we learn to manage in different ways through preventive practices before and during the tick season, as well as through daily practices, especially in areas where ticks are abundant. Many of these practices are preceded by traditions that have been adopted and maintained within families, and later further developed or replaced with new ones.

> My first tick encounter took place in Rosala [an island in the southwestern archipelago of Finland], where I stayed at my boyfriend's cabin in 2001. Even though I hadn't even been in the woods or grass, I had three ticks on my lower back. I spotted them while in the shower, after the trip to the cabin. My boyfriend habitually removed them with pliers; he was used to seeing ticks at his cabin. I had heard that you can get rid of ticks by putting a big pile of butter around them. Then the ticks have to dig themselves out of the skin. My boyfriend was able to tell me that that method is no longer recommended – the tick can "vomit" and then the bacteria is more likely transferred. At that time, I wasn't afraid of Lyme disease or brain fever, I just thought I should remove the ticks and that's it.[43]

41 SKS questionnaire answer number 092b.

42 See Lillbroända-Annala and Winberg, "Fästingen håller inget säkerhetsavstånd".

43 SKS questionnaire answer number 006.

In practices like tick removals and during the tick season when ticks are active, materiality plays an important role. Humans and companion animals become entangled with a range of preventive and protective products intended to keep ticks at a distance, to remove them, and even to kill them. As soon as tick season begins, a variety of products are placed in visible places in pharmacies and grocery stores. The products are kept on display throughout the season, along with advertisements reminding us of the risks of tick-borne diseases. Veterinary clinics and pet shops also actively promote protection for companion animals. In addition to Lyme disease and tick-borne encephalitis, which is extremely rare in pets but theoretically possible, dogs can also become infected with anaplasmosis when exposed to ticks.[44]

> I was diagnosed with Lyme disease in late autumn, when I was not really moving around the garden with bare legs, but instead "as instructed in long-sleeved clothes and boots," and I was no longer thinking that ticks were around. In this case, the tick had climbed along my arm from a raspberry bush. I noticed the tick when it had been attached for several days, so it was a real tick. At first, it seemed that there was no infection, but after a couple of weeks, a really feisty rash appeared on the skin, like a fungus that had grown on the skin. A doctor immediately determined, after seeing the rash, that it was Lyme disease. I took a three-week course of antibiotics, and luckily, there were no other symptoms.[45]

Seasonality in human – thing entanglements, as can be read from the previous quote, can be misleading as it urges us to protect ourselves from ticks during the season but not during the off-season, when ticks might still be active. The seasonality of tick-related products is a good example of materializations, i.e., what material does with us. When products are on display, we are encouraged to use repellents, tick collars on dogs, and tools for removing ticks, and when they are not on display, the opposite.

The products we use for tick encounters are connected with a certain use and skill: how to remove a tick in the right way and, many times, the alternative and more personal ways we tend to use the products. Confrontations with ticks can also bring about a sense of losing control over the body and the situation. Practices combined with material components reinforce the feeling of security and the feeling of control.

---

44 Sympaatti Oy, "Punkkien torjunta," https://www.sympaatti.fi/artikkelit/punkkien-torjunta, accessed April 15, 2021.

45 SKS questionnaire answer number 002.

> In the cupboard, I have a tick iron and disinfectant as well as a bandage, and in the bag, a sharp knife with a sharp tip. The tick iron does not work properly; it rarely works as you wish. The iron does not go properly between the skin and the tick, on its shoulders. It is usually the case that you notice the tick late and then you only get rid of the animal in two parts. Daily swimming trips, regular sauna baths, and tick checks here and there ensure a carefree feeling. Without ticks, it would be even more fun, as before.[46]

Socialization into a tick culture, as the recurring season could be called with all its practicalities and materializations, is influenced by everyday encounters between ticks and humans. These encounters or expected encounters are often described in terms of fear and worry, but some also express a casual coexistence, where the adoption of tick practices has a reassuring effect. It seems that the closer ticks come to human habitations, the more they make us worry.

> We have a real tick garden. There are a lot of ticks, and you can see them almost every day in the summer. It is not at all unusual for a tick to climb up the leg. Fortunately, most of the time, you can find the tick before it gets attached. Unfortunately, this is not always the case. In the family, we always do a tick check every summer evening. During the worst tick summers, ticks may be found daily. Fortunately, the tick check mainly prevents the transmission of borreliosis: the longer the tick is attached, the more likely it is to be caught, and if the tick is removed in time, there should be no problems. We have taken vaccinations against tick-borne encephalitis. However, there are not as many ticks every summer, and unfortunately during such summers, the dangers of ticks and doing tick checks may sometimes be missed. And even in April or late autumn, you may not always remember that ticks can attach then as well.[47]

5 Conclusions

In the relationship between humans and ticks, materializations and practices become important components in maintaining control and increasing the feeling of security. Thus, the practices and materiality associated with

46 SKS questionnaire answer number 070.

47 SKS questionnaire answer number 002.

preventing and removing ticks constitute a management strategy adapted, incorporated, and developed during the tick season. We become dependent upon these practices and the tick-related stuff. Therefore, a certain tick expertise – becoming *tick smart* – is legitimized through repetitions, experiences, and best practices. When examining these interactions and practices, a tick season lifestyle becomes visible. This lifestyle affects habits, body, materials, and space, which become important aspects of the human – tick and human – thing entanglements.

Knowledge of the health risks ticks can pose has had a profound effect on the respondents' attitudes toward ticks. The research material describes everyday experiences with ticks. The relationship with ticks is described in emotional terms, where contempt, worry and fear are commonly described feelings. For the majority of respondents, the risk of ticks has strongly influenced their relationship with nature, outdoor activities, and life in general. Many activities that the respondents previously enjoyed are now excluded due to worry and fear. That ticks are perceived to limit their lives means that the attitude toward ticks is colored by bitterness and resentment.

Therefore, many stories are filled with emotions and drama. Ticks are rarely presented in a positive light but instead are demonized to varying degrees and seen as the evil party in the human – tick relationship. The relatively few opposing voices speak of a coexistence where the adoption of tick practices has a reassuring effect. These respondents have a nonchalant and ambivalent attitude toward ticks, while most of them avoid tick-dense places and worry a lot about being infected by tick-borne diseases.

The health-promoting nature as we have come to know it takes on a different dimension in many answers when nature becomes dangerous, risky, and unattractive because of the tick.[48] Nature and ticks should be controlled by keeping lawns cut short, avoiding tick-dense areas, wearing tick-proof clothing, and doing regular tick checks on your own body and the bodies of others.

The connection between materiality and practices is strengthened, shaping the human – tick and human – thing entanglements. Our perception of ticks as threatening and risky is culturally constructed and culturally managed. We are socialized into a tick culture with tick-related practices for prevention and protection. These practices are preceded by traditions that have been adapted and practiced within the family and that find new material dimensions in the ever-growing range of products we are subjected to.

48 Lillbroända-Annala, "When Nature Becomes a Risk".

Bibliography

Archival Sources

Finnish Literature Society, SKS, questionnaire *Punkit tulevat!*, 2019.

Swedish Literature Society, SLS, questionnaire *Fästingen*, 2019.

Literature

Ahmed, Sara. *The Cultural Politics of Emotion*. Edinburgh: Edinburgh University Press, 2004.

Barad, Karen. "Posthumanist Performativity: Towards an Understanding of How Matter Comes to Matter." *Signs* 28, no. 3 (2003):801–831.

Barad, Karen. *Meeting the Universe Halfway: Quantum Physics and the Entanglement of Matter and Meaning*. Durham, NC: Duke University Press, 2007.

Beck, Ulrich. *Risk Society: Towards a New Modernity*. London: Sage, 1992.

Bennett, Jane. *Vibrant Matter: A Political Ecology of Things*. Durham, NC: Duke University Press, 2010.

Brown, Bill. "Thing Theory." In *Things*, edited by Bill Brown, 1–16. Chicago: University of Chicago Press, 2004.

Bryant, Levi, Nick Smicek, and Graham Harman (eds.). *The Speculative Turn: Continental Materialism and Realism*. Melbourne: RE Press, 2010.

Charles, Nickie and Charlotte Aull Davies. "My Family and Other Animals: Pets As Kin." In *Human and Other Animals*, edited by B. Carter and Nickie Charles, 72–73. London: Palgrave, 2011.

Coole, Diana, and Samantha Frost (eds.). *New Materialisms: Ontology, Agency, and Politics*. Durham: Duke University Press, 2010.

Damsholt, Tine and Dorthe Gert Simonsen. "Materialiseringer, processer, relationer og performativitet." In *Materialiseringer: Nye perspektiver på materialitet og kulturanalyse*, edited by T. Damsholt, D.G. Simonsen and C. Mordhorst, 9–38. Århus: Aarhus Universitetsforlag, 2009.

DeLanda, M. *A New Philosophy of Society: Assemblage Theory and Social Complexity.* London and New York: Continuum, 2006.

Deleuze, Gilles, and Félix Guattari. "Rhizome: Introduction." In *A Thousand Plateaus: Capitalism and Schizophrenia*, 3–28. London and New York: Continuum, 2007 [1980].

van Dooren, Thom, Eben Kirksey and Ursula Münster. "Multispecies Studies: Cultivating Arts of Attentiveness." *Environmental Humanities* 8, no. 1 (2016): 1–23. doi:10.1215/22011919-3527695.

Douglas, Mary. *Purity and Danger: An Analysis of Concepts of Pollution and Taboo.* London: Routledge and Kegan Paul, 1976 [1966].

Douglas, Mary. *Risk Accessibility According to the Social Sciences*. London: Routledge and Kegan Paul, 1986.

Douglas, Mary. *Risk and Blame. Essays in Cultural Theory*. London: Routledge, 1992.

Dyke, Ruth M. van (ed.). *Practicing Materiality*. Tucson: University of Arizona Press, 2015.

Ekström, Simon and Lars Kaijser. *Djur: Berörande möten och kulturella smärtpunkter*. Göteborg: Makadam, 2018.

Furedi, Frank. *Culture of Fear. Risk-taking and the Morality of Low Expectation*. United Kingdom: Continuum, 2002 [1997].

Gell, Alfred. "The Technology of Enchantment and the Enchantment of Technology." In *Anthropology, Art, and Aesthetics*, edited by J. Coote and A. Shelton, 40–67. Oxford: Clarendon Press, 1992.

Gell, Alfred. *Art and Agency: An Anthropological Theory*. Oxford: Clarendon Press, 1998.

Giddens, Anthony. *Modernitet och självidentitet: Självet och samhället i den senmoderna epoken*. Göteborg: Daidalos, 1999.

Hagelstam, Sonja and Sanna Lillbroända-Annala. "Människor och andra djur." *Budkavlen* 99 (2020): 1–10.

Hagström, Charlotte and Lena Marander-Eklund (eds.). *Frågelistan som källa och metod*. Lund: Studentlitteratur, 2005.

Haraway, Donna. *When Species Meet*. Minneapolis: University of Minnesota Press, 2008.

Hellmark Lindgren, Birgitta. *Pregnoscape: Den gravida kroppen som arena för motstridiga perspektiv på risk, kön och medicinsk teknik*. Uppsala: Department of Cultural Anthropology and Ethnology Institutionen för kulturantropologi och etnologi, 2006.

Helmreich, Stefan and Eben Kirksey. "The Emergence of Multispecies Ethnography." *Cultural Anthropology* 25, no. 4 (2010): 545–576.

Hodder, Ian. "The Entanglements of Humans and Things: A Long-Term View." *New Literary History* 45, no. 1 (2014): 19–36.

Hodder, Ian. *Entangled: An Archaeology of the Relationships Between Humans and Things*. Malden: Wiley-Blackwell, 2012.

Huldén, Lena. Uusien vektorivälitteisten tautien mahdollinen saapuminen Suomeen ilmastonmuutoksen ja ihmisten liikkuvuuden kylkiäisinä. Accessed September 15, 2023. https://julkaisut.valtioneuvosto.fi/handle/10024/163158.

Hytönen, Jukka, Jarmo Oksi, Jani Sormunen and Olli Vapalahti. 2021. "Puutiaisten elämä: Kuka pelkää punkkia?" Terveyskirjasto (2021). Accessed January 15, 2023. https://www.terveyskirjasto.fi/kpp00003/puutiaisten-elama.

Ingold, Tim. *The Perception of the Environment: Essays on Livelihood, Dwelling, and Skill*. London: Routledge, 2000.

Ingold, Tim. "Binding Against Boundaries: Entanglements of Life in an Open World," *Environment and Planning A* 40, no. 8 (2008): 1807–1808.

Ingold, Tim. "Anthropology Beyond Humanity." *Suomen Antropologi: Journal of the Finnish Anthropological Society* 38, no. 3 (2013): 5–23.

Ingold, Tim. *Making: Anthropology, Archaeology, Art, and Architecture*. London: Routledge, 2013.

Institutet för hälsa och välfärd TH. "Infektionssjukdomar och vaccinationer." Accessed April 6, 2021. https://thl.fi/sv/web/infektionssjukdomar-och-vaccinationer/sjukdomar-och-bekampning/sjukdomar-och-sjukdomsalstrare-a-o/fastingburen-hjarninflammation.

Latour, Bruno. *We Have Never Been Modern*. Cambridge: Harvard University Press, 1993.

Latour, Bruno. *Pandora's Hope: Essays on The Reality of Science Studies*. Cambridge: Harvard University Press, 1999.

Latour Bruno. *Reassembling the Social: An Introduction to Actor-Network Theory*. Oxford: Oxford University Press, 2005.

Lillbroända-Annala, Sanna and Oscar Winberg. "Fästingen håller inget säkerhetsavstånd.' Konkurrerande riskdiskurser om fästingar i media." *Trace Journal for Human-Animal Studies* 9 (2023): 58–80. doi: 10.23984/fjhas.121761.

Lillbroända-Annala, Sanna. "Fy, en fästing! Äckel i kroppsliga och affektiva möten mellan fästing, sällskapsdjur och människa." *Budkavlen* 101 (2022): 82–104. https://journal.fi/budkavlen/article/view/115761.

Lillbroända-Annala, Sanna. "When Nature Becomes a Risk: Solastalgia and Entanglement in Human-Tick Relations." *Cultural Analysis* 21, no. 2 (2023): 76–96. https://www.ocf.berkeley.edu/~culturalanalysis/volume21_2/vol21_2_Nature.html.

Luhmann, Niklas. *Risk: A Sociological Theory*. Berlin: Walter de Gruyter, 1993.

Lupton, Deborah. "Introduction: Risk and Sociocultural Theory." *Risk and Sociocultural Theory: New Directions and Perspectives*, edited by Deborah Lupton, 1–11. Cambridge: Cambridge University Press, 1999.

Marvin, Garry and Susan McHugh (eds.). *Routledge Handbook of Human-Animal Studies*. Abingdon: Routledge, 2014.

Miller, Daniel. *Material Culture and Mass Consumption*. Oxford: Blackwell, 1987.

Miller, Daniel. "Materiality: An Introduction." In *Materiality*, edited by Daniel Miller, 1–50. Durham: Duke University Press, 2005.

National Board on Research Integrity. Accessed August 4, 2023. https://tenk.fi/en/research-misconduct/responsible-conduct-research-rcr.

Ogden, Laura, Billy Hall and Kamiko Tanita. "Animals, Plants, People and Things: A Review of Multispecies Ethnography." *Environment and Society: Advances in Research* 4 (2013): 5–24.

Riis, Ole and Linda Woodhead. *A Sociology of Religious Emotion*. Oxford: Oxford University Press, 2010.

Räsänen, Tuomas and Nora Schuurman (eds.). *Kanssakulkijat: Monilajisten kohtaamisten jäljillä*. Helsinki: SKS, 2020.

Sympaatti Oy. "Punkkien torjunta." Accessed April 15, 2021. https://www.sympaatti.fi/artikkelit/punkkien-torjunta.

Trentmann, Frank. "Materiality in the Future of History: Things, Practices, and Politics." *Journal of British Studies* 48 (2009): 283–307.

Uusitalo, Ruut, Mika Siljander, Andreas Lindén, Jani J. Sormunen, Juha Aalto, Guy Hendrickx, Eva Kallio et al. "Predicting Habitat Suitability for Ixodes ricinus and Ixodes persulcatus Ticks in Finland." *Parasites & Vectors* 15, no. 310 (2022). doi:10.1186/s13071-022-05410-8.

Zöldi, Viktor, Topi Turunen, Olli Lyytikäinen and Jussi Sane. "Knowledge, Attitudes, and Practices Regarding Ticks and Tick-Borne Diseases, Finland." *Ticks and Tick-Borne Diseases* 8, no. 6 (2017): 872–877. doi:10.1016/j.ttbdis.2017.07.004.

PART 5

Multispecies Networks

CHAPTER 12

Fleas, Knowledge-Making, and the Epidemiology of Plague in British India: Perspectives from the Bombay Epidemic, 1905–1906

Emily Webster

1 Introduction

Between October 1905 and January 1906, in a cluster of houses behind the Parel Laboratory in Bombay, an unusual ecosystem unfolded. Monkeys, guinea pigs, and rats both wild and tame, were released inside perfect replicas of the traditional workers huts (or go-downs) of the city of Bombay. Inside these buildings, these experimental animals awaited a grim fate: some animals, or the fleas they carried with them, had been intentionally infected with virulent *Yersinia pestis,* the causative agent of bubonic plague – and a likely death sentence for its carriers.

These experiments, conducted by the Second Indian Plague Commission in 1905, were part of a larger set of knowledge-gathering activities that arose during the Third Plague Pandemic (1894–1950) designed to examine a key set of controversies on the etiology and epidemiology of plague: "How," as Sanitary Commissioner FHG Hutchinson succinctly inquired, "is plague infection carried from rat to rat, rat to man, or man to man?"[1] While some acceptance of the agency of rats in the plague had gained traction among the imperial scientific community as early as 1903, the question of *how* plague transmitted from rats to humans remained an unsettled question. Experiments by Antoine Yersin (1894), EF Hankin (1897–1898), George Nuttall (1897–1899), Ogata Masanori (1897) and Paul-Louis Simond (1898), all claimed success in instigating transmission through the rat-flea, *Xenopsylla cheopis,* but had achieved at best

1 Captain FHG Hutchinson, IMS, acting Deputy Sanitary Commissioner, Southern Registration District; To The Sanitary Commissioner of Bombay, dated Belgaum, 3rd August 1905. Proceedings, February 1906. Nos. 327–347. Measures for the Prevention of Plague. Calcutta Records 5. Government of India Home Department, Sanitary Branch. National Archives of India, Delhi.

mixed success when replicated.[2] This ambiguity, and widespread imperial competition to contribute to the knowledge-making practices around plague prompted, as medical anthropologist Christos Lynteris has noted, "an epistemologically intense, nuanced and complex era of plague research."[3] It was in this heightened context that the Government of India established a robust research program at the Parel Research Laboratory, Bombay, designed to settle the question of plague transmission dynamics.

Between 1898–1910, Indian Medical Service Officers of the Indian Plague Commission and Bombay Plague Committee drew on the local epidemic and information networks within the city to establish a series of long-term epidemiological and bacteriological experiments and surveillance mechanisms to observe, characterize, and define the boundaries of plague within the city.[4] Keeping with the theme of this edited volume, this chapter will explore how a subset of these experiments, concerning the role of the rat-flea, *Xenopsylla*

2 The Advisory Committee Appointed by the Secretary of State for India, the Royal Society, and the Lister Institute, "Reports on Plague Investigations in India," *The Journal of Hygiene* 6, no. 4 (1906), http://www.jstor.org/stable/3858976.

3 Christos Lynteris, "In Search of Lost Fleas: Reconsidering Paul-Louis Simond's Contribution to the Study of the Propagation of Plague," *Medical History* 66, no. 3 (2022), doi:10.1017/mdh.2022.19.

4 Christos Lynteris, "Pestis Minor: The History of a Contested Plague Pathology," *Bulletin of the History of Medicine* 93, no. 1 (2019); Nicholas H.A. Evans, "Blaming the Rat? Accounting for Plague in Colonial Indian Medicine," *Medicine Anthropology Theory* 5, no. 3 (2018), doi:10.17157/mat.5.3.371; No. 749 P, dated Bombay, the 12th August 1905. From Major WE Jennings, MD, DPH, IMS, Superintendent, Plague Operations, Bombay Presidency; To the Secretary to Government, General Department (Plague). Through The Surgeon-General with the Government of Bombay, pg. 145. National Archives of India, New Delhi, India; J.T.W Leslie, "Question whether it is possible or expedient to adopt generally any organized system for which the extermination of rats in areas which are infected with plague or threatened with an importation of the disease," May 1902, Calcutta Records 3, Sanitary Plague A Branch, Home Department, Government of India, nos. 114–116, pg. 6, National Archives of India; JA Turner, "Executive Health Officer's Report for Bombay, 1910–1925," IOR/V/25/840/24, India Office Records, AAS, BL; TS Weir, Reports of the Health Officer of Bombay, 1896–1909, IOR/V/25/840/23, Asian and African Studies Collection, British Library, London, United Kingdom; R.D. Saigol, IMS, "On Experiments Made to Determine the Jumping Power of Rats," July 10, Simla Records 3, Sanitary Plague A Branch, Home Department, nos. 70–73, pg. 9–10, National Archives of India; "Experiments to test the efficacy of M. Duclaux's cultures for the purpose of destroying rats," March 1901, Calcutta Records 1, Sanitary Plague A Branch, Home Department, Government of India, nos. 55–56, pg. 1–4, National Archives of India; J Dansyz, "Un Microbe Pathogene pour les Rats: Et son application a la destruction de ces animaux," in *Annales de L'Institut Pasteur* 14, no. 4 (1900); Cunningham, Captain J MD, IMS (Bombay Bacteriological Laboratory). *Scientific Memoirs by the Officers of the Medical and Sanitary Departments of the Government of India. The Destruction of Fleas by Exposure to the Suno* (Calcutta: Superintendent Government Printing, India, 1911). V/25/850/53. India Office Records, Asian and African Studies Collection, British Library.

cheopis, in plague transmission relied on multispecies relationships and ecologies within the city to develop and test the mechanisms of the rat-flea theory. Looking to a key set of experiments undertaken by the Commission in 1905–1906, it will examine how the urban ecology of Bombay and the highly specific, cyclic presentation of *Yersinia pestis* in the city informed the structure of experiments, allowing imperial medical officials to engineer these epizootics among "wild" rats, fleas, and experimental animals in ways that elucidated the local transmission pathways for plague. Looking to the published reports of the Second Plague Commission from the *Journal of Hygiene* and housed in archival repositories, as well as the sanitary records of the Government of Bombay, evidence suggests that imperial scientists treated the urban ecologies and the multiple species that lived within them as part of an "experimental system," and utilized the locally-specific presentation of plague to drive and design inquiry into the epizootic. While the ecological specificity of the epidemic in Bombay played a significant role in how these experiments were designed, however, the embeddedness of these experiments in a global, multi-scalar knowledge infrastructure meant that the findings of these studies were widely adopted and accepted as indicative of a more general plague etiology and epidemiology – with significant, long-lasting effects on the study of the disease.[5]

The role of colonial spaces as sites of experimentation and knowledge-production has been well-documented and theorized by historians of science and environment. Helen Tilley, Daniel Headrick, and others have demonstrated how, in the first half of the twentieth century, a practice of use of colonial sites as "living laboratories" emerged, in which scientific knowledge-gathering practices around human and nonhuman disease increasingly included controlled "field" experiments, designed to document, describe, and translate the mechanisms of disease transmission.[6] These colonial sites, including their infrastructures, the local and European experts working within them, and their

5 Helen Tilley, *Africa as a Living Laboratory* (Chicago: University of Chicago Press, 2011), 11–12. Tilley notes that while knowledge is "situated," there is too much circulation between metropole and colony, colony and colony, and between nation-states for any knowledge infrastructure to fall neatly into a single category, or be truly localized.

6 Tilley, *Africa as a Living Laboratory*, and Daniel R. Headrick, "Sleeping Sickness Epidemics and Colonial Responses in East and Central Africa, 1900–1940," ed. Philippe Büscher, *PLoS Neglected Tropical Diseases* 8, no. 4 (24 April 2014): e2772, doi:10.1371/journal.pntd.0002772. The persistence of field experimentation in colonial spaces contrasts with the divide that was (imperfectly) drawn between field and laboratory sciences over the 20th century. See Linda Nash, *Inescapable Ecologies: A History of Environment, Disease, and Knowledge* (Berkeley: University of California Press, 2006).

information systems, comprise a unique kind of "experimental system," – to borrow from historian of science Hans-Jörg Rheinberger – which occurred within highly specific environments, which created the conditions for highly impactful knowledge formation.[7]

Looking to the records of the Indian Plague Commission, however, we see an additional, critical aspect of plague's "experimental system": its reliance on material local ecology. Bombay, at the turn of the twentieth century, became a "laboratory" for testing the transmission dynamics of plague. The original relationship delineated by Simond between *Xenopsylla cheopis* and *Rattus* species were made robust; and early links between temperature, *Yersinia pestis* replication, and infectivity among fleas were forged, tested, and theorized.[8] Relying on extensive surveillance infrastructure and the unique, cyclic nature of the epidemic that had taken hold between 1896–1907, members of the Second Indian Plague Commission designed experiments that took into account the ecological conditions of the city and the predictable epizootic among "wild rats" to establish a rigorous course of inquiry into the relationship between rats, fleas, and humans in plague. Experiments conducted by the Commission were designed not to create controlled conditions in which to test plague transmission dynamics; rather, IMS officers embraced the permeability of their experimental systems to the uncontrolled, or "wild" environment. In doing so, these experiments demonstrated an important and often under-considered aspect of experimental systems: that material ecology played an important role in experimental systems and informed the kinds of questions and transmission pathways examined. These multi-species experiments, at once highly locally-driven and designed to test general theories of plague transmission, contributed to a broader etiology of plague that carried lasting impacts on the etiology of the disease.

2 Bombay as an Experimental System

The Third Plague Pandemic is estimated to have caused 15 million deaths globally, from its (hypothesized) emergence in Yunnan province, China, in 1854

7 Hans-Jörg Rheinberger, *Towards a History of Epistemic Things: Synthesizing Proteins in the Test Tube* (Stanford: Stanford University Press, 1997); Hans-Jörg Rheinberger, "Experimental Systems: Historiality, Narration, and Deconstruction," *Science in Context* 7, no. 1 (1994), doi:10.1017/S0269889700001599.

8 J.A. Turner, Esq., Executive Health Officer, Bombay Municipality, to the Municipal Commissioner for the City of Bombay. 8 May 1907. No. P. 1398., Home Department, Sanitary (Plague) Branch A, National Archives of India, New Delhi, India.

to its gradual decline in 1950. While the disease was global in scope, mortality was highly unevenly distributed; 10 of the 15 million deaths occurred in India, between 1896–1930.[9] Within India, no city suffered a higher mortality rate, nor longer-lived epidemic, than Bombay. The Bombay plague epidemic is estimated to have killed over 180,000 people in the city from its arrival in 1896 to its eventual disappearance in the 1930s. Death counts of the epidemic are often given in terms of human mortality, sources indicate that there were other residents of the city who suffered; that there was not one epidemic, but many nested epidemics. Scientific reports produced through the imperial government in Bombay showed millions of deaths, colony collapse, and significant behavioral changes among both *Rattus norvegicus* and *Rattus rattus* as they became infected with *Yersinia pestis*. Their epidemic was the result of another among *Xenopsylla cheopis*, or rat fleas, as the bacteria filled their stomachs with a biofilm and slowly starved them to death. The epizootic, as Plague Commissioner P.C.H. Snow notes, was as obvious as the human epidemic. "From the time the bubonic plague had established itself in Mandvi at the end of September 1896," he notes in his report. "Large numbers of rats were seen running about the streets and coming out of house connection pipes and drains in sickly or dying condition."[10]

As the first city outside of China to suffer cases of plague, Bombay garnered significant administrative and epidemiological attention.[11] The magnitude of the epidemic, and the threats it posed to administrative authority within the Empire's "Second City", made Bombay a concentrated site of scientific activity. Waldemar Haffkine (1860–1930), the renowned bacteriologist, was invited to establish a large bacteriological research laboratory dedicated to the study of plague nearly immediately after the first cases were discovered in the neighborhood of Mandvi in 1896.[12] By March 1897, the Bombay Plague Committee was formed to intervene on the burgeoning epidemic within the city, enacting

9 Myron Echenberg, *Plague Ports: The Global Urban Impact of Bubonic Plague, 1894–1901* (New York: New York University Press, 2007), pg. XII, 17, 50–51.

10 P.C.H. Snow, *Report on the Outbreak of Bubonic Plague in Bombay, 1896–1897* (Bombay: "Times of India" Steam Press, 1897), 9.V/27/856/7, Asian and African Studies Collection, British Library, London, United Kingdom.

11 Estimated number in Ira Klein, "Urban Development and Death: Bombay City, 1870–1914," *Modern Asian Studies* 20, no. 4 (1986): 729; and David Arnold, *Colonizing the Body: State Medicine and Epidemic Disease in Nineteenth-Century India* (Berkeley: University of California Press, 1993), 201.

12 W.B. Bannerman, "The Plague Research Laboratory of the Government of India, Parel, Bombay," *Proceedings of the Royal Society of Edinburgh* 24 (1904), doi:10.1017/S0370164600007781.

mass sanitizing campaigns, constructing plague camps, and establishing surveillance networks across the city.[13] In 1898, the first Indian Plague Commission, referred to as the Fraser commission after its leading physician, was established to pursue the etiology, epidemiology, and cause of the plague epidemic in India, beginning with Bombay; however, it was the Second Commission, established in 1905, that focused its attention most directly on drawing out the nuances and mechanisms of zoonotic transmission of plague – with Bombay as its major experimental site.[14]

There were several reasons that Bombay was chosen as a key site of investigation into the epidemiology and etiology of plague by the Second Indian Plague Commission, as elucidated in their 1906 report in *The Journal of Hygiene*. First, the city housed the laboratory and scientific infrastructure needed to conduct a variety of experiments on plague transmission. The first British Indian research laboratory, established in Bombay in 1884, was a site "in every way and well suited for the requirements of the Commission."[15] The medical laboratory housed at Grant Medical College served as a site of medical and bacteriological training for Indian doctors from its establishment in 1845, and housed Haffkine's initial experiments.[16] The city's municipal and health departments – and the pre-established Bombay Plague Committee – also facilitated surveillance and large-scale coordination of personnel for observation and experiments.

Most important to the Commission, however, was the unique ecology of plague in Bombay. In addition to the sheer magnitude of the epidemic and epizootic, Bombay suffered from plague cyclically, meaning outbreaks could be more or less predicted annually.[17] Seasonal prevalence was also "well-marked", and explaining cyclical seasonal patterns was considered to be a necessary condition for proving any hypothesis about plague transmission. Drawing on the clear relationship between the epidemic and epizootic in the city, the Commission expressed confidence that "[T]he general relationships of the epizootic and epidemic would be obtained from Bombay city, some errors being

13 Report of the Bombay Plague Committee Appointed by Government Resolution No. 1204/720P, on the Plague in Bombay, for the Period Extending from the 1st July 1897 to the 20th April 1896 (Bombay: Times of India Steam Press, 1898), IP/13/PC.5, Medical History of British India Collection, National Library of Scotland. Report of the Bombay Plague Committee, National Library of Scotland.

14 Evans, "Blaming the Rat?," 21–28.

15 "XXII. The Epidemiological Observations Made by the Commission in Bombay City," *The Journal of Hygiene* 7, no. 6 (1907): 725, doi:10.1017/S0022172400033684.

16 Lynteris, "Pestis Minor," 61; Pratik Chakrabarti, *Bacteriology in British* India (Cambridge: Cambridge University Press, 2011), 26–60.

17 Reports of the Indian Plague Commission, 530.

corrected by the very large number of plague rats and of human cases dealt with …"[18] Far from being considered a limitation of the city-as-experimental-site, its unique qualities were seen as integral to elucidating the relationship between rats, rat-fleas, environments, humans, and *Yersinia pestis*.

3 Urban Structure as Experimental System

Beginning in 1904, the Bombay Plague Committee began a program of rat capture and counting to monitor the epizootic within the city. Included in this process was the systematic trapping of rats from key locations around the city, combing of the rats for fleas, and dissection of rat and rat-flea for signs of *Yersinia pestis* infection.[19] These robust surveillance infrastructures, combined with existing experiments by members of the Government Laboratory (in particular WB Bannerman (1858–1924), director of the Plague Committee), had already raised a series of questions about the relationship between urban structure and ecology and the plague epizootic, and identified regions where urban structure may play a role in transmission.[20] Bannerman notes in his own 1906 experiments that different roof and housing structure types appeared to be more or less conducive to rats, observing,

> The structure of the houses in this country seems designed to favour the continued existence within them of the black rat … in Bombay the roofs of round country tiles and the curious shelf-like projections found in almost every room in the chawls, where firewood and dung cakes are stored, afford them ideal places for shelter and breeding.[21]

Indian Medical Service (IMS) officers also drew on the unique seasonality of the plague epizootic in the city to conduct studies on the role of climate on flea transmission. Disparities between Nuttall and Simond's findings in 1898 in the role of temperature in the activity of fleas inspired studies like those conducted in *The Fourth Progress Report of the Plague Research Commission*, in which "A large number of observations on the effect of temperature on the transmission of plague by fleas have been carried out in specially constructed

18 "Epidemiological Observations," 725.

19 "Epidemiological Observations," 728–752.

20 W.B. Bannerman, "Conditions Affecting the Origin and Spread of Plague," in "Measures for the Prevention of Plague," February 1906, Calcutta Records 5. Government of India Home Department. Sanitary Branch. Proceedings, February 1906. Nos. 327–347, 106. National Archives of India.

21 Bannerman, "Conditions Affecting the Origin and Spread of Plague," 106.

rooms both below and above the ordinary room temperature." While some were optimistic that the results of these experiments "have thrown considerable light on the problem of the seasonal prevalence of the disease" the Commission remained cautious about the conclusiveness of the findings.[22] It was precisely these questions that the Commission sough to settle and hoped that the cyclicality of Bombay's epidemic would facilitate.

In 1907, The Commission designed a series of ecological experiments to test the transmission pathways of plague in urban environments, grounded in the construction of environments that would foster an epizootic among rodent communities, using observations of the plague in the city to construct the experimental system. Experiments concerning "the relative importance of the Indian rat flea, *Xenopsylla cheopis*, and of actual close contact in the absence of fleas, in the dissemination of plague from animal to animal" highlight the focus on interconnected ecologies, porous experimental boundaries, and multispecies etiologies in these experiments.

The studies began with the construction of replica "go-downs" designed by Lt. Col. Bannerman and Captain Liston (1872–1950) (who had already begun to experiment with the transmission dynamics of the rat-flea in 1905) of the Bombay Bacteriological laboratory.

The body of the buildings were constructed to be "rat-proof", with 9-inch walls built of brick and mortar and concrete floors on top of a concrete slab. Inside, an "inspection chamber" made of wire netting about 3.25 by 3 feet wide was connected to wire netting covering the inside of the roof but separated from the rest of the hut, to prevent anything that might settle in the roof from entering the hut itself. The object of this structure was to allow wild rats to colonize the roofs, but to limit contact between "wild" rats and experimental animals, and thus controlling for the possibility of direct rat-to-rat transmission but allowing fleas to move from wild animals to experimental animals. (see Figure 13.1) [23]

Once identical huts had been constructed, an environment was engineered in which "wild" Bombay rats (and their resident fleas) would colonize the roofs. Drawing on observation (and presumably available materials) from the city, Bannerman and Liston instructed that there be three different types of roofing material across the six huts: the first two huts (1 and 2) were furnished with "country tile"; the second two (3 and 4) with "Mangalore tiles" (a red clay tile commonly used for roofing across India); and the third two (5 and 6) of corrugated steel.[24]

---

22 Bannerman, "Conditions Affecting the Origin and Spread of Plague," 106.

23 "Reports on the Plague Investigations in India," 450.

24 "Reports on the Plague Investigations in India," 451–453.

FIGURE 13.1 Photograph of the six "plague go-downs" constructed at the Plague Research Laboratory in Parel, Bombay, designed to mimic different urban structural conditions known to be more or less conducive to plague
SOURCE: "REPORTS ON PLAGUE INVESTIGATIONS IN INDIA." *THE JOURNAL OF HYGIENE* 6(1906): 450–451

Choices in roofing were based on ecological observations of rat density and plague mortality within the city: "in the case of go-downs Nos. 1 and 2," the report notes,

> the roofs of which offer good protection and shelter to the wild rat of Bombay, the flea supply is abundant and regular; in the case of go-downs Nos. 3 and 4, the roofs of which offer only poor protection to rats, the flea supply is more or less scanty; while in go-downs Nos. 5 and 6, the roofs of which are absolutely impervious to rats, no fleas should be able to gain access unless carried through the door on the experimental animals themselves, or by the attendant when feeding these animals.[25]

25 "Reports on the Plague Investigations in India," 453.

Once the requisite ecologies had been established in each hut, a series of experiments were designed to test the role of the rat-flea in transmission. In the first round of experiments, three guinea pigs were released in each hut for six days, then combed daily for fleas, to demonstrate the varied porousness of experimental environments to the broader urban ecology – and they noted that "the number of fleas varies in each instance with the accessibility of the roof to rats."[26] The experiments that followed focused on manipulating the number of fleas present in each go-down and the duration of exposure to fleas by experimental animals to establish the link more firmly between infected rat-fleas and epizootic plague. Some experiments relied on the existing epizootic to jump to the experimental guinea pigs, and observation of the trajectory of the epizootic among inoculated and uninoculated guinea pigs once it commenced. In the first round of experiments, several guinea pigs were inoculated with virulent *Yersinia pestis* and released to live alongside healthy guinea pigs in experimental huts 5 and 6 (devoid of fleas). In these experiments, it was found that the inoculated guinea pigs died, while uninoculated guinea pigs remained healthy. The experiment was then repeated in hut 2, where "rats had taken up their abode under the tiles" and thus "these go-downs were kept supplied with rat fleas, for as the insects left their hosts they fell down into the go-downs."[27] In these experiments, uninoculated animals died from plague in varying numbers; experiments were conducted during different periods, dependent on the pervasiveness of the epizootic within the city as a whole – and mortality rates among guinea pigs responded accordingly.[28] When conducted in November, during the cyclic increase in the urban epizootic, the experimental hut experienced an "epizootic of the most rapid description," with 115 fleas isolated from the last five animals to die of the disease – nearly ten times the typical number.[29]

The second set of experiments relied solely on transmission between experimental animals using fleas continuously replenished from the "wild" environment. Fleas isolated from other huts, in which epizootics were active among experimental guinea pigs, were then transferred to huts with healthy guinea pigs, and consistently added additional fleas over a number of days. The experiment showed that when the flea population was consistently replenished, the epizootic continued until all guinea pigs had died.[30]

---

26 "Reports on the Plague Investigations in India," 453.

27 "Reports on the Plague Investigations in India," 456.

28 "Reports on the Plague Investigations in India," 456–457.

29 "Reports on the Plague Investigations in India," 457.

30 "Reports on the Plague Investigations in India," 460.

While the controlled conditions were ultimately compromised – an "abundant supply of fleas" were found in go-down No. 6, for reasons unknown to experimenters – the existing hypotheses for why each would be useful in allowing or disallowing wild urban rats to colonize the space demonstrate a use of local urban ecology to experiment with and induce epizootics. In each case, the porousness of the experimental system was central to experimental design; *Xenopsylla cheopis* was isolated from the urban environment and conditions established for its breeding. What is more, Bannerman and Liston used the distinct seasonality of Bombay's epidemic to time their experiments. The first three experiments were undertaken when plague cases were sporadic among the city's wild rats (in June and July), and served as "control" experiments; the remaining three, tested against these original three, occurred in November, "during the period where the epizootic was just commencing."[31]

In keeping with Rheinberger's assertion that experiments often intentionally or unintentionally test multiple hypotheses, we might also look to how experiments controlled for, tested, and probed competing theories of plague transmission.[32] For example, in hut No. 1 and No. 3, a "certain amount of light" was allowed to penetrate through a small glass window in the tiles, and a small ventilation hole established in the housing, testing the role of sunlight and air flow on the spread of the microbe (in line with hygienic and miasmatic theories of disease). In multiple experiments, guinea pigs or chimpanzees were suspended in their own cages above the floor to avoid possible interaction with feces, urine, or infected soil in each hut, or set in cages in which fly paper was placed around the bottom six inches of the cage to prevent fleas jumping in (while simultaneously verifying existing findings from Egypt that fleas could not jump higher than six inches).[33] In another experiment, pregnant guinea pigs were removed from epizootic conditions and combed for fleas, and those already infected with plague were not found to transmit the disease to their offspring. Looking to these experiments, and the ways they used their environments to test and define the parameters of plague transmission, it is clear that the particular ecology of plague in Bombay had a direct effect on their conditions, observations, and outcomes; and the material environment in which the experiments were conducted was made inseparable from its findings.

---

31 "Reports on the Plague Investigations in India," 456.

32 Rheinberger, *Epistemic Things*, 76. Rheinberger argues that in experiments, "there at every step what is about to take shape creates unforeseen alternative directions for the next step to be taken."

33 "Reports on the Plague Investigations in India," 464–466.

This relationship becomes even more explicit in another set of studies which eschewed the controlled environments of the constructed huts altogether and instead loosed guinea pigs in houses recently impacted by plague. In a series of studies, guinea pigs were either allowed to run free in houses where plague had broken out or left in these houses in cages. The guinea pigs were then re-captured and combed for rat fleas. The Commission found that there were up to 40 rat-fleas on each of them, at least 40 percent of which were infected with the bacteria.[34] Perhaps most damningly, the same result occurred when guinea pigs were allowed to run free in houses that had been disinfected with sulfuric acid after the outbreak of plague, to much the same result – which indicated that British sanitary disinfection practices were not effective in removing fleas from housing and therefore were likely ineffective in preventing the re-emergence of plague.[35]

Engagement with the material urban environment in the design of rat-flea experiments carried significant implications for understanding of both plague etiology and understanding the legitimacy of plague control. The Commission engineered an environment in which the multispecies assemblages of the local plague epidemic could be tested, described, and (they believed) generalized. The resultant studies offered one of the most nuanced and detailed descriptions of plague dynamics to date, and provided support to existing theories of plague transmission while throwing others into question. Perhaps most strikingly, they undermined arguably the most controversial public health intervention enacted during the epidemic.[36]

4 Continuing Legacies: the Rat-Flea Hypothesis, Epidemiological Modeling, and Plague

From these experiments in which epizootics were manufactured within controlled environments – or observed in the broader urban ecology – a number of claims on the mechanisms of the transmission of plagues were posited and solidified.[37] As the results presented and the design of their experiments suggest, the transmission pathways elucidated by the Commission were quite ecologically specific. However, as historian Christos Lynteris argues in his

34 "Transmission of Plague by Fleas," Tables I–IV, in "Reports on the Plague Investigations in India," 482–483.

35 "Transmission of Plague by Fleas," Tables I–IV, in "Reports on the Plague Investigations in India," 481–482.

36 Evans, "Blaming the Rat?," 26–28; Arnold, *Colonizing the Body*, 200–239.

37 A.W. Bacot and C.J. Martin, "LXVII. Observations on the Mechanism of the Transmission of Plague by Fleas," *Journal of Hygiene* 13 (1914).

study of Paul-Louis Simond, the experiments also occurred in a different kind of "ecology" – what Anthropologist Charles Briggs calls a set of "ecologies of evidence."[38] In other words, these experiments were part of "broader assemblages" of networks of knowledge production that allowed certain types of evidence to be privileged, expanded upon, and mobilized over others.[39] Findings from the Second Plague Commission circulated widely across the Empire, and its experiments on the rat-flea cited widely in English-language medical journals throughout the 20th century. Articles in professional journals as widespread as *The British Medical Journal*, *the Journal of Hygiene*, and *the Journal of Infectious Diseases*, claimed that the study "established very conclusively the fact that the flea is the most important factor in the transmission of plague from rat to man."[40]

The rat-flea experiments conducted in Bombay were thus taken by imperial scientists to constitute a "representative" model for plague transmission – and one that definitively affirmed the rat-flea theory of disease. However, the findings conferred by the rat-flea experiments and the model of transmission they proposed, while often aligned with observations in other plague-cities, were constructed and affirmed under highly specific circumstances, which presented only a limited and partial view of plague dynamics. Following from the work of philosopher of science Nancy Cartwright, we might see the central slippage here as being in model interpretation: a representative model (a model that represented a particular phenomenon in the world) was taken by the wider scientific community to be an interpretive model (a model that could be linked to an abstract theory) in an instance where such a leap was dubious – if useful for integrating the study's results into existing "ecologies of evidence."[41] By relying heavily on the local ecology and its natural mechanisms to engineer a laboratory setting, what the experiments ultimately showed was

38 Charles L. Briggs, "Ecologies of Evidence in a Mysterious Epidemic," *Medicine Anthropology Theory* 3, no. 2 (2016), doi:10.17157/mat.3.2.430.

39 Briggs, "Ecologies of Evidence," 151.

40 Bacot and Martin, "Transmission of Plague by Fleas"; T.L. Anderson and J. Burton Cleland, "The Transmission of Plague," *The British Medical Journal* 1, no. 2414 (1907): 838; "Fleas and Plague: Recent Additions to Our Knowledge of the Mechanism by Which Fleas Probably Spread the Disease," *Public Health Reports (1896–1970)* 29, no. 19 (1914); Wheeler, C.M. and J.R. Douglas, "Sylvatic Plague Studies: V. The Determination of Vector Efficiency," *The Journal of Infectious Diseases* 77, no. 1 (1945), http://www.jstor.org/stable/30061611; Cole, LaMont C., "The Effect of Temperature on the Sex Ratio of Xenopsylla Cheopis Recovered from Live Rats," *Public Health Reports (1896–1970)* 60, no. 45 (1945), doi:10.2307/4585454.; W.M. Frazer, "Rats and Vermin and Their Role in the Spread of Disease," *The Journal of State Medicine (1912–1937)* 40, no. 12 (1932).

41 Stephan Hartmann, Carl Hoefer, and Luc Bovens (eds.), *Nancy Cartwright's Philosophy of Science*, Routledge Studies in the Philosophy of Science 3 (New York: Routledge, Taylor &

a detailed picture of the ecology of the Bombay epidemic itself and the role of rat-fleas within it.

While this model of plague transmission related to existing observations and proposed theoretical frameworks for plague transmission, including those listed above, historians of the plague (and indeed historical actors) have pointed out the ways that the Bombay epidemic was itself highly unusual. Features like housing type, urban structure, and rat ecology were highly specific to Bombay, and the epidemic itself, in its cyclicality, endemicity, and mortality rates did not easily map on to other plague-affected areas, and yet had a significant effect on transmission dynamics.[42] In being highly locally grounded, therefore, the study sacrificed its generalizability – but was nevertheless treated as generalizable evidence because of its adherence to basic mechanisms of scientific validity and seeming support of prevalent theoretical claims constructed by scientists across the British Empire. Meanwhile, the more robust, locally-specific results were largely ignored, and slum clearance and sanitation projects continued in spite of evidence of their ineffectiveness.[43]

The effects of this confluence of study design and broader imperial ecologies of knowledge carried long-term effects for the etiology of plague. Looking to several key studies in mathematical biology and epidemiology in the last twenty years, we can see how the cyclic nature of the epidemic in Bombay – and once again, fleas and findings on fleas from studies conducted during this time – are used to construct an ecology of plague that can be tested and verified. The fit of Kermack and McKendrick's SIR model (1927) with the data from the 1905–1906 Bombay epidemic has been referred to as the "most reproduced figure in books discussing mathematical epidemiology," and was taken as evidence of the model's strength as a predictor of epidemic structures.[44]

Francis Group, 2008), 24–25; Nancy Cartwright, *The Dappled World: A Study of the Boundaries of Science* (Cambridge: Cambridge University Press, 1999).

42 Emily Webster, "Plague in Bombay, 1896," in *Epidemic Urbanisms: Contagious Diseases in Global Cities*, ed. Mohammad Gharipour and Caitlin DeClerq (Bristol: Intellect Press, 2021); Ira Klein, "Urban Development and Death"; Prashant Kidambi, "'An Infection of Locality': Plague, Pythogenesis and the Poor in Bombay, c.1896–1905," *Urban History* 31, no. 2 (2004); Myron Echenberg, "Pestis Redux: The Initial Years of the Third Bubonic Plague Pandemic, 1894–1901," *Journal of World History* 13, no. 2 (2002); Arnold, *Colonizing the Body*, 200–239.

43 Sandeep Hazareesingh, "Colonial Modernism and the Flawed Paradigm of Urban Renewal: Uneven Development in Bombay, 1900–1925," *Urban History* 28, no. 2 (2001).

44 The SIR model estimates the trajectory of an epidemic within a population based on the number of susceptible (S), Infected (I), and Recovered (R) members of that population. It has become a cornerstone in epidemiological modelling since it was first posted in 1927.

However, this modeling structure was questioned by Bacaër in 2012, who pointed out the "remarkable seasonal pattern" of the epidemic undermined the original model, and Bacaër presents instead a seasonal model that explicitly used flea ecology described by the Commission to establish parameters with very different results.[45] Similarly, Keeling and Gilligan's widely-cited metapopulation model of zoonotic plague, relies on estimates of number of fleas per rat, flea life cycle and death rate, that are all derived from the structure of plague in India, and also cite the 1906–1907 Plague Commission explicitly in their studies.[46]

The observations taken from the 1906–1907 plague experiments were therefore used as evidence as to the epidemiology of plague generally – and yet, the study site itself was chosen because of its unique ecology. This tension has been pointed out by plague historians like Ann Carmichael, who posits the importance of marmots in the Alps as reservoirs of plague; and Matheus Alves Duarte da Silva, who examines the emergence of the concept of sylvatic plague and research on its circulation among wild rodents in the 20th century – a mechanism that remains highly uncertain and under-researched to this day.[47] These disparities and discontinuities carry real-time implications for plague research, as well; the ongoing outbreak of plague in Madagascar, which began in 1898 with the Third Plague Pandemic, is poorly understood in its ecological dynamics in part because it relies on some of these understudied transmission mechanisms.[48]

William Ogilvy Kermac and A.G. McKendrick, "A Contribution to the Mathematical Theory of Epidemics," *Proceedings of the Royal Society of London. Series A, Containing Papers of a Mathematical and Physical Character* 115, no. 772 (August 1927): 700–721, doi:10.1098/rspa.1927.0118.

45 Nicolas Bacaër, "The Model of Kermack and McKendrick for the Plague Epidemic in Bombay and the Type Reproduction Number with Seasonality," *Journal of Mathematical Biology*, 64 (2012), doi:10.1007/s00285-011-0417-5.

46 M.J. Keeling and C.A. Gilligan, "Bubonic Plague: A Metapopulation Model of a Zoonosis," *Proceedings of the Royal Society of London. Series B: Biological Sciences* 267 (2000).

47 Matheus Alves Duarte Da Silva, "Between Deserts and Jungles: The Emergence and Circulation of Sylvatic Plague (1920–1950)," *Medical Anthropology* 42, no. 4 (2023), doi:10.1080/01459740.2023.2189110.

48 Jennifer Alderson et al., "Factors Influencing the Re-Emergence of Plague in Madagascar," *Emerging Topics in Life Sciences* 4, no. 4 (2020), doi:10.1042/ETLS20200334; Voahangy Andrianaivoarimanana et al., "Understanding the Persistence of Plague Foci in Madagascar," *PLoS Neglected Tropical Diseases* 7, no. 11 (2013), doi:10.1371/journal.pntd.0002382.

5 Conclusion: Etiologies and Epidemiology of Time and Place

While much has been written on both experimental systems and field experimentation, this chapter has argued that there are three key insights to be gained from looking at the role of rat-flea ecologies in the design of field experiments in imperial epidemics. First, that by treating the urban space as a "experimental system" of its own – and attempting to utilize the locally-specific urban ecologies and porosity between controlled and uncontrolled environments to drive and design inquiry – the Indian Plague Commission was able to both "settle" major epistemic controversies and generate new hypotheses about plague transmission that were highly consequential. Second, the way that the specific ecology of plague in the city shaped the structure of experiments – with scientists utilizing both the cyclical nature of the epidemic and its sheer scale to both justify its use as an experimental site and to inform the structure of experiments – played a key role in the contributions of the Commission. Finally, that understanding this specificity allows us to think more critically about the multi-scalar imperial knowledge systems that allowed the findings of these studies to be widely adopted and accepted within broader plague etiology and epidemiology, despite continued emphasis on the unique ecology of the epidemic in the urban environment. Looking to the epizootics engineered by the Second Indian Plague Commission in Bombay, we can see how the ecological specificity of the epidemic in time and place allowed for particular experimental designs that themselves influenced the way knowledge was gathered, generalized, and legitimized – an epistemic lineage we are still discovering the ramifications of to this day.

Acknowledgements

Research for this project was funded by the National Science Foundation's Doctoral Dissertation Research Improvement Grant, the Social Science Research Council's International Dissertation Research Fellowship, and the Centre for South Asian Studies at the University of Chicago. Thanks to members of the Multispecies Insects Workshop, which met in Turku, Finland, in May 2022 and again in December 2022 for valuable insights and commentary on early drafts of this article, and to the participants of the Epizootics Beyond the Farm conference for suggestions on later iterations. Special thanks to Sanna Lillbroända and Christos Lynteris for their insights and recommendations.

Bibliography

"XXII. The Epidemiological Observations made by the Commission in Bombay City." *The Journal of Hygiene* 7, no. 6 (1907): 724–798. doi:10.1017/S0022172400033684.

The Advisory Committee Appointed by the Secretary of State for India, the Royal Society, and the Lister Institute. "Reports on Plague Investigations in India." *The Journal of Hygiene* 6, no. 4 (1906): 421–536. http://www.jstor.org/stable/3858976.

Alderson, Jennifer, Max Quastel, Emily Wilson and Duncan Bellamy. "Factors Influencing the Re-Emergence of Plague in Madagascar." *Emerging Topics in Life Sciences* 4, no. 4 (2020): 411–421. doi:10.1042/ETLS20200334.

Alves Duarte Da Silva, Matheus. "Between Deserts and Jungles: The Emergence and Circulation of Sylvatic Plague (1920–1950)." *Medical Anthropology* 42, no. 4 (2023): 325–339. doi:10.1080/01459740.2023.2189110.

Anderson, T.L. and J. Burton Cleland. "The Transmission of Plague." *The British Medical Journal* 1, no. 2414 (1907): 838.

Andrianaivoarimanana, Voahangy, Katharina Kreppel, Nohal Elissa, Jean-Marc Duplantier, Elisabeth Carniel, Minoarisoa Rajerison and Ronan Jambou. "Understanding the Persistence of Plague Foci in Madagascar." *PLoS Neglected Tropical Diseases* 7, no. 11 (2013): e2382. doi:10.1371/journal.pntd.0002382.

Arnold, David. *Colonizing the Body: State Medicine and Epidemic Disease in Nineteenth-Century India.* Berkeley: University of California Press, 1993.

Bacaër, Nicolas. "The Model of Kermack and McKendrick for the Plague Epidemic in Bombay and the Type Reproduction Number with Seasonality." *Journal of Mathematical Biology*, 64 (2012), 403–422. doi:10.1007/s00285-011-0417-5.

Bacot, A.W. and C.J. Martin. "LXVII. Observations on the Mechanism of the Transmission of Plague by Fleas." *Journal of Hygiene* 13 (1914): 423–439.

Bannerman, W.B. "The Spread of Plague in India." *Journal of Hygiene* (Lond) 6, no. 2 (1906): 179–211. doi:10.1017/s0022172400002813.

Bannerman, W.B. "The Plague Research Laboratory of the Government of India, Parel, Bombay." *Proceedings of the Royal Society of Edinburgh* 24 (1904): 113–144. doi:10.1017/S0370164600007781.

Briggs, Charles L. "Ecologies of Evidence in a Mysterious Epidemic." *Medicine Anthropology Theory* 3, no. 2 (2016). doi:10.17157/mat.3.2.430.

Cartwright, Nancy. *The Dappled World: A Study of the Boundaries of Science.* Cambridge: Cambridge University Press, 1999.

Chakrabarti, Pratik. *Bacteriology in British India.* Cambridge: Cambridge University Press, 2011.

Cole, LaMont C. "The Effect of Temperature on the Sex Ratio of Xenopsylla Cheopis Recovered from Live Rats." *Public Health Reports (1896–1970)* 60, no. 45 (1945): 1337–1342. doi:10.2307/4585454.

Cunningham, Captain J., MD. IMS (Bombay Bacteriological Laboratory). *Scientific Memoirs by the Officers of the Medical and Sanitary Departments of the Government of India. The Destruction of Fleas by Exposure to the Sun.* Calcutta: Superintendent Government Printing, India, 1911. V/25/850/53. India Office Records, Asian and African Studies Collection, British Library, London.

Dansyz, J. "Un Microbe Pathogene pour les Rats: Et son application a la destruction de ces Animaux." *Annales de L'Institut Pasteur* 14, no. 4 (1900): 193–204.

Echenberg, Myron. "Pestis Redux: The Initial Years of the Third Bubonic Plague Pandemic, 1894–1901." *Journal of World History* 13, no. 2 (2002): 429–449.

Echenberg, Myron. *Plague Ports: The Global Urban Impact of Bubonic Plague, 1894–1901.* New York: New York University Press, 2007.

Evans, Nicholas H.A. "Blaming the Rat? Accounting for Plague in Colonial Indian Medicine." *Medicine Anthropology Theory* 5, no. 3 (2018): 21–28. doi:10.17157/mat.5.3.371.

"Fleas and Plague: Recent Additions to Our Knowledge of the Mechanism by Which Fleas Probably Spread the Disease." *Public Health Reports (1896–1970)* 29, no. 19 (1914): 1133–1135.

Framing Animals as Epidemic Villains: Histories of Non-Human Disease Vectors, edited by Christos Lynteris. London: Palgrave MacMillan, 2019.

Frazer, W.M. "Rats and Vermin and Their Role in the Spread of Disease." *The Journal of State Medicine (1912–1937)* 40, no. 12 (1932): 717–726.

Hazareesingh, Sandeep. "Colonial Modernism and the Flawed Paradigm of Urban Renewal: Uneven Development in Bombay, 1900–1925." *Urban History* 28, no. 2 (2001): 235–255.

Headrick, Daniel R. "Sleeping Sickness Epidemics and Colonial Responses in East and Central Africa, 1900–1940," edited by Philippe Büscher, *PLoS Neglected Tropical Diseases* 8, no. 4 (24 April 2014): e2772. doi:10.1371/journal.pntd.0002772.

Keeling, M.J. and C.A. Gilligan. "Bubonic Plague: A Metapopulation Model of a Zoonosis." *Proceedings of the Royal Society of London. Series B: Biological Sciences* 267 (2000): 2219–2230.

Kermac, William Ogilvy and A.G. McKendrick. "A Contribution to the Mathematical Theory of Epidemics." *Proceedings of the Royal Society of London. Series A, Containing Papers of Mathematical and Physical Character* 115, no. 772 (August 1927): 700–721. doi:10.1098/rspa.1927.0118.

Klein, Ira. "Urban Development and Death: Bombay City, 1870–1914." *Modern Asian Studies* 20, no. 4 (1986): 725–754.

Klein, E. "Report of the Indian Plague Commission." *Nature* 65 (1902): 320–321.

Kidambi, Prashant. "'An Infection of Locality': Plague, Pythogenesis and the Poor in Bombay, c.1896–1905." *Urban History* 31, no. 2 (2004): 249–267.

Khumbar, Kiran. "The Minutes of the Indian Plague Commission." *Harvard Library Office for Scholarly Communication.* Accessed August 31, 2023. https://dash.harvard.edu/handle/1/37368749.

Lamb, George, and Plague Commission. *The Etiology and Epidemiology of Plague: A Summary of the Work of the Plague Commission.* 1908. Accessed November 29, 2023. https://wellcomecollection.org/works/kw68pvgs

Lynteris, Christos. "A 'Suitable Soil': Plague's Urban Breeding Grounds at the Dawn of the Third Pandemic." *Medical History* 61, no. 3 (2017): 343–357. doi: 10.1017/mdh.2017.32.

Lynteris, Christos. "In Search of Lost Fleas: Reconsidering Paul-Louis Simond's Contribution to the Study of the Propagation of Plague." *Medical Hist*ory 66, no. 3 (2022): 242–263. doi:10.1017/mdh.2022.19.

Lynteris, Christos. "Pestis Minor: The History of a Contested Plague Pathology." *Bulletin of the History of Medicine* 93, no. 1 (2019): 55–81.

Stephan Hartmann, Carl Hoefer, and Luc Bovens (eds.). *Nancy Cartwright's Philosophy of Science.* Routledge Studies in the Philosophy of Science 3. New York: Routledge, 2008.

Nash, Linda. *Inescapable Ecologies: A History of Environment, Disease, and Knowledge.* Berkeley: University of California Press, 2006.

Rheinberger, Hans-Jörg. *Towards a History of Epistemic Things: Synthesizing Proteins in the Test Tube.* Stanford: Stanford University Press, 1997.

Rheinberger, Hans-Jörg. "Experimental Systems: Historiality, Narration, and Deconstruction." *Science in Context* 7, no. 1 (1994): 65–81. doi:10.1017/S0269889700001599.

Sivaramakrishnan, Kavita. "Recasting Disease and Its Environment: Medical Practitioners, the Plague, and Politics in Colonial India, 1898–1910." In *Cultivating the Colonies: Colonial States and Their Environmental Legacies*, edited by Christina Folke, Karen Oslund, Niels Brimnes and Niklas Thode. Athens: Ohio University Press, 2011.

Snow, P.C.H. *Report on the Outbreak of Bubonic Plague in Bombay, 1896–1897.* Bombay: "Times of India" Steam Press, 1897. 9.V/27/856/7, Asian and African Studies Collection, British Library, London, United Kingdom.

Stenseth Nils Chr, Bakyt B. Atshabar, Mike Begon, Steven R. Belmain, Eric Bertherat, Elisabeth Carniel, Kenneth L. Gage, Herwig Leirs and Lila Rahalison. "Plague: Past, Present, and Future." *PLoS Medicine* 5, no. 1 (2008): e3. doi: 10.1371/journal.pmed.0050003.

Tilley, Helen. *Africa as a Living Laboratory.* Chicago: University of Chicago Press, 2011.

Wheeler, C.M., and J.R. Douglas. "Sylvatic Plague Studies: V. The Determination of Vector Efficiency." *The Journal of Infectious Diseases* 77, no. 1 (1945): 1–12. http://www.jstor.org/stable/30061611.

Webster, Emily. "Plague in Bombay, 1896." In *Epidemic Urbanisms: Contagious Diseases in Global Cities*, edited by Mohammad Gharipour and Caitlin DeClerq, 204–212. Bristol: Intellect Press, 2021.

CHAPTER 13

Epidemic Encounters: Mingling with Mosquitoes in Réunion and Mauritius

Karine Aasgaard Jansen

1 Introduction

> The chikungunya mosquito, it lives in the environment where we live, and thus we don't really notice it ... It's not a new mosquito. When I was a child, I lived in St. Rose where I often frequented the vanilla fields. And in those fields, there was only this type of mosquito, the white and black mosquito. I have always known the white and black mosquito.[1]

As illustrated by Mr. Noel's quote, in the small but densely populated Indian Ocean islands of Réunion and Mauritius, encounters with mosquitoes are everyday occurrences. This is particularly the case in impoverished, urban areas where residents not only share a close-knit environment with each other, but also with white and black striped *Aedes albopictus* and *aegypti* mosquitoes.[2] The female *Aedes* is a carrier of diseases such as dengue, yellow fever, zika, and chikungunya, and between 2005 and 2007, about 30 per cent of the islands' total populations were, for the first time, infected with chikungunya during an epidemic outbreak.[3]

In this chapter, I take a multispecies approach to discuss the entangling of human lives with that of the *Aedes* and how such interactions affected the diffusion and understanding of chikungunya in Réunion and Mauritius. Despite a surge of interest in multispecies encounters across the humanities and social

1 Mr. Noel, 56 years, interviewed 14 September, 2009, Réunion.

2 Due to its characteristic pattern, the *Aedes* is commonly known as the Asian tiger mosquito.

3 Philippe Renault et al., "L'épidemie de Chikungunya à la Réunion et à Mayotte, France, 2005–2006: Le Contexte et les Questions de Surveillance et d'Évaluation Posées," *Bulletin Épidémiologique hebdomaire* (*Institut de Veille Sanitaire INVS*) 38 (2008); François Taglioni and Jean-Sébastien Dehecq, "L'Environnement Socio-Spatial comme Facteur de l'Émergence des Maladies Infectieuses: Le Chikungunya dans l'Océan Indien, *EchoGéo* 9 (2009).

sciences, until now insects such as mosquitoes have largely been overlooked.[4] Instead, due to the risks that some mosquitoes pose to human health, they have primarily been investigated from the perspectives of natural science and biomedicine. However, to advance our knowledge on the encounters between people and mosquitoes that cause infection, I argue that we need to understand mosquitoes and other disease-carrying insects as co-participants rather than as mere vessels or vectors of disease.[5] This includes recognizing how so-called epidemic encounters reflect a set of multispecies networks or ongoing relations among species, objects and spaces.[6] While public health discourse conceives of the risk of arboviral disease in terms of proximity and exposure, risk also comes in form of the body's mingling with mosquitoes.[7] But where and how does this mingling occur? Furthermore, if you cannot avoid mosquitoes, how do you live with them? And what did people think when a familiar mosquito suddenly started spreading a previously unknown disease?

The chapter is based on longitudinal ethnographic fieldwork for a total of eight months in Réunion in 2009 and 2010, and for one year in Mauritius between 2017 and 2018. To answer the research questions above, I will first provide a comparative overview of the epidemic's course on the two islands. Second, I will outline the research background and the qualitative methods that inform this chapter. I will then map out the multispecies networks that facilitate interaction between mosquitoes and humans and, in turn, the spreading of chikungunya. This will be followed by a discussion of how public health efforts to control mosquito breeding grounds challenged residents' uses of these spaces and led them to question the biomedical etiology of chikungunya. Lastly, I will discuss why, in my opinion, a multispecies approach is essential to understand local disease etiologies and, in turn, to limit arboviral disease dispersion.

4 Anu Valtonen, Tarja Salmela and Outi Rantala, "Living with Mosquitoes," *Annals of Tourism Research* 83 (2020), doi:10.1016/j.annals.2020.102945.

5 Hanna Brown and Ann H. Kelly, "Material Proximities and Hotspots: Toward an Anthropology of Viral Hemorrhagic Fevers," *Medical Anthropology Quarterly* 28, no. 2 (2014).

6 Rohan Deb Roy, *Malarial Subjects: Empire, Medicine and Nonhumans in British India, 1820–1909* (Cambridge: Cambridge University Press, 2017).

7 Alex M. Nading, *Mosquito Trails: Ecology, Health, and the Politics of Entanglement* (Oakland: University of California Press, 2014).

2 Chikungunya: the 2005–2007 Outbreak in Réunion and Mauritius

Chikungunya can literally be translated from the Makonde or Kimakonde language spoken in today's southern Tanzania and northern Mozambique as "that which bends up."[8] It refers to an infected person's characteristically arched back resulting from symptoms of arthralgia, as the virus attacks the joints and leads to painful swelling and substantially reduced motor functions. These symptoms may last for only a few days or in some cases for several years. Other symptoms include fever, headaches, and rash.[9] Until late 2023, there was no effective vaccine against the chikungunya virus (CHIKV), but infection results in lifelong immunity.

While the 2005–2007 CHIKV epidemic was the first-ever documented occurrence of the disease in these Western Indian Ocean islands,[10] the circulation of viruses and vectors in the region is not new.[11] Disease, arboviral included, constitutes an integral part of the extensive network of maritime exchange and migration in the Indian Ocean World (IOW) described by historian David Arnold as a "disease zone."[12] Despite their remoteness, the small islands of Réunion and Mauritius are characterized by intense transoceanic movement and interconnectedness both by air and by sea.[13] As such, they also play a critical role as hubs for global disease dispersion.[14] As average temperatures rise worldwide, ecosystems become disrupted, and pathogens and their hosts spread to and thrive in geographical regions where they were previously unknown.[15] The CHIKV and the *Aedes* are no exceptions. While the epidemic started with small outbreaks on the Kenyan coast in June 2004, it then spread

8 Edward A. Alpers, "Chikungunya and Epidemic Disease in the Indian Ocean," in *Disease Dispersion and Impact in the Indian Ocean World*, ed. Gwyn Campbell and Eva-Maria Knoll (London: Palgrave Macmillan, 2020).

9 B. Lee Ligon, "Re-emergence of an Unusual Disease: The Chikungunya Epidemic," *Seminars in Pediatric Infectious Diseases* 17 (2006).

10 Gilles Pialoux et al., "Chikungunya, an Epidemic Arbovirus," *The Lancet Infectious Diseases* 7 (2007); Alpers, "Chikungunya and Epidemic Disease".

11 Anna Winterbottom and Facil Tesfaye (eds.), *Histories of Medicine in the Indian Ocean World, Volume 2: The Modern History* (New York: Palgrave Macmillan, 2016); Campbell and Knoll (eds.), *Disease Dispersion and Impact*.

12 David Arnold, "The Indian Ocean as a Disease Zone, 1500–1950," *South Asia* 14, no. 2 (1991).

13 Burkhard Schnepel and Edward A. Alpers (eds.), *Connectivity in Motion: Island Hubs in the Indian Ocean World* (Cham: Palgrave Macmillan, 2018).

14 Carla Mavian et al., "Islands as Hotspots for Emerging Mosquito-borne Viruses: A One-Health Perspective," *Viruses* 11, no. 1 (2019), doi:10.3390/v11010011.

15 Daniel R. Brooks, Eric P. Hoberg and Walter A. Boeger, *The Stockholm Paradigm. Climate Change and Emerging Disease* (Chicago: University of Chicago Press, 2019).

to the islands of Lamu, the Seychelles, Madagascar, the Comoros, Mayotte, Réunion and Mauritius, followed by India.[16] In addition, imported cases from exposed travelers returning from the affected areas were later identified in for example France, Italy, Hong Kong, USA, and Canada.[17]

Even though CHIKV and its carrier was identified in 1952,[18] chikungunya only became part of Réunion's and Mauritius's communicable disease surveillance systems following the 2005–2007 epidemic. Since the intensive public health interventions against malaria with the use of dichlorodiphenyltrichloroethane (DDT) in the early 1950s,[19] arboviral diseases have received very little attention on the islands.[20] The World Health Organization (WHO) declared for example Mauritius free from malaria in 1973 and Réunion in 1979.[21] Chikungunya had also previously been considered to be a benign disease. This changed, however, when chikungunya claimed its first fatality, a 10-year-old boy, in Réunion at the peak of the epidemic in January 2006. The boy was the first confirmed casualty due to CHIKV worldwide.[22] Moreover, at that point, more than 25,000 new cases were registered in Réunion during the last week of January, and 45,000 cases in the first week of February.[23]

In total, 266,000 cases of chikungunya were registered in Réunion by the end of the epidemic in April 2007. That is nearly 30 per cent of the island's total population of 802,000 as of 2007.[24] In addition, 250 people died from the disease.[25] Mauritius only confirmed 11,472 cases of chikungunya during the epidemic. In contrast to Réunion, these numbers were based on laboratory tests. However, studies of crude death rates (CDRs), that is the number of deaths occurring among the population of a given geographical area during a

16 Taglioni and Dehecq, "L'Environnement Socio-Spatial," 4.

17 Alpers, "Chikungunya and Epidemic Disease".

18 Pialoux et al., "Chikungunya"; Alpers, "Chikungunya and Epidemic Disease".

19 In contrast to CHIKV, malaria is a parasite that the Anopheles mosquito spreads.

20 Karine Aasgaard Jansen, "Tropical Disease and the Making of France in Réunion," in *Histories of Medicine in the Indian Ocean World,* ed. Anna Winterbottom and Facil Tesfaye (New York: Palgrave Macmillan, 2016).

21 World Health Organization, "Countries and Territories Certified Malaria-Free by WHO," last updated January 12, 2024, https://www.who.int/teams/global-malaria-programme/elimination/countries-and-territories-certified-malaria-free-by-who.

22 P. Leyral, "Les Autorités Sanitaires de l'Île le Reconnaisent Officielment: Un Enfant est Mort Directement du Chikungunya," *Le Journal de l'Île de de la Réunion* (*JIR*), February 4, 2006.

23 Taglioni and Dehecq, "L'Environnement Socio-Spatial," 15.

24 J.-L. Rallu, "Populations et Sociétés: Populations et Développement dans l'Outre-Mer de l'Union Européenne," *Bulletin Mensuel d'Information de l'Institut National d'Études Démographiques* 456 (2009).

25 Phillipe Renault et al., "L'épidemie de Chikungunya".

set year, corresponds to 743 excess deaths during the peak months of the epidemic in 2005 and 2006.[26] This amounts to 400,000 affected cases, and 32% of the total population of about 1.2 million at the time.[27]

While the media in Réunion accused Mauritius of underestimating the number of affected cases, the Mauritian media accused Réunion of overestimating them. Both claims probably hold some truth. While Mauritius is a small nation-state that is largely financially dependent on international tourism, Réunion is a French overseas department (DOM). Despite the fact that Réunion is located only 226 kilometers from Mauritius, it is a fully integrated part of France and the European Union (EU). If Mauritius had declared the chikungunya epidemic a "public health crisis" like Réunion did, it would have most likely had a severe negative impact on the country's overall economy. In Réunion, it led instead to a financial relief package totaling 91 million Euros from the French government toward research and sanitary improvements.[28] Still, the increase in aid only occurred at the same time as it was discovered that the *Aedes* mosquito also exists in parts of southern France and that CHIKV could, and in fact did, spread to the mainland. By then it had already been a year since the first case had been registered in Réunion.[29]

3 Research Background and Methods

Most research on the 2005–2007 Western Indian Ocean chikungunya epidemic has been done by epidemiologists and entomologists. There are therefore few studies of chikungunya that are based on qualitative and not quantitative methods. However, as a medical anthropologist I am interested in how epidemics are experienced, explained, and handled at a local level.[30] This entails an understanding of epidemics as plagues characterized by stigmatization,

---

26 Sanjay Beesoon et al., "Chikungunya fever, Mauritius, 2006," *Emerg Infect Dis.* 14, no. 2 (2008), doi:10.3201/eid1402.071024.

27 Taglioni and Dehecq, "L'Environnement Socio-Spatial," 15; Michel Watin, "La Médiatisation de l'Épidémie de Chikungunya à Maurice et à La Réunion (2005–2006)," in *La Santé dans l'Espace Public*, ed. Hélène Romeyer (Rennes: Presses de l'EHESP, 2010).

28 Anon, "M. de Villepin Promet 76 million d'Euros Contre le Chikungunya," *Le Monde*, February 28, 2006.

29 Jansen, Karine Aasgaard, "The Printed Press's Representations of the 2005–2007 Chikungunya Epidemic in Réunion: Political Polemics and (Post)Colonial Disease," *Journal of African Media Studies* 4, no. 2 (2012).

30 Ann H. Kelly, Frédérik Kéck and Christos Lynteris (eds.), *The Anthropology of Epidemics* (New York: Routledge, 2019).

shame, blame, and fear of contagion.[31] To Herring and Swedlund,[32] the distinction between epidemic and plague is not primarily historically anchored, but rather displays two different ways of thinking about infectious disease.[33] To investigate how people's entanglements with mosquitoes have informed their experiences with and conceptualizations of chikungunya, I have therefore employed various qualitative methods during ethnographic fieldwork on both islands. The main methods that inform this chapter are participant observation, semi-structured interviews, and multispecies ethnography.

The principal approach to data collection was the core anthropological method of participant observation. Participant observation entails the gathering of ethnographic data based upon the researcher's participation in everyday life and provides a good indication of social life as it unfolds in contexts. It also encourages willingness to follow unexpected elements as they present themselves in the field. This study draws for example on a range of informal conversations about chikungunya with numerous people in various social settings that I later wrote down in my field diary at the end of each day. The conversations were typically held outside, especially in Réunion. They also frequently took place while carrying out various household chores and gardening activities such as doing laundry, watering plants, picking and shelling beans, or having coffee on the overbuilt terrace (*varang*) of a traditional Réunionese Creole house (*kaz creole*).

To complement my fieldnotes, I also conducted semi-structured and recorded interviews in the local lingua franca Creole or French with 15 key research participants in Réunion and 11 in Mauritius. Most of these interviews took place in their homes. All these research participants had been infected with chikungunya during the 2005–2007 epidemic. In addition, I interviewed and had several meetings and discussions on a running basis with public health experts, epidemiologists, and entomologists in both islands. These interviews were conducted on the basis of their professional roles, centered around issues of public vector-control, and normally took place at their offices during work hours.

Upon the discovery of three new cases of CHIKV in Réunion in August 2009, I also joined a team of field agents from the Regional Department of Health and Social Affairs (*Direction Régionale des Affaires Sanitaires et Sociales*

31 Cf. D. Ann Herring and Alan S. Swedlund (eds.), *Plagues and Epidemics: Infected Spaces Past and Present* (Oxford: Berg, 2010).

32 Herring and Swedlund (eds.), *Plagues and Epidemics*.

33 Cf. Charles E. Rosenberg, *Explaining Epidemics and Other Studies in the History of Medicine* (New York: Cambridge University Press, 1992).

[DRASS]) for a week to conduct daily garden inspections for mosquito breeding grounds.[34] This was a precautionary measure used to prevent the disease from spreading any further. Learning from natural scientists' and public health practitioners' observational tools and procedures in the field[35] allowed me to observe the field agents' interactions with the public and simultaneously learn how to pay attention to mosquitoes and their larvae, as well as to people with respect to their interactions with mosquitoes. For example, "to trail mosquitoes,"[36] I chose to settle in hot and humid coastal areas known to provide favorable breeding conditions for the *Aedes*. Most of my ethnographic fieldwork was carried out in the town of St. Pierre (approximate population of 26,000) in Réunion and in Pointe aux Sables, a suburb to Mauritius's capital city of Port Louis (approximate population of 118,815). Drawing on multispecies ethnography, I also focused my research on ecological spaces or so-called "contact zones" where residents' lives overlap (biologically, socially, and politically) with those of *Aedes* mosquitoes.[37] These spaces included gardens, backyards, or so-called wastelands. While anthropological and ethnographical research has traditionally been human-centered, multispecies ethnographers study the host of organisms whose lives and deaths are linked to human social worlds.[38] Multispecies ethnography thus goes beyond anthropology in challenging and re-negotiating traditional boundaries between nature and culture within the humanities and social sciences. Instead of perceiving nonhumans as objects, they are rather taken as intrinsic parts of people's ongoing relations and interactions with their surrounding environment. Seeking out areas with an abundance of mosquitoes became a way to learn how to live with them, as my research participants did. It also allowed me to adapt a more ethical approach to mosquitoes as fellow earthbound creatures, rather than thinking of them merely as annoying others to be avoided or killed.[39]

34 DRASS is known today as *Agence Régional de Santé La Réunion* (ARS).

35 Heather Ann Swanson, "Methods for Multispecies Anthropology: Thinking with Salmon Otolithes and Scales," *Social Analysis* 61, no. 2 (2017).

36 Nading, *Mosquito trails*.

37 Donna J. Haraway, *When Species Meet* (Minneapolis: Minnesota University Press, 2008), 244; Alex Aisher and Vinitra Damodaran, "Introduction: Human-Nature Interactions through a Multispecies Lens," *Conservation and Society* 14, no. 4 (2016).

38 Eben S. Kirksey and Stefan Helmreich, "The Emergence of Multispecies Ethnography," *Cultural Anthropology* 25 no. 4 (2010): 545.

39 Valtonen et al., "Living with Mosquitoes".

4 Mingling with Mosquitoes in Gardens and Wastelands

As Hugh Raffles points out, mosquitoes are some of our closest cohabitants.[40] They eat our food, feed on our blood, and share our homes and ecosystems. While mosquitoes are unable to verbally communicate with humans, they are more than capable of making their presence known to us, and vice versa. Mosquitoes seek out our skin by detecting our exhalation of carbon dioxide and sensing our odor and body heat. Sometimes we can also hear a mosquito approach us by the buzzing sound caused by the high frequency of its wing-beats. If we do not detect and kill it, a female mosquito will then penetrate one of our blood vessels with her razor-sharp system of six thin, needle like mouthparts known as a proboscis.[41] To allow her to fill up, she drips saliva into the bite. Her saliva keeps our blood from coagulating, but also causes us to develop the itchy welts we commonly associate with mosquito bites. Moreover, her potent saliva blocks out our immune system and can serve as a conduit for the transferring of viruses such as CHIKV, or parasites such as malaria.

A bite constitutes thus an intra-action between mosquitoes and people.[42] For a brief moment, the female mosquito's body is merged with that of a human, allowing for the creation of new life as feeding on our blood enables the female mosquito to lay eggs.[43] The *Aedes* breeds in stagnant water, which is gathered in both natural and artificial reservoirs. Natural containers are for instance rain ponds and plant axils, while artificial ones include gutters, flowerpots, bottles, plastic food containers and discarded car tires.[44] The *Aedes* thrives particularly in artificial reservoirs created by urban spaces where access to adequate sanitation and clean water is limited. As argued by Raude and Setbon, "the relationship between the socioeconomic position and the risk of chikungunya disease is relatively well established."[45] Due to their conglomeration of social housing, squatter settlements and overall low socio-economic profile,[46] parts of both St. Pierre and Pointe aux Sables have a reputation for being so-called problem

40 Hugh Raffles, *Insectopedia* (New York: Pantheon Books, 2010).

41 In contrast, male mosquitoes feed on nectar and fruit juices only.

42 Karen Barad, *Meeting the Universe Halfway: Quantum Physics and the Entanglement of Matter and Meaning* (Durham, NC: Duke University Press, 2007).

43 Valtonen et al., "Living with Mosquitoes".

44 Dorothy Bonn, "How did Chikungunya reach the Indian Ocean?," *The Lancet Infectious Diseases* 6, no. 9 (2006).

45 Jocelyn Raude and Michel Setbon, "The Role of Environmental and Individual Factors in the Social Epidemiology of Chikungunya Disease on Mayotte Island," *Health & Place* 15, no. 3 (2009): 690.

46 Cf. Nading, *Mosquito Trails*.

neighborhoods (*quartiers chauds*). Then again, as I will return to, some people may be exposed to mosquitoes because they have no running water and must collect it haphazardly, whereas others are at risk because they have plenty of water with which to water the plants on their balconies.[47]

Everyday encounters between residents and mosquitoes are also enhanced by the *Aedes's* activeness during the day, and local living patterns. As previously mentioned, many research participants tended to spend a lot of their free time outdoors. This was especially the case in Réunion, where the garden (*jardin creole*) functions as an outdoor extension of the house (*kaz creole*) and acts as a stage for the unfolding of everyday domestic life.[48] A *jardin creole* is composed of two compartments: the front yard (*kour devant*) with an overbuilt terrace (*varang*) and the backyard (*kour derriere*). While these spaces have distinct functions in terms of comfort and practicality, a Réunionese household transgresses dichotomous understandings of home as a bounded entity that demarcates the inside from the outside. It is thus not only a space shared with human household members or "companion species" such as pets,[49] but also with plants and insects like the *Aedes*. For example, in the *kour devant* people plant flowers in pots instead of in the ground. Depending on the household's available space and income, the *kour devant* may consist of everything from a couple to hundreds of potted flowers. The saucers that are used to collect excessive water are favored breeding grounds for the *Aedes* mosquito. These saucers are not the only contact zone in a *jardin creole* however, as the *kour derriere* commonly houses the household's poultry, dog or bird cages, and various sheds. The *kour derrier* is private, meaning it is only meant for the household's members, and is often used to wash and dry clothes, to clean rice and lentils in the outdoor kitchen, and to store gardening tools or other useful objects – and sometimes various discarded items like car tires. Both the flowerpots in the *kour devant,* and the accumulation of items in the *kour derriere,* makes thus for a flourishing environment for mosquitoes.[50]

47 Arachu Castro, Yasmin Khawja and James Johnston, "Social Inequalities and Dengue Transmission in Latin America," in *Plagues and Epidemics: Infected Spaces Past and Present*, ed. D. Ann Herring and Alan C. Swedlund (Oxford: Berg, 2010).

48 Éliane Wolff, *Quartiers de Vie: Approche Ethnologique des Populations Défavorisées de l'Île de la Réunion* (Paris: Meridiens Klincksieck, 1991).

49 Haraway, *When Species Meet.*

50 Karine Aasgaard Jansen, "Challenging Chikungunya: Resistance to Public Health Measures and Etiology during the 2005–2007 Epidemic in Réunion," in *Disease Dispersion and Impact in the Indian Ocean World*, ed. Gwyn Campbell and Eva-Maria Knoll (London: Palgrave Macmillan, 2020).

In Pointe aux Sables, people live somewhat differently than in St. Pierre. While some of my research participants do indeed have gardens, the *kaz creole* is a feature particular to Réunion. One of the major factors to *Aedes* proliferation in Pointe aux Sables, is therefore not potted plants and their saucers, but rather so-called wastelands scattered across the neighborhood. Although the wastelands are primarily used as dumping grounds, they are also sometimes used for keeping poultry or dogs, storage, impromptu social gatherings or even squatting. As such, they resemble to a large degree a communal version of a Réunionese *kour derrier* and serve an important social role in the community. This can be illustrated further with the case of a public semi-open washhouse (*lavoir*) in a neighborhood inhabited by several of my informants in St. Pierre. This is also where the first case of chikungunya in Réunion was identified. A *lavoir* consists of a long row of large double washbasins set in stone. In addition to being used to do laundry, the *lavoir* also functioned as a neighborhood meeting place. While the women who lived in the nearby houses would come together to do their laundry and chat, men often sat in small groups on the grassy slope next to the *lavoir*, drinking rum and eating take-away food, playing cards, dominos, or the guitar, and watching the women work. As much as the *lavoir* itself contributed toward the accumulation of stagnant water in ponds, so did some of the men's leftover rubbish such as bottles, bottle caps, and food containers.[51]

Taken together, these gardens, wastelands, and their array of objects, critters, and fauna constitutes thus a living multispecies network, or a "natureculture" community,[52] consisting of mosquitoes (and a multiple of other insects and arachnids), domestic animals, birds, plants, microbes, and people. They also show how humans cannot be separated from non-humans, and how, over time, we have come to be deeply entwined and entangled in each other's lives.[53] The smallest component of this multispecies network, a virus, is even unique in the sense that it can only survive by entering and reproducing within the cells of another living being.[54] This natureculture community makes thus, in the words of Eben Kirksey, for an array of "emergent ecologies" where new symbiotic assemblages of lifeforms can live and flourish.[55] It also provides an

51 Jansen, "Challenging Chikungunya".

52 Haraway, *When Species Meet.*

53 Agustín Fuentes, "Naturalcultural Encounters in Bali: Monkeys, Temples, Tourists, and Ethnoprimatology," *Cultural Anthropology* 25, no. 4 (2010).

54 Eben S. Kirksey, "Welcome to the Virosphere," *E-flux Journal* 130 (2022).

55 Eben S. Kirksey, *Emergent Ecologies* (Durham: Duke University Press, 2015).

ideal opportunistic ecological space for the female *Aedes* in which she can feed on both nectar and blood, and breed in a wide range of water containers.

5 Virus or Vector? Mosquito Control and Local Disease Etiologies

Mingling with mosquitoes in contact zones also creates pathogenic possibility, however. Chikungunya transmits after all when humans interact with, and are bitten by, the *Aedes* in these shared emergent ecologies. In limiting the spreading of arboviral diseases, it is therefore critical to separate humans from potentially disease-carrying mosquitoes. Consequently, public health interventions aimed at vector control are often characterized by laborious efforts to disentangle the entanglements that bind humans and mosquitoes together.[56] While there are different preventive strategies for integrated vector management such as mosquito surveillance and larvae elimination, the urgency of the 2005–2007 chikungunya outbreak required swift action in the form of fogging. Fogging is a technique used for killing mosquitoes that involves using a fine spray of pesticide directed by a blower. In Réunion and Mauritius, fogging is usually known as "demosquitofication" (*demostification*).

During the epidemic in Réunion, these demosquitofications were mostly conducted by French mainland (*metropolitain*) soldiers working together with local DRASS agents. As outsiders to the island, the soldiers were unaware of or ignored the cultural spatial boundaries in the *jardin creole*. In their search for mosquitoes, they ventured into residents' backyards (*kour derriere*), which are usually considered off-limits for non-household members.[57] This led to some tensions between the French public health authorities and Réunionese residents.[58] One of my key research participants, 72-year-old Gabrielle, risked for example a substantial fine for denying the soldiers access to her *jardin* to conduct a demosquitofication. She and several others also declined their offers of free anti-repellents or refused to use any products to protect themselves against the *Aedes* such as mosquito nets.

The situation in Mauritius was slightly different. In targeting the *Aedes* and their breeding grounds, fogging was mostly carried out in wastelands. Like in Réunion, people were also urged to get rid of all potential sources for the

56 Ann H. Kelly and Javier Lezaun, "Urban Mosquitoes, Situational Publics, and the Pursuit of Interspecies Separation in Dar es Salaam," *American Ethnologist* 41, no. 2 (2014).

57 For further reading on the connection between arboviral disease, vector control and colonialism in Réunion, see for example Jansen, "Tropical Disease".

58 Jansen, "Challenging Chikungunya".

accumulation of stagnant water close to their homes. However, according to 45-year-old Philippe, the Ministry of Health (MOH) never actually declared an epidemic. In his opinion, this did not mean that the Mauritian public health authorities did not take adequate measures to get rid of mosquitoes, but that they tried to keep their efforts discreet. Philippe believed that this was a necessary precaution to avoid causing irrevocable damage to people's livelihoods in the island's high-end tourism sector. For this reason, some also suspected that fogging was primarily carried out at tourist resorts and at night only. For example, 59-year-old Florence claimed that no demosquitofications took place in her neighborhood during the epidemic even though she lived in an area that was highly affected by chikungunya.

What several research participants in Réunion and Mauritius had in common is that many of them questioned the biomedical etiology of chikungunya, that is whether mosquitoes were vectors for CHIKV. Many research participants doubted for example that mosquitoes caused chikungunya because they had neither seen nor felt a mosquito bite them before they fell ill. Others questioned how something so small as a mosquito could cause such excruciating pain. Moreover, while research participants commonly presented mosquitoes to me as a familiar environmental feature of the islands, chikungunya was often rather depicted as something alien and aerial. For example, 41-year-old Nathalie from Réunion claimed that since mosquitoes had always existed in Réunion, it made no sense that they would now spread a new disease. Instead, she was convinced that chikungunya was an airborne and not a vector-borne disease.[59] 72-year-old Marianne from Mauritius also argued that chikungunya spread through air since there had always been lots of mosquitoes and sources of stagnant water in her neighborhood. Besides, for as long as she could remember, she had taken precautions to protect herself from getting bitten by not using saucers for her potted plants, burning mosquito repelling coils and using anti-mosquito creams. Yet, regardless of such everyday encounters with mosquitoes throughout her entire life, Marianne still fell severely ill with chikungunya.

By not distinguishing between vector and virus, Gabrielle, Nathalie, Marianne, and several other research participants in Réunion and Mauritius made sense of chikungunya as a new disease to the islands. These complementary etiologies were not caused by a lack of access to public health information, but rather by a reluctance to accept such biomedical explanations for disease

59 Karine Aasgaard Jansen, "The 2005–2007 Chikungunya Epidemic in Réunion: Ambiguous Etiologies, Memories and Meaning-Making," *Medical Anthropology* 32, no. 2 (2013).

transmission.[60] As medical historian Charles E. Rosenberg argued, there are two fundamental styles of etiological reasoning: while "configuration" is holistic and emphasizes contextual, environmental features in explaining the spreading of infectious disease, "contamination" represents the turn towards germ theory in the 1870s. According to Rosenberg, in most cases people employ these in combination.[61] To think of chikungunya as an airborne rather than a vector-borne disease also allowed for the continuation of normal, everyday life on the islands. This included mingling with mosquitoes in shared, ecological spaces such as gardens and wastelands.

6 Conclusion: Why a Multispecies Approach?

In his analysis of how residents of Ciudad Sandino in Nicaragua share their lives with dengue-carrying mosquitoes, Alex Nading argues that in studying the diffusion of dengue, it is both difficult but also analytically unproductive to separate humans from non-humans such as mosquitoes because their lives are so intertwined and entangled.[62] I agree. While such disentanglements may be necessary, they do not always take local uses and management of contact zones into consideration. While wastelands may traditionally have been considered neglected spaces of uncultivated land on the margins of society,[63] as shown by multispecies ethnographers among others, wastelands are also valuable spaces for conserving urban biodiversity.[64] This includes insects such as the *Aedes* mosquito. By not taking the biomedical etiology of chikungunya at face value, research participants in Réunion and Mauritius thereby challenged public health interventions that threatened to negatively impact their familiar surroundings and interactions with mosquitoes.[65] In the words of Priscilla Wald, "contagion is more than an epidemiological fact."[66] Yet, epidemiological

60 Jansen, "The 2005–2007 Chikungunya Epidemic".

61 Rosenberg, *Explaining Epidemics.*

62 Nading, *Mosquito Trails.*

63 Rémi Beau, Franceska Di Pietro and Amélie Robert, "Anthropocene Wastelands: From the Margins to the Center", in *Urban Wastelands: A Form of Urban Nature*, ed. Franceska Di Pietro and Amélie Robert (Cham: Springer International Publishing, 2021).

64 Anna L. Tsing, *The Mushroom at the End of the World: On the Possibility of Life in Capitalist Ruins* (New Jersey: Princeton University Press, 2015); Jamie Lorimer, *Wildlife in the Anthropocene: Conservation After Nature* (Minneapolis: University of Minnesota Press, 2015).

65 Jansen, "Challenging Chikungunya".

66 Priscilla Wald, *Contagious: Cultures, Carriers, and the Outbreak Narrative* (Durham: Duke University Press, 2008).

research often misses social dimensions of disease diffusion and control.[67] Epidemiological research that informs public health measures typically looks at disease as a microbial invasion. This entails that the so-called solution to the problem of arboviral diseases is often presented as improved medical education and care,[68] and has resulted in standardized global public health measures in the fight against arboviral diseases.[69] The problem, however, is that these often fail when adapted to diverse local contexts largely unknown to the promoters of the interventions.[70]

Given how Gabrielle and Nathalie did not protect themselves against chikungunya infection, I am not claiming that mingling with mosquitoes is unproblematic. Nevertheless, as I have tried to show in this chapter, by not accounting for the where and how of multispecies encounters in the design and execution of vector control, such design and execution are also likely to be less effective. Epidemic encounters entail living with and sharing a wide range of multispecies networks. To limit the spread of arboviral diseases, all these complex entanglements between humans, insects and environments need to be considered.

Bibliography

Aisher, Alex and Vinitra Damodaran. "Introduction: Human-Nature Interactions through a Multispecies Lens." *Conservation and Society* 14, no. 4 (2016): 293–304.

Alpers, Edward A. "Chikungunya and Epidemic Disease in the Indian Ocean." In *Disease Dispersion and Impact in the Indian Ocean World*, edited by Gwyn Campbell and Eva-Maria Knoll, 211–237. London: Palgrave Macmillan, 2020.

Anon. "M. de Villepin Promet 76 million d'Euros Contre le Chikungunya." *Le Monde*, February 28, 2006.

Arnold, David. "The Indian Ocean as a Disease Zone, 1500–1950." *South Asia* 14, no. 2 (1991): 1–21.

Barad, Karen. *Meeting the Universe Halfway: Quantum Physics and the Entanglement of Matter and Meaning*. Durham: Duke University Press, 2007.

---

67 James A. Trostle, *Epidemiology and Culture* (New York: Cambridge University Press, 2005).

68 Meike Wolf, "Is There Really Such a Thing as 'One Health'? Thinking About a More than Human World from the Perspective of Cultural Anthropology," *Social Science & Medicine* 129 (2015): 8.

69 Randall M. Packard, *A History of Global Health: Interventions into the Lives of Other People* (Baltimore: John Hopkins University Press, 2016).

70 Leonore Manderson, "Applying Medical Anthropology in the Control of Infectious Disease," *Tropical Medicine and International Health* 3, no. 12 (1998).

Beau, Rémi, Franceska Di Pietro and Amélie Robert. "Anthropocene Wastelands: From the Margins to the Center." In *Urban Wastelands: A Form of Urban Nature,* edited by Franceska Di Pietro and Amélie Robert, 361–377. Cham: Springer International Publishing, 2021.

Beesoon, Sanjay, Ellen Funkhouset, Navaratnam Kotea, Andrew Spielman and Rebecka M. Robich. "Chikungunya fever, Mauritius, 2006." *Emerg Infect Dis.* 14, no. 2 (2008): 337–338. doi:10.3201/eid1402.071024.

Bonn, Dorothy. "How did Chikungunya reach the Indian Ocean?" *The Lancet Infectious Diseases* 6, no. 9 (2006): 543.

Brooks, Daniel R., Eric P. Hoberg and Walter A. Boeger. *The Stockholm Paradigm. Climate Change and Emerging Disease.* Chicago: University of Chicago Press, 2019.

Brown, Hanna and Ann H. Kelly. "Material Proximities and Hotspots: Toward an Anthropology of Viral Hemorrhagic Fevers." *Medical Anthropology Quarterly* 28, no. 2 (2014): 280–303.

Campbell, Gwyn and Eva-Maria Knoll (eds.). *Disease Dispersion and Impact in the Indian Ocean World.* London: Palgrave Macmillan, 2020.

Castro, Arachu, Yasmin Khawja and James Johnston. "Social Inequalities and Dengue Transmission in Latin America." In *Plagues and Epidemics: Infected Spaces Past and Present,* edited by D. Ann Herring and Alan C. Swedlund, 231–251. Oxford: Berg, 2010.

Deb Roy, Rohan. *Malarial Subjects: Empire, Medicine and Nonhumans in British India, 1820–1909.* Cambridge: Cambridge University Press, 2017.

Fuentes, Agustín. "Naturalcultural Encounters in Bali: Monkeys, Temples, Tourists, and Ethnoprimatology." *Cultural Anthropology* 25, no. 4 (2010): 600–624.

Haraway, Donna J. *When Species Meet.* Minneapolis: Minnesota University Press, 2008.

Herring, D. Ann and Alan S. Swedlund. *Plagues and Epidemics: Infected Spaces Past and Present.* Oxford: Berg, 2010.

Jansen, Karine Aasgaard. "The Printed Press's Representations of the 2005–2007 Chikungunya Epidemic in Réunion: Political Polemics and (Post)Colonial Disease." *Journal of African Media Studies* 4, no. 2 (2012): 227–242.

Jansen, Karine Aasgaard. "The 2005–2007 Chikungunya Epidemic in Réunion: Ambiguous Etiologies, Memories and Meaning-Making." *Medical Anthropology* 32, no. 2 (2013): 174–189.

Jansen, Karine Aasgaard. "Tropical Disease and the Making of France in Réunion." In *Histories of Medicine in the Indian Ocean World, Volume 2: The Modern Period,* edited by Anna Winterbottom and Facil Tesfaye, 147–171. New York: Palgrave Macmillan, 2016.

Jansen, Karine Aasgaard. "Challenging Chikungunya: Resistance to Public Health Measures and Etiology during the 2005–2007 Epidemic in Réunion." In *Disease Dispersion and Impact in the Indian Ocean World,* edited by Gwyn Campbell and Eva-Maria Knoll, 237–255. London: Palgrave Macmillan, 2020.

Kirksey, Eben S. and Stefan Helmreich. "The Emergence of Multispecies Ethnography." *Cultural Anthropology* 25 no. 4 (2010): 545–576.

Kirksey, Eben S. "Welcome to the Virosphere." *E-flux Journal* 130 (2022).

Kirksey, Eben S. *Emergent Ecologies*. Durham, NC: Duke University Press, 2015.

Kelly, Ann H. and Javier Lezaun. "Urban Mosquitoes, Situational Publics, and the Pursuit of Interspecies Separation in Dar es Salaam." *American Ethnologist* 41, no. 2 (2014): 368–383.

Kelly, Ann H., Frédérik Kéck and Christos Lynteris (eds.). *The Anthropology of Epidemics*. New York: Routledge, 2019.

Leyral, P. "Les Autorités Sanitaires de l'Île le Reconnaisent Officielment: Un Enfant est Mort Directement du Chikungunya." *Le Journal de l'Île de de la Réunion* (*JIR*), February 4, 2006.

Ligon, B. Lee. "Re-emergence of an Unusual Disease: The Chikungunya Epidemic." *Seminars in Pediatric Infectious Diseases* 17 (2006): 99–104.

Lorimer, Jamie. *Wildlife in the Anthropocene: Conservation After Nature*. Minneapolis: University of Minnesota Press, 2015.

Mavian, Carla, Melissa Dulcey, Olga Munoz, Marco Salemi, Amy Y. Vittor and Ilaria Capua. "Islands as Hotspots for Emerging Mosquito-borne Viruses: A One-Health Perspective." *Viruses* 11, no. 1 (2019): 11. doi:10.3390/v11010011.

Manderson, Leonore. "Applying Medical Anthropology in the Control of Infectious Disease." *Tropical Medicine and International Health* 3, no. 12 (1998): 1020–1027.

Nading, Alex M. *Mosquito Trails: Ecology, Health, and the Politics of Entanglement*. Oakland: University of California Press, 2014.

Packard, Randall M. *A History of Global Health: Interventions into the Lives of Other People*. Baltimore: John Hopkins University Press, 2016.

Pialoux, Gilles, Bernard A. Gaüzere, Stephan Jauréguiberry and Michel Strobel. "Chikungunya, an Epidemic Arbovirus." *The Lancet Infectious Diseases* 7 (2007): 319.

Raffles Hugh. *Insectopedia*. New York: Pantheon Books, 2010.

Rallu, J.-L. "Populations et Sociétés: Populations et Développement dans l'Outre-Mer de l'Union Européenne." *Bulletin Mensuel d'Information de l'Institut National d'Études Démographiques* 456 (2009): 1–4.

Raude, Jocelyn and Michel Setbon. "The Role of Environmental and Individual Factors in the Social Epidemiology of Chikungunya Disease on Mayotte Island." *Health & Place* 15, no. 3 (2009): 690.

Renault, Philippe, Daouda Sissoko, Martine Ledrans, Vincent Pierre and Gilles Brücker. "L'épidemie de Chikungunya à la Réunion et à Mayotte, France, 2005–2006: Le Contexte et les Questions de Surveillance et d'Évaluation Posées." *Bulletin Épidémiologique Hebdomaire* (Institut de Veille Sanitaire INVS) 38 (2008): 39–40.

Rosenberg, Charles E. *Explaining Epidemics and Other Studies in the History of Medicine*. New York: Cambridge University Press, 1992.

Schnepel, Burkhard and Edward A. Alpers (eds.). *Connectivity in Motion: Island Hubs in the Indian Ocean World*. Cham: Palgrave Macmillan, 2018.

Swanson, Heather Ann. "Methods for Multispecies Anthropology: Thinking with Salmon Otolithes and Scales." *Social Analysis* 61, no. 2 (2017): 81–99.

Taglioni, François and Jean-Sébastien Dehecq. "L'Environnement Socio-Spatial comme Facteur de l'Émergence des Maladies Infectieuses: Le Chikungunya dans l'Océan Indien." *EchoGéo* 9 (2009): 1–42.

Trostle, James A. *Epidemiology and Culture*. New York: Cambridge University Press, 2005.

Tsing, Anna L. *The Mushroom at the End of the World: On the Possibility of Life in Capitalist Ruins*. New Jersey: Princeton University Press, 2015.

Valtonen, Anu, Tarja Salmela and Outi Rantala. "Living with Mosquitoes." *Annals of Tourism Research* 83 (2020). doi:10.1016/j.annals.2020.102945.

Wald, Priscilla. *Contagious: Cultures, Carriers, and the Outbreak Narrative*. Durham: Duke University Press, 2008.

Watin, Michel. "La Médiatisation de l'Épidémie de Chikungunya à Maurice et à La Réunion (2005–2006)." In *La Santé dans l'Espace Public*, edited by Hélène Romeyer, 133–149. Rennes: Presses de l'EHESP, 2010.

Winterbottom, Anna and Facil Tesfaye (eds.). *Histories of Medicine in the Indian Ocean World, Volume 2: The Modern Period*. New York: Palgrave Macmillan, 2016.

Wolf, Meike. "Is There Really Such a Thing as 'One Health'? Thinking About a More than Human World from the Perspective of Cultural Anthropology." *Social Science & Medicine* 129 (2015): 5–11.

Wolff, Éliane. *Quartiers de Vie: Approche Ethnologique des Populations Défavorisées de l'Île de la Réunion*. Paris: Meridiens Klincksieck, 1991.

World Health Organization. "Countries and Territories Certified Malaria-Free by WHO." Last updated January 12, 2024. https://www.who.int/teams/global-malaria-programme/elimination/countries-and-territories-certified-malaria-free-by-who.

CHAPTER 14

Humans, Ticks, and the Conflict over the Cervids

Heta Lähdesmäki and Tuomas Räsänen

1 Introduction

In recent decades, the convergence of anthropogenic environmental changes has made life easier for ticks in northern climes, such as Finland. Consequently, their prevalence has been increasing and their habitat has extended northward. Ticks need rodents and mammals as host species during their life cycle. The warming and shortening of winters, in particular, has not only improved the proliferation of ticks but also supported an explosion in the deer population. In Finland, this consists of the white-tailed deer (*Odocoileus virginianus*), the roe deer (*Capreolus capreolus*) and a much smaller population of fallow deer (*Dama dama*). A combined deer population of more than 150,000 provides ticks with an abundant reservoir of blood. In the context of Finland, ticks, and the diseases they spread – mainly borreliosis and Tick-borne encephalitis (TBE) – have altered the perceptions of humans to nature and outdoor activities. Many are cautious, if not afraid, of hiking in forests and fields. Checking one's body for ticks and sometimes removing them has become standard procedure after spending time in vegetated terrain.[1] Concern and fears about ticks has soured the attitudes of many Finns toward deer, and particularly the White-tailed deer, and caused conflicts between humans regarding the different ways of using and enjoying nature.

In this chapter, we will analyze how the proliferation of ticks has affected human attitudes towards cervids. We will focus particularly on the human relationship to White-tailed deer, and to a lesser extent to Roe deer. They are both non-native species in Finland that only came into the country during the twentieth century. However, from the point of view of thorough analysis it is also helpful to make some comparisons to elk (*Alces alces*), with which humans have shared the natural terrain of Finland for millennia.

1 Sanna Lillbroända-Annala, "'Fästingeländet,' Praktiker och materialisering mellan människor, fästingar och natur," *Laboratorium för folk och kultur* 8 (2021), accessed September 29, 2023, https://bragelaboratorium.com/2021/11/11/fastingelandet-praktiker-och-materialisering-mellan-manniskor-fastingar-och-natur/. See also chapter by Lillbroända-Annala in this book.

There are two common species of ticks in Finland: the castor bean tick (*Ixodes Ricinus*) and the taiga tick (*Ixodes Persulcatus*). The former has probably inhabited Finland for as long as their host species, that is, since the ice retreated after the latest glaciation period. We have no way of knowing how common ticks were in earlier times, but certainly the scarcity of cervids throughout the twentieth century, caused by excessive hunting,[2] limited their number. Ticks have only begun to thrive in Finland as the population of cervids has multiplied in recent decades.

In this chapter we will examine how knowledge of ticks and the diseases they spread have affected human attitudes toward deer. We will analyze these networks of humans, ticks and cervids, and the conflicts that have arisen from the functioning of these networks, from the perspective of the concepts of belonging and alienness/invasiveness. Animal species are often perceived as either belonging to a site or being out-of-place. The terms people use about animals reveal how their belonging is perceived: native species are perceived to occupy their natural range, whereas species that have traveled between countries or continents with human help are called invasive species and animals that do so without human help are alien species. We argue that the discrepancy in attitudes towards different ticks' host species stems from the fact that the elk is a native species in Finland, whereas the White-tailed deer suffers from the intensified and culturally-accepted hostility held towards alien species.[3]

Our study material consists of four different types of sources. First, we will use newspaper articles published from the 1990s until 2020 that discuss ticks and cervids. Many of these newspapers were published in Finnish coastal areas, where the tick problem is most pressing, but the materials also include articles from national newspapers, such as *Maaseudun Tulevaisuus* and *Hufvudstadsbladet*. These materials have been gathered using the digital newspaper archive managed by the National Library of Finland. Second, we will use the materials produced by the questionnaire for the public conducted by our research project, in which people have recollected and explained their encounters, experiences and thoughts about ticks. Third, we have also used semi-structured thematic interviews, eight of which were conducted by Heta

2 Tuire Nygrén, *Suomen hirvikannan säätely – biologiaa ja luonnonvarapolitiikkaa*, Ph.D. diss. (University of Joensuu, 2009), 18–19.

3 Peter Coates, *American Perceptions of Immigrant and Invasive Species: Strangers on the Land* (Berkeley: University of California Press, 2007); Owain Jones and Paul Cloke, *Tree Cultures: The Place of Trees and Trees in Their Place* (Oxford: Berg, 2002); Dolly Jørgensen, "Migrant Muskoxen and the Naturalization of National Identity in Scandinavia," in *The Historical Animal*, ed. Susan Nance (Syracuse: Syracuse University Press, 2015); Heta Lähdesmäki, *Susien paikat. Ihminen ja susi 1900-luvun Suomessa*, Ph.D. diss. (University of Jyväskylä, 2020).

Lähdesmäki between 2019–2023 on Seili Island in the Archipelago Sea, and ten were conducted by students from the University of Turku between 2019 and 2022 on Ruissalo Island in Turku and elsewhere in the Archipelago Sea. Lastly, we also use reports and surveys produced by the Finnish Wildlife Agency and the Finnish Hunters' Association.[4]

2 Histories Separated: Deer and Ticks

In 2004, Juha Kairikko, a longtime executive director of the Finnish Hunters' Association, wrote that in the current ecopolitical climate it would be inconceivable to introduce a large mammal species, such as the White-tailed deer, into Finland from other continents.[5] In recent decades, non-native species, whether deliberately introduced into Finland or inadvertently as stowaways, have become one of the most serious environmental problems in the country.[6] In the not so distant past, things were seen rather differently. The White-tailed deer was first introduced into Finnish nature in 1934 from the United States, when a group of Finnish immigrants living in Minnesota organized a transfer of seven fawns to Finland as a gift to the hunters of their ancestral homeland. The deer were first kept in an enclosure, but one of the fawns escaped and the remaining six were later released. A series of reintroductions occurred in the 1940s and 1950s. By the late 1950s, the population was considered large enough to be hunted.[7] The deer population remained relatively low until the late twentieth century: only exceeding 10,000 individuals by the mid-1970s and 40,000 by 2000, with the majority of the species being

4 As sources, we used a report derived from a survey of 721 hunters conducted by the Finnish Wildlife Agency in 2018 and a report stemming from a survey conducted by the Finnish Hunters' Association in 2022 that contains responses from 333 hunters. Antti Rinne and Mikko Toivola, *Valkohäntäpeura Varsinais-Suomessa – kannanhoidon kysely valkohäntäpeuran metsästäjille. Raportti* (Varsinais-Suomi: Suomen Riistakeskus, 2018); The Finnish Hunters' Association, "Valkohäntäpeuran jäsenkysely tukee vahvasti Liiton linjauksia," December 15, 2022, accessed October 8, 2023, https://metsastajaliitto.fi/uutiset/valkohantapeuran-jasenkysely-tukee-vahvasti-liiton-linjauksia. Lähdesmäki's research was supported by the Academy of Finland (project no. 341118) and Kone Foundation (project *Elämän saari*).

5 Juha K. Kairikko and Jaakko Ruola, *Valkohäntäpeura* (Jyväskylä: Suomen metsästäjäliitto, 2004), 42.

6 See, for example, David Pimentel, *Biological Invasions: Economic and Environmental Costs of Alien Plant, Animal, and Microbe Species* (Boca Raton: CRC Press, 2011).

7 Kairikko and Ruola, *Valkohäntäpeura*, 42–45, 60–65; Petri Nummi, *Suomeen istutetut riistaeläimet* (Helsinki: Helsingin yliopisto, 1988), 26–29, 30–32; Heikki Lehikoinen, *Tuo hiisi hirviäsi: Metsästyksen kulttuurihistoria Suomessa* (Helsinki: Teos, 2007), 74.

confined to the south-western region of Finland, which was the mildest area of the country.[8]

There is little to indicate any major problems or controversies from the human point of view in these early decades of coexistence. At the time when the deer were introduced into Finnish nature, some had voiced concerns about the potential damage the species could inflict on agriculture or forests. Others were merely suspicious of a new species being introduced into Finland. All these concerns seemed to have been in vain. As Juha Kairikko puts it in his historical overview of the species, the transfer "brought a fine game species to Finland," and as for non-hunting locals, they were mostly "thrilled with these beautiful cervids." In some senses, Finns were fortunate as the White-tailed deer introduced into Finland were not infected with brainworm (*Parelaphostrongylus tenuis*), a parasite that is harmless to deer, but can be fatal to elk.[9] Such an eventuality would have made coexistence less harmonious. Interestingly, it was not the brainworm that changed the relationship between humans and deer, but a pathogen-carrying arachnid.

From the point of view of humans, the histories of deer and ticks, however, have only crossed paths in the twenty-first century. In their history of White-tailed deer in Finland, for example, Kairikko and Ruola did not mention ticks at all. Humans living in the Finnish countryside were fully aware of ticks, and older generations of contemporary Finns often recollect how ticks were thought to dwell in places where alders grew. From these trees they were able to attach themselves to the skin of passersby. Many people recollected being cautioned as children not to play beneath alders.[10]

Many of these early preconceptions about ticks have proven to be false in scientific studies. Ticks do not specifically favor alders, but thrive in many types of habitats, including urban green spaces.[11] What is more, they definitely do not lurk on trees. Yet, an idea that has retained its strength, although in changing form, is that ticks are somehow special among Finnish arthropods. There are other species in Finland, for example, that suck human blood or

8 Lehikoinen, *Tuo hiisi hirviäsi*, 75.

9 Kairikko and Ruola, *Valkohäntäpeura*, 42, 56. For the deer population and its development, see the Finnish Wildlife Agency, "LUKE: Valkohäntäpeurakannan kasvu pysähtyi," March 18, 2022, accessed September 19, 2023, https://riista.fi/luke-valkohantapeurakannan-kasvu-pysahtyi-2/.

10 See, for example, The Finnish Literature Society (FLS), questionnaire 2019, "Punkit tulevat," SKS 017, 027, 030, 046, 060; Interviews TKU/A/19/124; EF 7.9.2021.

11 See, for example, Jani Sormunen et al., "Enhanced Threat of Tick-Borne Infections within Cities?: Assessing Public Health Risks Due to Ticks in Urban Green Spaces in Helsinki, Finland," *Zoonoses and Public Health* 67, no. 7 (2020), doi:10.1111/zph.12767.

are able to give nasty stings, such as mosquitoes and wasps. But these hardly generate a similar aura of danger as ticks, which have elicited warnings about certain environments. An air of mystery still persists as to how they find their way onto human skin, as well as a sense of revulsion about how they penetrate inside the human body. Wasps and mosquitoes are not thrown into stoves, as sometimes happens to ticks, as if by doing so humans are able to expel evil.[12]

3 Histories Entangled: the Fear of Ticks and the Alienation of Deer

The fear of ticks, ubiquitous in present-day Finland, and the idea of these arachnids being the most dangerous animal in Finland can be traced to the mid-2000s. This was the time when the number of media stories about ticks exploded, accompanied by scientists speculating that these arachnids had been moving northward and increased in number due to climate change.[13] Apparently both factors – changes in eco-systemic and communicative spaces – strengthened each other. Subsequently, the coverage in the media was full of stories citing medical experts and educated Finns about how borreliosis could cause permanent disabilities if untreated over time. Even more disturbingly, an even more harmful subtype of TBE virus also began to spread on the Finnish mainland. This was a disease that could cause lifelong misery and in rare cases also death.[14] And these were not just scandalous stories circulated by tabloid publications. Antti Vaheri, a professor of virology at the University of Helsinki, stated the following when questioned by a journalist about the true danger of ticks: "A tick is the most dangerous animal in Finland. It can disable or even kill a human being."[15] When this idea gradually percolated through to the whole society, it is no wonder that the fear of ticks also transformed attitudes toward deer.

---

12 FLS, "Punkit tulevat," SKS 055, 063; Sanna Lillbroända-Annala, "Fy, en fästing! Äckel i kroppsliga och affektiva möten mellan fästing, sällskapsdjur och människa," *Budkavlen* 101 (2022).

13 See, for example, Liina Kjellberg, "Puutiainen odottaa heinikossa," *Maaseudun Tulevaisuus* (hereafter *MT*), July 15, 2015; "Punkkien aiheuttamat taudit yleistyneet," *MT*, May 19, 2006; Sanna Lillbroända-Annala and Oscar Winberg, "'Fästingen håller inget säkerhetsavstånd': Konkurrerande riskdiskurser om fästingar i media," *TRACE ∴ Journal for Human-Animal Studies* 9 (2023).

14 See, for example, "Puutiaisaivokuume lisääntynyt Manner-Suomessa," *MT*, January 19, 2007; Marjut Weman, "Varo punkkeja," *MT*, June 18, 2007; Shahin Doagu, "Puutiaisaivokuume on lisääntyvä terveysriski," *MT*, April 24, 2009.

15 Tarja Halla, "Kuinka vaarallinen punkki on?" *MT*, May 31, 2010.

As the historian Peter Coates has pointed out, for example, native species are seen as part of their own ecosystem, that is, to belong to the areas where they exist, while species classified as invasive, or alien, are often viewed as intruders and not belonging to their new environment. Moreover, animals (and plants) can also be seen to possess a national identity. Indeed, ideas of nationality have influenced our understanding of the non-human world of nature.[16]

Finns initially connected the deer to the southwestern landscape when they appeared in Finland and called them *Laukonpeura*, "the deer of Laukko," according to the place where they were first introduced. According to an article published in the popular newspaper *Ilta-Sanomat* in 1952:

> Local residents protect these animals with proper affection, here and there they are downright happy to see that the deer of Laukko has settled in the backwoods of our farms. The deer seem to have won over local peoples' hearts as no-known acts of violence have been committed against them.[17]

This indicates that the White-tailed deer was not initially seen as an alien species in Finland. On the contrary, the newspaper article from 1952 suggests they were welcomed as a game animal that was suited to their new environment. People are sometimes able to quickly include new species to the national fauna. Historian Dolly Jørgensen has written about how muskoxen were first introduced to Norway and then came to Sweden on their own, where the species "quickly became understood as a central element in the mountains, probably because of its novelty and distinctive appearance."[18]

However, bestowing a sense of belonging on a species comes with the possibility of denial if the relationship between the granter and grantee turns problematic. In our interviews, conducted between 2019 and 2023, many people living or spending time in the Archipelago Sea area described White-tailed deer as an introduced or alien species that had been brought to Finland by humans.[19] We do not claim that ticks were the only factor behind the about-face in people's mind toward White-tailed deer. A significant factor in the minds of the interviewees concerned their sheer number, as the deer began

16 Coates, *American Perceptions* (2007), 3.

17 "Laukonpeurat ja talvi," *Ilta-Sanomat*, 12 March, 1952, 8.

18 Jørgensen, "Migrant Muskoxen," 196.

19 See, for example, interviews TKU/A/22/89-lit; TKU/A/22/87-lit; TKU/A/22/67-lit; TKU /A/22/71-lit; H. V. 9.9.2021; EF 7.9.2021.

to appear "everywhere, under every window,"[20] as one interviewee put it. This ensured a collision between the interests of humans and the deer. This has also been the case with other "problem animals," and even with species that are not seen as alien, such as pigeons.[21]

First, the exploding numbers of deer made them a potential cause of traffic accidents. If you could spot "ten to twenty deer on both sides of the road,"[22] it obviously increased the sense of risk for drivers. In 2017 alone, for example, there were over 1600 traffic collisions caused by White-tailed deer in southwestern Finland.[23] No wonder people started to feel that "the White-tailed deer population is dangerously large."[24] A second problem concerned gardening. Invasive alien species transform ecosystems as they affect native species through competition, predation, hybridization, and disease. Alien species are therefore often considered synonymous with danger, negative change and damage to both the ecosystem and to humans and they have cause unease and fear among the human population.[25] This was exactly what many Finns felt toward the White-tailed deer. Interviewees described how White-tailed deer ate whatever humans tried to cultivate, whether that be ornamental or edible plants, and also endangered plants and young trees in the natural environment.[26]

However, we argue that ticks have greatly amplified the alienation of deer, which has consequently led to conflicts between humans about the place of deer in Finnish nature. Many have connected the dots between the simultaneous rapid increase in the populations of both ticks and deer.[27] As one resident put it in *Paraisten kuulutukset*, the local newspaper for the Turku Archipelago: "In my first fifty years in Parainen, I had never seen a tick, nor did I know anyone who had. But I hadn't seen a deer either. Now, in the last ten years, not a single summer has gone by without me or a family member having one or

20 Interview I.V.6.7.2020.

21 For example, as they have become more numerous, city pigeons' presence has been seen more as a negative thing. See, for example, Colin Jerolmack, "How Pigeons Became Rats: The Cultural-Spatial Logic of Problem Animals," *Social Problems* 55, no. 1 (2008), doi:10.1525/sp.2008.55.1.72.

22 Interview TKU/A/19/108.

23 Rinne and Toivola, *Valkohäntäpeura Varsinais-Suomessa*, 2.

24 Teija Uitto, "Peurakanta on päässyt vaarallisen suureksi," *Vakka: Vakka-Suomen sanomat*, July 21, 2017.

25 Coates, *American Perceptions*, 1.

26 See, for example, interviews E.F.7.9.2021; H. V. 9.9.2021; TKU/A/22/67-lit; TKU/A/22/70-lit.

27 See, for example, interviews TKU/A/19/124; TKU/A/22/66-lit; TKU/A/22/67-lit; H. V. 9.9.2021; E.F.7.9.2021.

more ticks [on us]. And the deer are watching overhead with the biggest eyes."[28] Others were familiar with ticks, but, according to them, they were previously seen mostly seen in animal.[29] Now, however, with the presence of White-tailed deer it was possible to repeatedly find many ticks in humans, "thanks to the deer."[30]

People felt that deer brought ticks close to humans as the former animal roamed in peoples' yards and gardens. One interviewee worried about the impact of White-tailed deer: "once there were seven [deer] at the same time in the yard and if there are some thousand ticks in each of them and they shake them all around in the area then there are [many of] them [ticks]."[31] The connection between ticks and deer was seen as a huge public health problem by many, and they quickly began to question the usefulness of White-tailed deer and even their right to exist in Finland. Some wondered why we needed White-tailed deer, which brought ticks and diseases with them when there were other game animals in the country.[32] The above-mentioned newspaper article in *Paraisten kuulutukset* explicitly pointed out to the alienness of the White-tailed deer when criticizing the abundance of the deer: "[The White-tailed deer] is an alien species that should be controlled in the same way as lupine, which is also beautiful [...] Let's start a large-scale hunt and pick off the deer, and the ticks at the same time, from the nature of Parainen."[33]

According to many studies, the prevalence of tick-borne diseases is connected to local White-tailed deer populations.[34] These findings were also discussed in the local media.[35] In Finland, the distribution of ticks has been

28 Jussi Vuorinen, "Sudet, kauriit ja punkit," *Paraisten kuulutukset*, January 14, 2016.

29 See, for example, interview TKU/A/22/66-lit.

30 Interview H. V. 9.9.2021.

31 Interview TKU/A/19/124.

32 Interview H. V. 9.9.2021. The role of deer in spreading ticks and tick-borne diseases was acknowledged also by respondents for the survey conducted in 2013 to villagers near the Ekenäs Archipelago. See Milla Niemi and Madeleine Nyman, *Valkohäntäpeuran ekologiset ja sosiaaliset vaikutukset Tammisaaren saariston kansallispuistossa ja sen lähialueilla* (Vantaa: Metsähallitus, 2013), 23, 28, 32, 35–36, accessed October 23, 2023, https://julkaisut.metsa.fi/assets/pdf/lp/Asarja/a204.pdf.

33 Vuorinen, "Sudet, kauriit ja punkit". The garden lupine (*Lupinus polyphyllus*) is among the most worrisome invasive species in Finnish nature.

34 D.C. Duffy et al., "Ixodes Scapularis (Acari: Ixodidae) Deer Tick Mesoscale Populations in Natural Areas: Effects of Deer, Area, and Location," *Journal of Medical Entomology* 31, no. 1 (1994), doi:10.1093/jmedent/31.1.152; Niemi and Nyman, *Valkohäntäpeuran*, 23; Juho Matala et al., *Hirvieläinten vaikutuksia yhteiskuntaan, elinkeinoihin ja ekosysteemiin: Synteesiraportti* (Helsinki: Luonnonvarakeskus, 2021), 111.

35 See, for example, Kaj Sundqvist, "Rådjur och förekomst av fästingar," Åbo Underrättelser, October 10, 2017; Annika Rentola, "Fästingen är en risk också för turisten i Europa,"

observed as expanding northwards, coinciding with the spread of the White-tailed and Roe deer populations.[36] The White-tailed deer is not the only host species for ticks. Besides other cervids, smaller animals, such as moles and hares, provide a blood reservoir for ticks in different stages of their life cycle. Our interviewees, however, never raised (with one exception) any other animals as being responsible for the tick problem.[37] As for our questionnaire, when other wild animals were mentioned, the interviewees often pointed the finger at other species that were considered aliens, such as Roe deer and Common raccoon dogs.[38] Other host species are therefore regularly overlooked. This is not due to a lack of knowledge, since experts have repeatedly educated the public about the issue in newspapers and in other forms of the media. Thus, there must be another explanation for why White-tailed deer are perceived as being the principal culprit in Finland. We argue that this stems from the tendency of humans to categorize animals into those who belong and those who do not.

The White-tailed and Roe deer are both alien species in Finland, but there is one crucial difference in their alienness: Roe deer have come to Finland from Sweden with their own hooves,[39] while the much more populous White-tailed deer were imported by humans from the United States. The elk, which is another cervid that hosts ticks, was mentioned even more rarely and never as pointedly. It is a native species and the most iconic of cervids in the Finnish imagology of nature. As such, it is deeply entrenched in people's minds as an integral part of local nature. Such a sentiment was evidenced by the responses

Hufvudstadsbladet, April 28, 2018; Mikael Piippo, "Fästingarna uppskattar mild vinter," *Hufvudstadsbladet*, November 27, 2018; "Rådjur och harar bakom ökning av TBE," Åbo Underrättelser, August 30, 2018.

36 Matala et al., *Hirvieläinten vaikutuksia yhteiskuntaan*, 89.

37 One interviewee blamed the Common raccoon dog.

38 For other responses critical of the White-tailed deer and references to other animals, see, for example, FLS, "Punkit tulevat," SKS 006, 044, 060, 063, 064, 071, 074, 077, 081, N004, N048. See also, interview B.W. 8.9.2021. Similar results have been found in an earlier survey on the White-tailed deer. People living near the Ekenäs Archipelago National Park also wanted to reduce the White-tailed deer population, although many still thought that the species formed a valuable part of the diversity in the surrounding areas. Niemi and Nyman, *Valkohäntäpeuran*, 33, 34. It is noteworthy, though, that this survey was conducted in 2013, when the tick problem was discussed far less frequently than it would be a few years later.

39 The history of the Roe deer in Finland is somewhat uncertain, but according to some theories they have existed in the country since prehistoric times, but disappeared during the early modern age, only to reappear in recent decades from Sweden across Tornio River. Once across the river they then fanned out to most of the country. See, Lehikoinen, *Tuo hiisi hirviäsi* (2007), 73.

of our interviewees. The elk was hardly ever out-of-place in Finnish nature (except perhaps when running onto roads and causing car accidents) and had every right to exist alongside humans,[40] despite spreading ticks and causing enormous financial damage to forestry and landowners.[41] Finns clearly categorized cervids according to their belonging. Indeed, even though elks were as big a threat to human health as other cervids, they were given absolution. In contrast, by invading Finland the White-tailed deer had brought the snake into paradise. But they were not the only guilty party in Eden, as hunters were also perceived by many Finns as the devil's henchmen.

4 Conflicting Relationships

Many people viewed deer hunting and the decimation of the deer population as the best and quickest solution to the ever-growing tick problem.[42] In newspapers this stance found its staunchest supporter in Margareta Gustafsson, a parasitologist from Åbo Akademi University, who relentlessly advocated for the extermination of the deer population on the grounds that it was a harmful and invasive species that risked causing "nationwide epidemics." She also accused hunters of harboring disdain for human health for the sake of having an abundant stock of deer for them to enjoy.[43] However, it was far from being a scientifically-proven fact that deer were the main factor behind the tick problem. Researchers of tick ecology, although admitting that deer did spread ticks and contributed to their population growth, argued that they constituted just one reason among many for the proliferation of the ticks. They also pointed to other host species, as well as climate change. This latter factor not only favored ticks, but also deer.[44] To overcome this uncertainty, Gustafsson suggested an

40 See, for example, interviews E.F.7.9.2021; H. V. 9.9.2021; TKU/A/22/67-lit.

41 Jere Nieminen, *Hirviä ja ihmisiä: Hirven yhteiskunnallisen läsnäolon hallinta 2000-luvun alussa*, Ph.D. diss. (Tampere: Tampere University Press, 2015).

42 Interviews TKU/A/22/66-lit; H. V. 9.9.2021; JH 29.11.2019; I.V. 6.7.2020; E.F.7.9.2021; H. V. 9.9.2021; TKU/A/22/84-lit.

43 Margareta Gustafsson, "Själö – en fästingrik och farlig turistattraktion," *Åbo Underrättelser*, May 27, 2017; Margareta Gustafsson, "Hjortdjuren som fästingbussar," *Hufvudstadsbladet*, July 9, 2017; Margareta Gustafsson, "Fästingar försvinner om man tar kål på gnagare och vitsvanshjort," *Hufvudstadsbladet*, June 2, 2018; Margareta Gustafsson and Anneli Jalkanen, "Risk för TBE- och borreliaepidemier," *Vasabladet*, June 3, 2018; Margareta Gustafsson and Anneli Jalkanen, "Fästingsmittarna och fästingspridarna – gnagare och hjortjur," Åbo Underrättelser, June 6, 2018.

44 Jonna Hongell, "Klimatet gynnar fästingar," *Vasabladet*, May 28, 2017; Mikael Piippo, "Fästingar uppskattar mild vinter," *Hufvudstadsbladet*, November 27, 2018.

experiment be undertaken whereby the environmental administration should choose an island on which there would be a total cull of the deer population in order to determine whether this affected the tick population.[45]

Many hunters, however, strongly disagreed with the idea of reducing the deer population. They also downplayed the role of deer as a source of the tick problem. For hunters, the White-tailed deer was the most important game animal in the southern part of Finland.[46] In their target program for 2019–2027, The Finnish Hunters' Association acknowledged that there were too many deer in the country, but neither they nor the respondents of the 2022 survey for hunters were in favor of any legislative changes.[47] The hunters argued that even though the deer were an alien species they posed more benefits than harm to the community, thus further complicating the situation.[48] This seems to have also been the view of many game management officials in Finland, as the White-tailed deer has yet to be added to the national list of harmful invasive alien species. The deer are categorized as an "established invasive species" and a game animal that cannot be freely hunted, unlike species designated in the harmful invasive species category, as defined in Finnish law.[49]

---

45 "Pargasiternas majmöte," *Paraisten kuulutukset*, May 28, 2015; Margareta Gustafsson and Anneli Jalkanen, "Risk för TBE- och borreliaepidemier," *Vasabladet*, June 3, 2018.

46 According to the 2018 survey, most hunters perceived the White-tailed deer as the most important game animal in Southwestern Finland. Rinne and Toivola, *Valkohäntäpeura Varsinais-Suomessa*, 2, 3, 7. Bow hunters from Eastern Finland, who hunted on Seili Island, were an exception. Initiated by the university and Metsähallitus, the hunters tried to reduce the deer population or at least prevent it from growing. They hunted according to the quotas determined by the game management association but seem to have understood local peoples' views about less deer on the island. Interview with hunters, November 27, 2019.

47 The Finnish Hunters' Association, "Metsästäjäliiton tavoiteohjelma 2019–2027," 20, accessed October 8, 2023, https://metsastajaliitto.fi/sites/default/files/2019-12/Tavoiteohjelma-Metsastajaliitto.pdf. The majority of the 2022 survey respondents felt that the White-tailed deer population was tolerable, but 40 percent felt that the population was too big. Less than 15 percent of respondents thought that changing the hunting permit system would be a solution to cutting the deer population. Rinne and Toivola, *Valkohäntäpeura Varsinais-Suomessa*, 5, 7.

48 According to the 2022 survey by The Finnish Hunters' Association, hunters did not consider White-tailed deer to be an invasive alien species, but an alien species that was also a useful game animal and did not wish the animal's status would change. The Finnish Hunters' Association, "Valkohäntäpeuran jäsenkysely". See also, The Finnish Hunters' Association, "Metsästäjäliiton tavoiteohjelma 2019–2027," 9.

49 Hunting Act 615/1993, 5 §; The Natural Resources Institute Finland, Invasive Alien Species, "Valkohänt*äkauris* (*valkohäntäpeura*)," accessed November 9, 2023, https://vieraslajit.fi/lajit/MX.47629. Finnish Biodiversity Information Facility, "White-tailed Deer – Odocoileus virginianus," accessed November 9, 2023, https://laji.fi/en/taxon/MX.47629.

In the material we analyzed, hunters did not mention ticks as a problem that derived from deer, but they did cite other harmful effects. Local hunting officials, for example, listed other reasons for the explosion of the tick population, including climate change and an increase in ticks' nutrition. This nutritional factor meant that there were plenty of other host species for ticks to feed from. The hunting officials did not explain the logic of how other animals could have contributed to the rapidly increasing number of ticks, since deer were the only host species in which the population had grown. But it seems plausible that there were other reasons for their reluctance to admit that deer were a problem. These hunting advocates reasoned that the deer population had already reduced by lynx and that it would not be able to cope with more intensive hunting.[50]

In a dispute over the role of the White-tailed deer in the growing tick problem, ordinary people seem to have mostly taken an anti-deer stance. Many people increasingly felt that the way hunting was conducted – by making sure that there were sufficient number of deer for future hunters – was a problem.[51] One interviewee expressed the opinion that decision-makers belittled the problem and allowed hunters to hunt deer, while at the same time not addressing the fact that tens of thousands of people were becoming seriously ill from ticks.[52] In our questionnaire, a number of respondents expressed frustration at White-tailed deer and saw them as the main reason for ticks being so life-threateningly ubiquitous. Some argued in a straightforward manner that the White-tailed deer was an invading species that had no place in Finnish nature. One put it as follows: "Alien species introduced into Finnish nature, including those imported for hunting, should be completely removed from our nature." Another respondent argued that the White-tailed deer was the main villain and hoped that all of them would be terminated, since "there are enough elk for hunters."[53] In short, there has been and still is an unresolved conflict of interests between hunters and other users of nature in Finland. For hunters, deer were not a part of the problem, while many felt that they were, alongside the hunters themselves. Non-hunters argued that if hunters were to admit that

50 Emelie Melin, "Rädjursjakt ingen lösning på fästingplågen," *Österbottens Tidning*, July 2, 2013; Paulina Ek, "Rådjursjakt inget alternativ," *Syd-Österbotten*, July 11, 2013.

51 Interview TKU/A/22/87-lit. See also interview H. V. 9.9.2021.

52 Interview H. V. 9.9.2021.

53 FLS, "Punkit tulevat," SKS 007, 071. See also FLS, "Punkit tulevat," SKS 007, 017.

deer were a problem, they would have a hard time in justifying their ambition to maintain as big a deer population as possible for their beloved hobby.[54]

5 Conclusion

In recent years, there have been between 2000 and 6000 reported Lyme Borreliosis cases annually in Finland, and between 69 to 148 cases of TBE, from which most are severe cases with a mortality rate of approximately 1 percent.[55] In 2021, researchers at the Natural Resources Institute Finland argued that it is estimated that the total economic burden of tick-borne diseases in Finland is around €10 million per year. Hence, it is easy to see why some are speaking of a national epidemic, particularly vis-à-vis borreliosis. However, there are no easy solutions available to lessen the impact of this epidemic. All people can do is to resort to pre-emptive measures and, in the event of infection, to undertake a heavy course of antibiotics.

The fear of ticks has spoiled the relationship between humans and White-tailed deer. The species was admired by most for several decades after its introduction into Finland. Yet, it is now perceived as a villain and an unwelcomed alien. Demands to cull the deer population have grown louder. Moreover, the reluctance of hunters to act accordingly has created a conflict between them and non-hunting local residents. These interspecific conflicts have also led to a

54 This is not just our interpretation, but is also a view shared by many conservationists in Finland as well as some writers in our source material. Besides the above-listed statements by Margareta Gustafsson, see also Anne-Mari Packalén-Reinikainen, "Valkohäntäpeura on ongelmallisin riistaeläin," *MT*, February 21, 2014; Veli-Jussi Jalkanen, "Hirvieläinvahingot ovat tietoinen ja kallis valinta," *MT*, October 5, 2015; Maarit Gockel, "Valkohäntäpeurojen haittoja vähätellään," *MT*, July 28, 2017; Linnea Nordling, "Metsästyslupien määrä pohjaa kannan aliarvioon," *MT*, September 22, 2017; Leena Kauppila, "Villisikojen ja peurojen talviruokinta lopetettava," *MT*, September 28, 2018. As for the view of conservationists, see, for example, Liisa Hulkko, "Ilves tarvitsi asianajajan," *Luonnonsuojelija* 49, no. 4 (2023); Oona Lohilahti, "Ilvesten kaatoluvat järkyttivät," *Luonnonsuojelija* 49, no. 4 (2023).

55 The Finnish Institute for Health and Welfare, "Lyme borreliosis register surveillance," accessed November 9, 2023, https://sampo.thl.fi/pivot/prod/en/infestat/borre/fact_infestat_borre?&row=hcdmunicipality2020-572572&column=weeks-546468&filter=measure-546834; The Finnish Institute for Health and Welfare, "Puutiaisaivotulehduksenesiintyvyys Suomessa," accessed November 9, 2023. https://thl.fi/fi/web/infektiotaudit-ja-rokotukset/taudit-ja-torjunta/taudit-ja-taudinaiheuttajat-a-o/puutiaisaivotulehdus/puutiaisaivotulehduksen-esiintyvyys-suomessa. Matala et al., *Hirvieläinten vaikutuksia yhteiskuntaan*, 111.

heightened sense of mistrust of officials, who, according to some local people, do not take the deer and tick issue seriously enough.

Hunting, however, is hardly a perfect solution to counter threat posed by ticks. Those who advocate hunting as a solution embrace the idea of the old days when everything was supposedly in its rightful place. But there is no going back. The ecosystem in which deer thrive is not the same as it used to be, and neither is the climate. There is no proof that the removal of deer would reduce the number of ticks. A more philosophical objection stems from the fact that not only is the environment changing, but so too is human society. Jørgensen points out that many locals in Norway and Sweden view the muskoxen as an important symbol of the area. In other words, despite its original alienness, it has become integrated into their societies.[56] The White-tailed deer is also an alien species to Finland. Are we, then, to deprive muskoxen and White-tailed deer – or any species – of its belonging, whenever problems occur? This would be a highly anthropocentric stance, especially since the problems are of human origin: White-tailed deer did not ask to be transferred to Finland, nor did they bring ticks with them. Although deer do not contract the same diseases as humans, they are most likely also plagued by the blood-sucking parasites. In other words, they are not the cause of the problem but are co-sufferers alongside humans.

Bibliography

Archival Sources

Finnish Literature Society, SKS, questionnaire *Punkit tulevat!*, 2019.

Collection of Folkloristics and Study of Religion (TKU). Archives of the School of History, ulture and Arts Studies, University of Turku.

Newspapers and Periodicals

Hufvudstadsbladet

Ilta-Sanomat

Maaseudun Tulevaisuus

Paraisten kuulutukset

Syd-Österbotten

Vakka: Vakka-Suomen sanomat

Vasabladet

56 Jørgensen, "Migrant muskoxen," 198–199.

Åbo Underrättelser
Österbottens Tidning

Other Sources

Hunting Act 615/1993
Seili interviews, in Lähdesmäki's possession.

Literature

Coates, Peter. *American Perceptions of Immigrant and Invasive Species: Strangers on the Land.* Berkeley: University of California Press, 2007.

Duffy, D.C., S.R. Campbell, D. Clark, C. DiMotta, and S. Gurney. "Ixodes Scapularis (Acari: Ixodidae) Deer Tick Mesoscale Populations in Natural Areas: Effects of Deer, Area, and Location." *Journal of Medical Entomology* 31, no. 1 (1994): 152–158. doi:10.1093/jmedent/31.1.152.

The Finnish Biodiversity Information Facility. "White-tailed Deer – Odocoileus virginianus." Accessed November 9, 2023. https://laji.fi/en/taxon/MX.47629.

The Finnish Hunters' Association. "Metsästäjäliiton tavoiteohjelma 2019–2027." Accessed October 8, 2023. https://metsastajaliitto.fi/sites/default/files/2019-12/Tavoiteohjelma-Metsastajaliitto.pdf.

The Finnish Hunters' Association. "Valkohäntäpeuran jäsenkysely tukee vahvasti Liiton linjauksia", December 15, 2022. Accessed October 8, 2023. https://metsastajaliitto.fi/uutiset/valkohantapeuran-jasenkysely-tukee-vahvasti-liiton-linjauksia.

The Finnish Institute for Health and Welfare. "Lyme borreliosis register surveillance." Accessed November 9, 2023. https://sampo.thl.fi/pivot/prod/en/infestat/borre/fact_infestat_borre?&row=hcdmunicipality2020-572572&column=weeks-546468&filter=measure-546834.

The Finnish Institute for Health and Welfare. "Puutiaisaivotulehduksen esiintyvyys Suomessa." Accessed November 9, 2023. https://thl.fi/fi/web/infektiotaudit-ja-rokotukset/taudit-ja-torjunta/taudit-ja-taudinaiheuttajat-a-o/puutiaisaivotulehdus/puutiaisaivotulehduksen-esiintyvyys-suomessa.

The Finnish Wildlife Agency. "LUKE: Valkohäntäpeurakannan kasvu pysähtyi," March 18, 2022. Accessed September 19, 2023. https://riista.fi/luke-valkohantapeurakannan-kasvu-pysahtyi-2/.

Hulkko, Liisa. "Ilves tarvitsi asianajajan." *Luonnonsuojelija* 49, no. 4 (2023): 3.

Ilvesviita, Pirjo. *Paaluraudoista kotkansuojeluun: suomalainen metsästyspolitiikka 1865–1993*. Rovaniemi: Lapin yliopisto, 2005.

Jerolmack, Colin. "How Pigeons Became Rats: The Cultural-Spatial Logic of Problem Animals." *Social Problems* 55, no. 1 (2008): 72–94. doi:10.1525/sp.2008.55.1.72.

Jones, Owain, and Paul Cloke. *Tree Cultures: The Place of Trees and Trees in Their Place.* Oxford: Berg, 2002.

Jørgensen, Dolly. "Migrant Muskoxen and the Naturalization of National Identity in Scandinavia." In *The Historical Animal*, edited by Susan Nance, 184–201. Syracuse: Syracuse University Press, 2015.

Kairikko, Juha and Jaakko Ruola. *Valkohäntäpeura*. Jyväskylä: Suomen metsästäjäliitto, 2004.

Lehikoinen, Heikki. *Tuo hiisi hirviäsi. Metsästyksen kulttuurihistoria Suomessa*. Helsinki: Teos, 2007.

Lillbroända-Annala, Sanna. "Fy, en fästing! Äckel i kroppsliga och affektiva möten mellan fästing, sällskapsdjur och människa." *Budkavlen* 101 (2002): 82–104.

Lillbroända-Annala, Sanna. "'Fästingeländet,' Praktiker och materialisering mellan människor, fästingar och natur." *Laboratorium för folk och kultur* 8 (2021). Accessed September 29, 2023. https://bragelaboratorium.com/2021/11/11/fastingelandet-praktiker-och-materialisering-mellan-manniskor-fastingar-och-natur/.

Lillbroända-Annala, Sanna, and Oscar Winberg. "'Fästingen håller inget säkerhetsavstånd': Konkurrerande riskdiskurser om fästingar i media." *TRACE ∴ Journal for Human-Animal Studies* 9 (2023): 58–80. doi: 10.23984/fjhas.121761.

Lohilahti, Oona. "Ilvesten kaatoluvat järkyttivät." *Luonnonsuojelija* 49, no. 4 (2023): 7.

Lähdesmäki, Heta. *Susien paikat. Ihminen ja susi 1900-luvun Suomessa*. Nykykulttuurin tutkimuskeskuksen julkaisuja 127. Ph.D. diss. University of Jyväskylä, 2020.

Matala, Juho, Ari Nikula, Jani Pellikka, Sami Aikio, Jukka Forsman, Heikki Henttonen and Katja Holmala. *Hirvieläinten vaikutuksia yhteiskuntaan, elinkeinoihin ja ekosysteemiin: Synteesiraportti*. Helsinki: Luonnonvarakeskus, 2021.

Natural Resources Institute Finland. Invasive Alien Species. "Valkohäntäkauris (valkohäntäpeura)." Accessed November 9, 2023. https://vieraslajit.fi/lajit/MX.47629.

Niemi, Milla, and Madeleine Nyman. *Valkohäntäpeuran ekologiset ja sosiaaliset vaikutukset Tammisaaren saariston kansallispuistossa ja sen lähialueilla*. Vantaa: Metsähallitus, 2013. Accessed October 23, 2023. https://julkaisut.metsa.fi/assets/pdf/lp/Asarja/a204.pdf.

Nieminen, Jere. *Hirviä ja ihmisiä: Hirven yhteiskunnallisen läsnäolon hallinta 2000-luvun alussa*. Ph.D. diss. Tampere: Tampere University Press, 2015.

Nummi, Petri. *Suomeen istutetut riistaeläimet*. Helsinki: Helsingin yliopisto, 1988.

Nygrén, Tuire. *Suomen hirvikannan säätely – biologiaa ja luonnonvarapolitiikkaa*. Ph.D. diss. University of Joensuu, 2009.

Ojalammi, Sanna. "Susi, tunteet ja tila." *Maaseudun uusi aika* no. 1/2015.

Pál, Viktor, Tuomas Räsänen, and Mikko Saikku. "Introduction to the Environmental Histories of Finland." In *Green Development of Greenwashing?: Environmental Histories of Finland*, edited by Viktor Pál, Tuomas Räsänen, and Mikko Saikku, 1–8. Winwick: Whitehorse Press, 2023.

Philo, Chris and Chris Wilbert, "Animal spaces, beastly places. An introduction." In *Animal Spaces, Beastly Places. New Geographies of Human-Animal Relations*, edited by Chris Philo and Chris Wilbert, 1–35. London: Routledge, 2007.

Pimentel, David. *Biological Invasions: Economic and Environmental Costs of Alien Plant, Animal, and Microbe Species*. Boca Raton: CRC Press, 2011.

Rinne, Antti and Mikko Toivola. *Valkohäntäpeura Varsinais-Suomessa – kannanhoidon kysely valkohäntäpeuran metsästäjille. Raportti*. Varsinais-Suomi: Suomen Riistakeskus, 2018.

Sormunen, Jani Jukka, Niko Kulha, Tero Klemola, Satu Mäkelä, Ella-Maria Vesilahti, and Eero Juhani Vesterinen. "Enhanced Threat of Tick-borne Infections Within Cities? Assessing Public Health Risks Due to Ticks in Urban Green Spaces in Helsinki, Finland." *Zoonoses and Public Health* 67, no. 7 (2020): 823–839. doi:10.1111/zph.12767.

Teperi, Jouko. *Sudet Suomen rintamaiden ihmisten uhkana 1800-luvulla*. Helsinki: Suomen Historiallinen Seura, 1977.

Index

www.ingramcontent.com/pod-product-compliance
Lightning Source LLC
Chambersburg PA
CBHW071229230926
51849CB00001B/61